ESTIMATOR'S ELECTRICAL
MAN-HOUR MANUAL

T H I R D
E D I T I O N

Man-Hour Manuals and Other Books
by John S. Page

Conceptual Cost Estimating Manual

Cost Estimating Manual for Pipelines and
  Marine Structures

Estimator's Electrical Man-Hour Manual/3rd Edition

Estimator's Equipment Installation
  Man-Hour Manual/3rd Edition

Estimator's General Construction
  Man-Hour Manual/2nd Edition

Estimator's Man-Hour Manual on Heating,
  Air Conditioning, Ventilating, and Plumbing/2nd Edition

Estimator's Piping Man-Hour Manual/5th Edition

John S. Page has wide experience in cost and labor estimating, having worked for some of the largest construction firms in the world. He has made and assembled numerous types of estimates including lump-sum, hard-priced, and scope, and has conducted many time and method studies in the field and in fabricating shops. Mr. Page has a B.S. in civil engineering from the University of Arkansas and received the Award of Merit from the American Association of Cost Engineers in recognition of outstanding service and cost engineering.

ESTIMATOR'S ELECTRICAL MAN-HOUR MANUAL

T H I R D

E D I T I O N

JOHN S. PAGE

G|P
P| Gulf Professional Publishing
An Imprint of Elsevier

To all electrical estimators who have
spent many hours burning midnight
oil and scratching heads trying to
estimate reasonable labor units,
I dedicate this manual.

Originally published by Gulf Professional Publishing,
Houston, TX.

For information, please contact:
Manager of Special Sales
Butterworth–Heinemann
An Imprint of Elsevier
225 Wildwood Avenue
Woburn, MA 01801–2041
Tel: 781-904-2500
Fax: 781-904-2620
For information on all Butterworth–Heinemann publications
available, contact our World Wide Web home page at:
http://www.bh.com

Library of Congress Cataloging-in-Publication Data

Page, John S.
    Estimator's electrical man-hour manual / John S. Page. — 3rd ed.
      p. cm.
    Includes bibliographical references and index.

    ISBN-13: 978-0-88415-228-6 ISBN-10: 0-88415-228-6 (alk. paper)

    1. Electric engineering—Estimates. I. Title.
  TK435.P3 1999
  621.319′24′0299—dc21   99-18584
  ISBN-13: 978-0-88415-228-6   CIP
  ISBN-10: 0-88415-228-6

Transferred to Digital Printing 2009

CONTENTS

Section 2—BRANCH, SERVICE, AND FEEDER WIRING

Section 3—WIRING DEVICES—SWITCHES, OUTLETS, AND RECEPTACLES

Section 4—
SURFACE METAL RACEWAY AND BRANCH BUSWAY

Section 5—LIGHTING FIXTURES

Section 6—UNDERFLOOR DUCT

Section 7—BUS DUCT

Section 8—ELECTRIC HEATING AND VENTILATING

Section 9—PANELBOARDS AND ACCESSORIES

Section 10—SWITCHBOARDS AND ACCESSORIES

Section 11—
SWITCHES, STARTERS, CONTROLS, AND GUTTERS

Section 12—MOTOR CONTROL CENTERS

Section 13—MOTOR CONTROLS AND MOTORS

Section 14—POWER TRANSFORMERS

Section 15—OUTSIDE OVERHEAD SYSTEMS

Section 16—OUTSIDE UNDERGROUND SYSTEMS

Section 17—
COMMUNICATIONS AND SIGNAL SYSTEMS

Section 18—
ELECTRICAL INSTRUMENT INSTALLATION

Section 19—
ANCHORS, FASTENERS, HANGERS AND SUPPORTS

Section 20—
DEMOLITION, EXCAVATION AND CONCRETE

Section 21—TECHNICAL INFORMATION

PREFACE

This third edition is fully updated with the addition of a new section on electrical instrument installation consisting of eleven new tables.

The labor units, which are expressed in manhours throughout this manual, are for assisting the estimator in estimating electrical installation labor cost for an individual item of work or total project direct cost.

The thousands of manhour units that follow, for the most part, are averages of many projects constructed in the Gulf Coast area. Most of these projects were petrochemical related.

After extensive time and method studies and production evaluation of individual electricians, it was determined that the average performing productivity for the Gulf Coast area was equal to 70%. The manhours throughout this manual are based on this percentage.

To correctly apply these manhours to a particular project, a productivity factor giving consideration to location and conditions should be established for application against these manhours. The reader should therefore carefully note the introduction on the following pages, because it outlines a method for obtaining such a productivity factor.

The following basic manhour units provide a separate time allowance for a particular labor operation under specific conditions and circumstances all in accordance with the notes as appears on the individual table pages.

To determine the direct labor dollar value of a project a composite labor rate should be established and applied against the various estimated manhours for the individual units or in total, whichever is desired. The introduction that follows outlines a method to establish the composite rate.

The Human Factor in Estimating

In this high-tech world of sophisticated software packages, including several for labor and cost estimating, you might wonder what a collection of manhour tables offers that a computer program does not. The answer is the *human factor*. In preparing a complete estimate for a refinery, petrochemical, or other heavy industrial project one often confronts 12–18 major accounts, and each account has 5–100 or more sub-accounts, depending on the project and its engineering design. While it would seem that such numerous variables provide the perfect opportunity for computerized algorithmic solution, accurate, cost-effective, realistic estimating is still largely a function of human insight and expertise. Each project has unique aspects that still require the seasoned consideration of an experienced professional, such as general economy, projects supervision, labor relations, job conditions, construction equipment, and weather, to name a few.

Computers are wonderful tools. They can solve problems as no human can, but I do not believe construction estimating is their forté. I have reviewed several construction estimating software packages and have yet to find one that I would completely rely on. Construction estimating is an art, a science, and a craft, and I recommend that it be done by those who understand and appreciate all three of these facets. This manual is intended for those individuals.

John S. Page

INTRODUCTION
Production and Composite Rate

It is common practice to estimate labor operations in manhour units. However, it is obvious that a unit manhour and its dollar value would not be the same for all projects in all locations. Therefore, the following is offered for establishing productivity factors and composite rates for application against the following manhour units.

There should be sound reasoning and understanding to back up a monetary unit before it is applied to an item for labor value. The best reasoning is manhours based on what we call productivity efficiency coupled with production elements.

After comparison of many projects, constructed under varied conditions, we have found that production elements can be grouped into six different classifications and that production percentages can be classified into five different categories.

The six different classifications of production elements are:

1. General Economy
2. Project Supervision
3. Labor Relations
4. Job Conditions
5. Equipment
6. Weather

The five ranges of productivity efficiency percentages are:

Type	Range
1. Very low	10– 40%
2. Low	41– 60%
3. Average	61– 80%
4. Very good	81– 90%
5. Excellent	91–100%

Since there is such a wide range between the productivity percentages, let us attempt to evaluate each of the six elements, giving an example with each, and see just how a true productivity percentage can be obtained.

1. *General Economy:* This is nothing more than the state of the nation or area in which your project is to be constructed. The things that should be reviewed and evaluated under this category are:

a. Business trends and outlooks
b. Construction volume
c. The employment situation

Let us assume that after giving due consideration to these items you find them to be very good or excellent. This sounds good, but actually it means that your productivity range will be very low. This is due to the fact that with business being excellent the top supervision and craftsmen will be mostly employed and all that you will have to draw from will be inexperienced personnel. Because of this, in all probability, it will tend to create bad relationships between owner representatives, contract supervisors, and the various craftsmen, thus making very unfavorable job conditions. On the other hand, after giving consideration to this element you may find the general economy to be of a fairly good average. Should this be the case, you should find that productivity efficiency tends to rise. This is due to the fact that under normal conditions there are enough good supervisors and craftsmen to go around and they are satisfied, thus creating good job conditions and understanding for all concerned. We have found, in the past, that general economy of the nation or area where your project is to be constructed, sets off a chain reaction to the other five elements. We, therefore, suggest that very careful consideration be given this item.

As an example, to show how a final productivity efficiency percentage can be arrived at, let us say that we are estimating a project in a given area and after careful consideration of this element, we find it to be of a high average. Since it is of a high average, but by no means excellent, we estimate our productivity percentage at 75%.

2. *Project Supervision:* What is the caliber of your supervision? Are they well-seasoned and experienced? What can you afford to pay them? What supply do you have to draw from? Things that should be looked at and evaluated under this element are:

a. Experience
b. Supply
c. Pay

Like general economy this too must be carefully analyzed. If business is normal, in all probability, you will be able to obtain good supervision, but if business is excellent the chances are that you will have a poor lot to draw from. Should the contractor try to cut overhead by the use of cheap supervision he will usually wind up doing a very poor job. This usually results in a dissatisfied client, a loss of profit, and a loss of future work. This, like the attachment of the fee for a project, is a problem over which the estimator has no control. It must be left to management. All the estimator can do is evaluate and estimate his project accordingly.

To follow through with our example, after careful analyis of the three (3) items listed under this element, let us say that we have found our supervision will be normal for the project involved and we arrive at an estimated productivity rate of 70%.

3. *Labor Conditions:* Does your organization possess a good labor relations man? Are there experienced first class satisfied craftsmen in the area where your project is to

be located? Like project supervision, things that should be analyzed under this element are:

a. Experience
b. Supply
c. Pay

A check in the general area where your project is to be located should be made to determine if the proper experienced craftsmen are available locally, or will you have to rely on travelers to fill your needs. Can and will your organization pay the prevailing wage rates?

For our example, let us say that for our project we have found our labor relations to be fair but feel that they could be a little better and that we will have to rely partially on travelers. Since this is the case, we arrive at an efficiency rating of 65%.

4. *Job Conditions:* What is the scope of your project and just what work is involved in the job? Will the schedule be tight and hard to meet, or will you have ample time to complete the project? What kind of shape or condition is the site in? Is it low and mucky and hard to drain, or is it high and dry and easy to drain? Will you be working around a plant already in production? Will there be tie-ins, making it necessary to shut down various systems of the plant? What will be the relationship between production personnel and construction personnel? Will most of your operations be manual, or mechanized? What kind of material procurement will you have? There are many items that could be considered here, dependent on the project; however, we feel that the most important items that should be analyzed under this element are as follows:

a. Scope of work
b. Site conditions
c. Material procurement
d. Manual and mechanized operations

By a site visitation and discussion with owner representatives, coupled with careful study and analysis of the plans and specifications, you should be able to correctly estimate a productivity percentage for this item.

For our example, let us say that the project we are estimating is a completely new plant and that we have ample time to complete the project but the site location is low and muddy. Therefore, after evaluation we estimate a productivity rating of only 60%.

5. *Equipment:* Do you have ample equipment to complete your project? Just what kind of shape is it in and will you have good maintenance and repair help? The main items to study under this element are:

a. Usability
b. Condition
c. Maintenance and repair

This should be the simplest of all elements to analyze. Every estimator should know what type and kind of equipment his company has, as well as what kind of mechanical shape it is in. If equipment is to be obtained on a rental basis then the estimator should know the agency he intends to use as to whether they will furnish good equipment and good maintenance.

Let us assume, for our example, that our company equipment is in very good shape, that we have an ample supply to draw from and that we have average mechanics. Since this is the case we estimate a productivity percentage of 70%.

6. *Weather:* Check the past weather conditions for the area in which your project is to be located. During the months that your company will be constructing, what are the weather predictions based on these past reports? Will there be much rain or snow? Will it be hot and mucky or cold and damp? The main items to check and analyze here are as follows:

    a. Past weather reports
    b. Rain or snow
    c. Hot or cold

This is one of the worst of all elements to be considered. At best all you have is a guess. However, by giving due consideration to the items as outlined under this element, your guess will at least be based on past occurrences.

For our example, let us assume that the weather is about half good and half bad during the period that our project is to be constructed. We must then assume a productivity range of 50% for this element.

We have now considered and analyzed all six elements and in the examples for each individual elements have arrived at a productivity efficiency percentage. Let us now group these percentages together and arrive at a total percentage.

Item	Productivity Percentage
1. General economy	75
2. Production supervision	70
3. Labor relations	65
4. Job conditions	60
5. Equipment	70
6. Weather	50
Total	390%

Since there are six elements involved, we must now divide the total percentage by the number of elements to arrive at an average percentage of productivity.

$$390\% \div 6 = 65\% \text{ average productivity efficiency}$$

At this point we caution the estimator. This example has been included as a guide to show one method that may be used to arrive at a productivity percentage. The preceding elements can and must be considered for each individual project. By so doing, coupled with the proper manhour tables that follow, a good labor value estimate can be properly executed for any place in the world, regardless of its geographical location and whether it be today or twenty years from now.

Composite Rate

Next, we must consider the composite rate. In order to correctly arrive at a total direct labor cost, using the manhours as appear in the following tables, this must be done.

Most organizations consider field personnel with a rating of superintendent or greater as a part of job overhead, and that of general foreman or lower as direct job labor cost. The direct manhours as appear on the following pages have been determined on this basis. Therefore, a composite rate should be used when converting the manhours to direct labor dollars.

Again, the estimator must consider labor conditions in the area where the project is to be located. He must ask himself how many men will he be allowed to use in a crew, can he use crews with mixed crafts, and how many crews of he various crafts will he need.

Following is an example that may be used to obtain a composite rate:

From the total required estimated manhours we assume that the project we are estimating can be completed on schedule with the use of four 10-man crews, and that a general foreman will be required for every two crews.

Rate of craft in the given area:

General Foreman . . . . $22.00 per hour
Foreman . . . . $21.50 per hour
Working Steward. . . . . $21.10 per hour
Journeyman. $21.00 per hour
Fourth-year Apprentice . . . . $14.00 per hour

Note: General Foreman and Foreman are dead weight since they do not work with their tools, however, they must be considered and charged to the composite crew.

Crew for Composite Rate:

One General Foreman. . . . . 4 hours @ $22.00 $88.00
One Foreman. . . . . 8 hours @ $21.50 $172.00
One Working Stewart. . . . . 8 hours @ $21.10 $168.80
Seven Journeymen. . . . . 8 hours @ $21.00 $1,176.00
Two Apprentices . . . . 8 hours @ $14.00 $224.00

Total for 80 hours $1,828.80

$1,828.80 ÷ 80 = $22.86 Composite Manhour Rate for 100% Time

It is well to note that this time that as stated in the preface the manhours are based on an average productivity of 70%. Therefore, the composite rate of $22.86 as compiled becomes equal to 70% productivity.

Let us now assume that we have evaluated a certain project to be bid and find it to be of a low average with an overall productivity rating of only 65%. This means a loss in productive manhours or dollars paid per manhour. Therefore, the manhours or composite rate per manhour needs an adjustment as follows:

Manhour Adjustment:

$$\frac{70\%}{65\%} = 1.08 \times \text{Manhours} = \text{Productive Manhours}$$

or

$$\frac{70\%}{65\%} = 1.08 \times \$22.86 = \$24.68 \text{ Composite Rate for 65\% Productivity}$$

Simply by multiplying the number of manhours estimated for a given block or item of work by the arrived at composite rate, a total estimated direct labor cost, in dollar value can be easily and accurately obtained.

It is our express desire and sincere hope that the foregoing will enable the ordinary electrical estimator to turn out a better labor estimate and assist in the elimination of much guesswork.

Section 1

BRANCH, SERVICE, AND FEEDER ROUGH-IN

This section includes manhour time frames covering the installation of boxes, fittings, conduits, etc. required for roughing in electrical lighting and power systems for a process or industrial plant.

Throughout the tables, consideration has been given to all necessary labor for operations such as unloading, handling, and installing, that one might expect to encounter for this type of work. The tables include time allowances for general foreman through apprentice.

The manhour tables that follow are average of many projects of varied nature and include time allowances for particular operations or blocks of work in accordance with the notes there-on.

Before applying the manhour units in this or any section, we caution the estimator to be thoroughly familiar with the Introduction of this manual.

SQUARE BOXES, COVERS, AND RINGS

Walls, Flat Slabs, and Concrete Pan Form Construction
(Concealed Installation)

MANHOURS EACH

Item	Size (Inches)	Manhours for Height to			
		10'	15'	20'	25'
Square Type					
Box	4	.41	.45	.48	.51
Box Extension	4	.28	.31	.33	.35
Plaster Ring	4	.15	.17	.18	.19
Cover — 1- or 2-Device	4	.14	.15	.16	.17
Tile Cover — 1- or 2-Device	4	.19	.21	.22	.24
Offset Cover — 1-Device	4	.29	.32	.34	.36
Cover Surface Mtg. Device	4	.29	.32	.34	.36
Blank Cover	4	.15	.17	.18	.19
Cover with K.O.	4	.15	.17	.18	.19
Swivel Hanger Cover	4	.41	.45	.48	.51
Box	4-11/16	.48	.53	.56	.60
Box Extension	4-11/16	.32	.35	.38	.40
Plaster Ring	4-11/16	.23	.25	.27	.29
Cover — 1- or 2-Device	4-11/16	.23	.25	.27	.29
Tile Cover — 1- or 2-Device	4-11/16	.29	.32	.34	.36
Cover Surface Mtg. Device	4-11/16	.45	.50	.54	.57
Blank Cover	4-11/16	.24	.26	.28	.30
Cover With K.O.	4-11/16	.24	.26	.28	.30

Wall And Ceiling Surfaces (Exposed Installation)

MANHOURS EACH

Item	Size (Inches)	Manhours for Height to			
		10'	15'	20'	25'
Square Type					
Box	4	.33	.36	.39	41
Box Extension	4	.28	.31	.33	.35
Plaster Rings	4	.15	.17	.18	.19
Cover — 1- or 2-Device	4	.14	.15	.16	.17
Tile Cover — 1- or 2-Device	4	.19	.21	.22	.24
Offset Cover — 1-Device	4	.29	.32	.34	.36
Cover Surface Mtg. Device	4	.29	.32	.34	.36
Blank Cover	4	.15	.17	.18	.19
Cover With K.O.	4	.15	.17	.18	.19
Swivel Hanger Cover		.41	45	48	.51
Box	4-11/16	.38	.42	.45	.47
Box Extension	4-11/16	.32	.35	.38	.40
Plaster Ring	4-11/16	.23	.25	.27	.29
Cover — 1- or 2-Device	4-11/16	.23	.25	.27	.29
Tile Cover — 1- or 2-Device	4-11/16	.29	.32	.34	.36
Cover Surface Mtg. Device	4-11/16	.45	.50	.54	.57
Blank Cover	4-11/16	.24	.26	.28	.30
Cover With K.O.	4-11/16	.24	.26	.28	.30

Manhours include checking out of job storage, handling, job hauling, layout and installation items as outlined.

Manhours exclude installation of special fasteners and anchors, fittings, conduits, raceway, and scaffolding. See respective tables for these time frames.

For individual heights greater than 25 feet increase manhours 10% for each additional 5 feet or fraction thereof.

SQUARE BOXES, COVERS, AND RINGS

Concrete Block, Brick, or Hollow Tile Walls

(Concealed Installation)

MANHOURS EACH

Item	Size (Inches)	Manhours for Height to			
		10'	15'	20'	25'
Square Type					
Box	4	.47	.52	.55	.59
Box Extension	4	.29	.32	.34	.36
Plaster Ring	4	.16	.18	.19	.20
Cover — 1- or 2-Device	4	.15	.17	.18	.19
Tile Cover — 1- or 2-Device	4	.20	.22	.24	.25
Offset Cover — 1-Device	4	.30	.33	.35	.37
Cover Surface Mtg. Device	4	.16	.18	.19	.20
Blank Cover	4	.16	.18	.19	.20
Cover with K.O.	4	.43	.47	.51	.54
Swivel Hanger Cover	4				
Box	4-11/16	.54	.59	.64	.67
Box Extension	4-11/16	.34	.37	.40	.42
Plastic Ring	4-11/16	.25	.28	.29	.31
Cover — 1- or 2-Device	4-11/16	.25	.28	.29	.31
Tile Cover — 1- or 2-Device	4-11/16	.31	.34	.36	.39
Cover Surface Mtg. Device	4-11/16	.47	.52	.55	.59
Blank Cover	4-11/16	.26	.29	.31	.32
Cover with K.O.	4-11/16	.26	.29	.31	.32

Wood Frame Construction (Concealed Installation)

MANHOURS EACH

Item	Size (Inches)	Manhours for Height to			
		10'	15'	20'	25'
Square Type					
Box	4	.25	.28	.29	.31
Box Extension	4	.18	.20	.21	.22
Plaster Ring	4	.12	.13	.14	.15
Cover — 1- or 2-Device	4	.12	.13	.14	.15
Tile Cover — 1- or 2-Device	4	.15	.17	.18	.19
Offset Cover — 1-Device	4	.20	.22	.24	.25
Cover Surface Mtg. Device	4	.20	.22	.24	.25
Blank Cover	4	.12	.13	.14	.15
Cover with K.O.	4	.12	.13	.14	.15
Swivel Hanger Cover	4	.25	.28	.29	.31
Box	4-11/16	.29	.32	.34	.36
Box Extension	4-11/16	.26	.29	.31	.32
Plaster Ring	4-11/16	.20	.22	.24	.25
Cover — 1- or 2-Device	4-11/16	.20	.22	.24	.25
Tile Cover — 1- or 2-Device	4-11/16	.26	.29	.31	.32
Cover Surface Mtg. Device	4-11/16	.35	.39	.41	.44
Blank Cover	4-11/16	.20	.22	.24	.25
Cover with K.O.	4-11/16	.20	.22	.24	.25

Manhours include checking out of job storage, handling, job hauling, layout, and installing items as outlined.

Manhours exclude installation of special fasteners and anchors, fittings, conduit, raceway, and scaffolding. See respective tables for these time frames.

For individual heights greater than 25 feet increase manhours 10% for each additional 5 feet or fraction thereof.

SQUARE BOXES, COVERS, AND RINGS
Suspended Ceiling Construction
(Concealed Installation)

MANHOURS EACH

Item	Size (Inches)	Manhours for Height to			
		10'	15'	20'	25'
Square Type					
Box	4	.29	.32	.34	.36
Box Extension	4	.22	.24	.26	.27
Plaster Ring	4	.13	.14	.15	.16
Cover — 1- or 2-Device	4	.12	.13	.14	.15
Tile Cover — 1- or 2-Device	4	.16	.18	.19	.20
Offset Cover — 1-Device	4	.24	.26	.28	.30
Cover Surface Mtg. Device	4	.24	.26	.28	.30
Blank Cover	4	.13	.14	.15	.16
Cover with K.O.	4	.13	.14	.15	.16
Swivel Hanger Cover	4	.29	.32	.34	.36
Box	4-11/16	.34	.37	40	.42
Box Extension	4-11/16	.30	.33	.35	.37
Plaster Ring	4-11/16	.20	.22	.24	.25
Cover — 1- or 2-Device	4-11/16	.20	.22	.24	.25
Tile Cover — 1- or 2-Device	4-11/16	.25	.28	.29	.31
Cover Surface Mtg. Device	4-11/16	.31	.34	.36	.39
Blank Cover	4-11/16	.20	.22	.24	.25
Cover with K.O.	4-11/16	.20	.22	.24	.25

Manhours include checking out of job storage, handling, job hauling, layout, and installing items as outlined.

Manhours exclude installing special fasteners and anchors, fittings, conduit, raceway, and scaffolding. See respective tables for these time frames.

For individual heights greater than 25 feet increase manhours 10% for each additional 5 feet or fraction thereof.

OCTAGON AND SPECIAL BOXES, ROUND COVERS, AND RINGS

Flat Concrete Slab Construction (Concealed Installation)

MANHOURS EACH

Item	Size (Inches)	Manhours for Height to			
		10′	15′	20′	25′
Octagon Type					
Box	4	.48	.53	.56	.60
Hung Ceiling Box	4	.45	.50	.53	.56
Box Extension	4	.27	.30	.32	.34
Round Type					
Plaster Ring	4	.15	.17	.18	.19
Device Cover	4	.15	.17	.18	.19
Blank Cover	4	.15	.17	.18	.19
Cover with K.O.	4	.15	.17	.18	.19
Switch or Receptacle Cover	4	.15	.17	.18	.19
Swivel Hanger Cover	4	.51	.56	.60	.64
Special Boxes					
Concrete Box	4	.32	.35	.38	.40
Concrete Box Plate	4	.15	.17	.18	.19

Concrete Pan Form Construction (Concealed Installation)

MANHOURS EACH

Item	Size (Inches)	Manhours for Height to			
		10′	15′	20′	25′
Octagon Type					
Box	4	.60	.66	.71	.75
Hung Ceiling Box	4	.56	.62	.66	.70
Box Extension	4	.34	.37	.40	.42
Round Type					
Plaster Ring	4	.16	.18	.19	.20
Device Cover	4	.16	.18	.19	.20
Blank Cover	4	.16	.18	.19	.20
Cover with K.O.	4	.16	.18	.19	.20
Switch or Receptacle Cover	4	.16	.18	.19	.20
Swivel Hanger Cover	4	.64	.70	.75	.79
Special Boxes					
Concrete Box	4	.44	.48	.52	.55
Concrete Box Plate	4	.16	.18	.19	.20

Manhours include checking out of job storage, handling, job hauling, layout and installing items as outlined.

Manhours exclude installing special fasteners and anchors, fittings, conduit, raceway, and scaffolding. See respective tables for these time frames.

For individual heights greater than 25 feet increase manhours 10% for each additional 5 feet or fraction thereof.

OCTAGON AND SPECIAL BOXES, ROUND COVERS, AND RINGS

Concrete Wall Construction

(Concealed Installation)

MANHOURS EACH

Item	Size (Inches)	Manhours for Height to			
		10'	15'	20'	25'
Octagon Type					
Box	4	.50	.55	.59	.62
Hung Ceiling Box	4	.47	.52	.55	.59
Box Extension	4	.29	.32	.34	.36
Round Type					
Plaster Ring	4	.15	.17	.18	.19
Device Cover	4	.15	.17	.18	.19
Blank Cover	4	.15	.17	.18	.19
Cover with K.O.	4	.15	.17	.18	.19
Switch or Receptacle Cover	4	.15	.17	.18	.19
Swivel Hanger Cover	4	.15	.58	.62	.66
Special Boxes					
Concrete Box	4	.34	.37	.40	.42
Concrete Box Plate	4	.15	.17	.18	.19

Concrete Block, Brick, and Hollow Tile Walls

(Concealed Installation)

MANHOURS EACH

Item	Size (Inches)	Manhours for Height to			
		10'	15'	20'	25'
Octagon Type					
Box	4	.53	.58	.62	.66
Hung Ceiling Box	4	.50	.55	.59	.62
Box Extension	4	.30	.33	.35	.37
Round Type					
Plaster Ring	4	.16	.18	.19	.20
Device Cover	4	.16	.18	.19	.20
Blank Cover	4	.16	.18	.19	.20
Cover with K.O.	4	.16	.18	.19	.20
Switch or Receptacle Cover	4	.16	.18	.19	.20
Swivel Hanger Cover	4	.56	.62	.66	.70

Manhours include checking out of job storage, handling, job hauling, layout, and installing items as outlined.

Manhours exclude installing special fasteners and anchors, fittings, conduit, raceway, and scaffolding. See respective tables for these time frames.

For individual heights greater than 25 feet increase above manhours 10% for each additional 5 feet or fraction thereof.

OCTAGON BOXES, ROUND COVERS, AND RINGS
Wall and Ceiling Construction
(Exposed Installation)

MANHOURS EACH

Item	Size (Inches)	Manhours for Height to			
		10'	15'	20'	25'
Octagon Type					
Box	4	.43	.47	.51	.54
Hung Ceiling Box	4	.41	.45	.48	.51
Box Extension	4	.24	.26	.28	.30
Round Type					
Plaster Ring	4	.14	.15	.16	.17
Device Cover	4	.14	.15	.16	.17
Blank Cover	4	.14	.15	.16	.17
Cover with K.O.	4	.14	.15	.16	.17
Switch or Receptacle Cover	4	.14	.15	.16	.17
Swivel Hanger Cover	4	.46	.51	.54	.57

Wood Frame Construction
(Concealed Installation)

MANHOURS EACH

Item	Size (Inches)	Manhours for Height to			
		10'	15'	20'	25'
Octagon Type					
Box	4	.41	.45	.48	.51
Hunt Ceiling Box	4	.38	.42	.45	.47
Box Extension	4	.23	.25	.27	.29
Round Type					
Plaster Ring	4	.13	.14	.15	.16
Device Cover	4	.13	.14	.15	.16
Blank Cover	4	.13	.14	.15	.16
Cover with K.O.	4	.13	.14	.15	.16
Switch or Receptacle Cover	4	.13	.14	.15	.16
Swivel Hanger Cover	4	.43	47	.51	.54

Manhours include checking out of job storage, handling, job hauling, layout, and installing items as outlined.

Manhours exclude installing special fasteners and anchors, fittings, conduit, raceway, and scaffolding. See respective tables for these time frames.

For individual heights greater than 25 feet increase manhours 10% for each additional 5 feet or fraction thereof.

OCTAGON BOXES, ROUND COVERS, AND RINGS
Suspended Ceiling Construction
(Concealed Installation)

MANHOURS EACH

Item	Size (Inches)	Manhours for Height to			
		10'	15'	20'	25'
Octagon Type					
Box	4	.50	.55	.59	.62
Hung Ceiling Box	4	.47	.52	.55	.59
Box Extension	4	.28	.31	.33	.35
Round Type					
Plaster Ring	4	.16	.18	.19	.20
Device Cover	4	.16	.18	.19	.20
Blank Cover	4	.16	.18	.19	.20
Cover with K.O.	4	.16	.18	.19	.20
Switch or Receptacle Cover	4	.16	.18	.19	.20
Swivel Hanger Cover	4	.53	.58	.62	.66

Manhours include checking out of job storage, handling, job hauling, layout, and installing items as outlined.

Manhours exclude installing special fasteners and anchors, fittings, conduit, raceway, and scaffolding. See respective tables for these time frames.

For individual heights greater than 25 feet increase manhours 10% for each additional 5 feet or fraction thereof.

KNOCKOUT TYPE OUTLET GANG BOXES AND COVERS

Flat Concrete Slab Construction

(Concealed Installation)

MANHOURS EACH

Item	Manhours for Height to			
	10'	15'	20'	25'
Square Boxes				
2-Gang	0.48	0.53	0.56	0.60
3-Gang	0.85	0.94	1.00	1.06
4-Gang	0.85	0.94	1.00	1.06
5-Gang	1.10	1.21	1.29	1.37
6-Gang	1.38	1.52	1.62	1.72
7-Gang	1.82	2.00	2.14	2.27
8-Gang	2.31	2.54	2.72	2.88
Device Covers				
2-Gang	0.23	0.25	0.27	0.29
3-Gang	0.32	0.35	0.38	0.40
4-Gang	0.32	0.35	0.38	0.40
5-Gang	0.32	0.35	0.38	0.40
6-Gang	0.48	0.53	0.56	0.60
7-Gang	0.48	0.53	0.56	0.60
8-Gang	0.48	0.53	0.56	0.60
Surface Mounting Cover Device				
2-Gang	0.32	0.35	0.38	0.40
3-Gang	0.32	0.35	0.38	0.40
4-Gang	0.47	0.52	0.55	0.59
5-Gang	0.47	0.52	0.55	0.59
6-Gang	0.66	0.73	0.78	0.82
Blank Covers				
2-Gang	0.23	0.25	0.27	0.29
3-Gang	0.23	0.25	0.27	0.29
4-Gang	0.32	0.35	0.38	0.40
5-Gang	0.32	0.35	0.38	0.40
6-Gang	0.48	0.53	0.56	0.60
7-Gang	0.48	0.53	0.56	0.60
8-Gang	0.48	0.53	0.56	0.60

Manhours include checking out of job storage, handling, job hauling, layout, and installing items as outlined.

Manhours exclude installing special fasteners and anchors, fittings, conduit, raceway, and scaffolding. See respective tables for these time frames.

For individual heights greater than 25 feet increase manhours 10% for each additional 5 feet or fraction thereof.

KNOCKOUT TYPE OUTLET GANG BOXES AND COVERS

Concrete Pan Form Construction (Concealed Installation)

MANHOURS EACH

Item	Manhours for Height to			
	10'	15'	20'	25'
Square Boxes				
2-Gang	0.60	0.66	0.71	0.75
3-Gang	1.06	1.17	1.25	1.32
4-Gang	1.06	1.17	1.25	1.32
5-Gang	1.37	1.51	1.61	1.71
6-Gang	1.70	1.87	2.00	2.12
7-Gang	2.25	2.48	2.65	2.81
8-Gang	2.85	3.14	3.35	3.56
Device Covers				
2-Gang	0.26	0.29	0.31	0.32
3-Gang	0.37	0.41	0.44	0.46
4-Gang	0.37	0.41	0.44	0.46
5-Gang	0.37	0.41	0.44	0.46
6-Gang	0.55	0.61	0.65	0.69
7-Gang	0.55	0.61	0.65	0.69
8-Gang	0.55	0.61	0.65	0.69
Surface Mounting Cover Device				
2-Gang	0.37	0.41	0.44	0.46
3-Gang	0.37	0.41	0.44	0.46
4-Gang	0.54	0.59	0.64	0.67
5-Gang	0.54	0.59	0.64	0.67
6-Gang	0.76	0.84	0.89	0.95
Blank Covers				
2-Gang	0.26	0.29	0.31	0.32
3-Gang	0.26	0.29	0.31	0.32
4-Gang	0.37	0.41	0.44	0.46
5-Gang	0.37	0.41	0.44	0.46
6-Gang	0.48	0.53	0.56	0.60
7-Gang	0.48	0.53	0.56	0.60
8-Gang	0.48	0.53	0.56	0.60

Manhours include checking out of job storage, handling, job hauling, layout, and installing items as outlined.

Manhours exclude installing special fasteners and anchors, fittings, conduit, raceway, and scaffolding. See respective tables for these time frames.

For individual heights greater than 25 feet increase manhours 10% for each additional 5 feet or fraction thereof.

KNOCKOUT TYPE OUTLET GANG BOXES AND COVERS

Concrete Wall Construction

(Concealed Installation)

MANHOURS EACH

Item	Manhours for Height to			
	10'	15'	20'	25'
Square Boxes				
2-Gang	0.50	0.55	0.59	0.62
3-Gang	0.88	0.97	1.04	1.10
4-Gang	0.88	0.97	1.04	1.10
5-Gang	1.14	1.25	1.34	1.42
6-Gang	1.44	1.58	1.69	1.80
7-Gang	1.89	2.08	2.22	2.36
8-Gang	2.40	2.64	2.82	2.99
Device Covers				
2-Gang	0.23	0.25	0.27	0.29
3-Gang	0.32	0.35	0.38	0.40
4-Gang	0.32	0.35	0.38	0.40
5-Gang	0.32	0.35	0.38	0.40
6-Gang	0.48	0.53	0.56	0.60
7-Gang	0.48	0.53	0.56	0.60
8-Gang	0.48	0.53	0.56	0.60
Surface Mounting Cover Device				
2-Gang	0.32	0.35	0.38	0.40
3-Gang	0.32	0.35	0.38	0.40
4-Gang	0.47	0.52	0.55	0.59
5-Gang	0.47	0.52	0.55	0.59
6-Gang	0.66	0.73	0.78	0.82
Blank Covers				
2-Gang	0.23	0.25	0.27	0.29
3-Gang	0.23	0.25	0.27	0.29
4-Gang	0.32	0.35	0.38	0.40
5-Gang	0.32	0.35	0.38	0.40
6-Gang	0.48	0.53	0.56	0.60
7-Gang	0.48	0.53	0.56	0.60
8-Gang	0.48	0.53	0.56	0.60

Manhours include checking out of job storage, handling, job hauling, layout, and installing items as outlined.

Manhours exclude installing special fasteners and anchors, fittings, conduit, raceway, and scaffolding. See respective tables for these time frames.

For individual heights greater than 25 feet increase manhours 10% for each additional 5 feet or fraction thereof.

KNOCKOUT TYPE OUTLET GANG BOXES AND COVERS

Concrete Block, Brick, and Hollow Tile Walls

(Concealed Installation)

MANHOURS EACH

Item	Manhours for Height to			
	10'	15'	20'	25'
Square Boxes				
2-Gang	0.53	0.58	0.62	0.66
3-Gang	0.94	1.03	1.11	1.17
4-Gang	0.94	1.03	1.11	1.17
5-Gang	1.21	1.33	1.42	1.51
6-Gang	1.52	1.67	1.79	1.90
7-Gang	2.00	2.20	2.35	2.50
8-Gang	2.50	2.75	2.94	3.12
Device Covers				
2-Gang	0.24	0.26	0.28	0.30
3-Gang	0.34	0.37	0.40	0.42
4-Gang	0.34	0.37	0.40	0.42
5-Gang	0.34	0.37	0.40	0.42
6-Gang	0.50	0.55	0.59	0.62
7-Gang	0.50	0.55	0.59	0.62
8-Gang	0.50	0.55	0.59	0.62
Surface Mounting Cover Device				
2-Gang	0.34	0.37	0.40	0.42
3-Gang	0.34	0.37	0.40	0.42
4-Gang	0.49	0.54	0.58	0.61
5-Gang	0.49	0.54	0.58	0.61
6-Gang	0.69	0.76	0.81	0.86
Blank Covers				
2-Gang	0.24	0.26	0.28	0.30
3-Gang	0.24	0.26	0.28	0.30
4-Gang	0.34	0.37	0.40	0.42
5-Gang	0.34	0.37	0.40	0.42
6-Gang	0.50	0.55	0.59	0.62
7-Gang	0.50	0.55	0.59	0.62
8-Gang	0.50	0.55	0.59	0.62

Manhours include checking out of job storage, handling, job hauling, layout, and installing items as outlined.

Manhours exclude installing special fasteners and anchors, fittings, conduit, raceway, and scaffolding. See respective tables for these time frames.

For individual heights greater than 25 feet increase manhours 10% for each additional 5 feet or fraction thereof.

KNOCKOUT TYPE OUTLET GANG BOXES AND COVERS

Wall and Ceiling Construction

(Exposed Installation)

MANHOURS EACH

Item	Manhours for Height to			
	10'	15'	20'	25'
Square Boxes				
2-Gang	0.43	0.47	0.51	0.54
3-Gang	0.77	0.85	0.91	0.96
4-Gang	0.77	0.85	0.91	0.96
5-Gang	0.99	1.09	1.17	1.28
6-Gang	1.24	1.36	1.46	1.55
7-Gang	1.64	1.80	1.93	2.05
8-Gang	2.08	2.29	2.45	2.59
Device Covers				
2-Gang	0.22	0.24	0.26	0.28
3-Gang	0.30	0.33	0.35	0.37
4-Gang	0.30	0.33	0.35	0.37
5-Gang	0.30	0.33	0.35	0.37
6-Gang	0.46	0.51	0.54	0.57
7-Gang	0.46	0.51	0.54	0.57
8-Gang	0.46	0.51	0.54	0.57
Surface Mounting Cover Device				
2-Gang	0.30	0.33	0.35	0.37
3-Gang	0.30	0.33	0.35	0.37
4-Gang	0.45	0.50	0.53	0.56
5-Gang	0.45	0.50	0.53	0.56
6-Gang	0.63	0.69	0.74	0.79
Blank Covers				
2-Gang	0.22	0.24	0.26	0.28
3-Gang	0.22	0.24	0.26	0.28
4-Gang	0.30	0.33	0.35	0.37
5-Gang	0.30	0.33	0.35	0.37
6-Gang	0.48	0.53	0.56	0.60
7-Gang	0.48	0.53	0.56	0.60
8-Gang	0.48	0.53	0.56	0.60

Manhours include checking out of job storage, handling, job hauling, layout, and installing items as outlined.

Manhours exclude installing special fasteners and anchors, fittings, conduit, raceway, and scaffolding. See respective tables for these time frames.

For individual heights greater than 25 feet increase manhours 10% for each additional 5 feet or fraction thereof.

KNOCKOUT TYPE OUTLET GANG BOXES AND COVERS

Wood Frame Construction

(Concealed Installation)

MANHOURS EACH

Item	Manhours for Height to			
	10′	15′	20′	25′
Square Boxes				
2-Gang	0.41	0.45	0.48	0.51
3-Gang	0.72	0.79	0.85	0.90
4-Gang	0.72	0.79	0.85	0.90
5-Gang	0.94	1.03	1.11	1.17
6-Gang	1.17	1.29	1.38	1.46
7-Gang	1.55	1.71	1.28	1.93
8-Gang	1.96	2.16	2.31	2.45
Device Covers				
2-Gang	0.21	0.23	0.25	0.26
3-Gang	0.29	0.32	0.34	0.36
4-Gang	0.29	0.32	0.34	0.36
5-Gang	0.29	0.32	0.34	0.36
6-Gang	0.43	0.47	0.51	0.54
7-Gang	0.43	0.47	0.51	0.54
8-Gang	0.43	0.47	0.51	0.54
Surface Mounting Cover Device				
2-Gang	0.29	0.32	0.34	0.36
3-Gang	0.29	0.32	0.34	0.36
4-Gang	0.42	0.46	0.49	0.52
5-Gang	0.42	0.46	0.49	0.52
6-Gang	0.59	0.65	0.69	0.74
Blank Covers				
2-Gang	0.21	0.23	0.25	0.26
3-Gang	0.21	0.23	0.25	0.26
4-Gang	0.29	0.32	0.34	0.36
5-Gang	0.29	0.32	0.34	0.36
6-Gang	0.43	0.47	0.51	0.54
7-Gang	0.43	0.47	0.51	0.54
8-Gang	0.43	0.47	0.51	0.54

Manhours include checking out of job storage, handling, job hauling, layout, and installing items as outlined.

Manhours exclude installing special fasteners and anchors, fittings, conduit, raceway, and scaffolding. See respective tables for these time frames.

For individual heights greater than 25 feet increase manhours 10% for each additional 5 feet or fraction thereof.

KNOCKOUT TYPE OUTLET GANG BOXES AND COVERS

Suspended Ceiling Construction

(Concealed Installation)

MANHOURS EACH

Item	Manhours for Height to			
	10'	15'	20'	25'
Square Boxes				
2-Gang	0.50	0.55	0.59	0.62
3-Gang	0.88	0.97	1.04	1.10
4-Gang	0.88	0.97	1.04	1.10
5-Gang	1.14	1.25	1.34	1.42
6-Gang	1.44	1.58	1.69	1.80
7-Gang	1.89	2.08	2.22	2.36
8-Gang	2.40	2.64	2.82	2.99
Device Covers				
2-Gang	0.23	0.25	0.27	0.29
3-Gang	0.32	0.35	0.38	0.40
4-Gang	0.32	0.35	0.38	0.40
5-Gang	0.32	0.35	0.38	0.40
6-Gang	0.48	0.53	0.56	0.60
7-Gang	0.48	0.53	0.56	0.60
8-Gang	0.48	0.53	0.56	0.60
Surface Mounting Cover Device				
2-Gang	0.32	0.35	0.38	0.40
3-Gang	0.32	0.35	0.38	0.40
4-Gang	0.47	0.52	0.55	0.59
5-Gang	0.47	0.52	0.55	0.59
6-Gang	0.66	0.73	0.78	0.82
Blank Covers				
2-Gang	0.23	0.25	0.27	0.29
3-Gang	0.23	0.25	0.27	0.29
4-Gang	0.32	0.35	0.38	0.40
5-Gang	0.32	0.35	0.38	0.40
6-Gang	0.48	0.53	0.56	0.60
7-Gang	0.48	0.53	0.56	0.60
8-Gang	0.48	0.53	0.56	0.60

Manhours include checking out of job storage, handling, job hauling, layout, and installing items as outlined.

Manhours exclude installing special fasteners and anchors, fittings, conduit, raceway, and scaffolding. See respective tables for these time frames.

For individual heights greater than 25 feet increase manhours 10% for each additional 5 feet or fraction thereof.

HANDY AND SECTIONAL SWITCH BOXES
Concrete Wall Construction
(Concealed Installation)

MANHOURS EACH

Item	Manhours for Height to			
	10'	15'	20'	25'
Handy Type				
Box	0.41	0.45	0.48	0.51
Box Extension	0.27	0.30	0.32	0.34
Box Cover	0.25	0.17	0.18	0.19
Sectional Switch Box				
Box with K.O.	0.41	0.45	0.48	0.51
Box with Clamps	0.31	0.34	0.36	0.39

Concrete Block, Brick, and Hollow Tile Walls
(Concealed Installation)

MANHOURS EACH

Item	Manhours for Height to			
	10'	15'	20'	25'
Handy Type				
Box	0.45	0.50	0.53	0.56
Box Extension	0.30	0.33	0.35	0.37
Box Cover	0.17	0.18	0.19	0.20
Sectional Switch Box				
Box with K.O.	0.45	0.50	0.53	0.56
Box with Clamps	0.34	0.38	0.40	0.43

Manhours include checking out of job storage, handling, job hauling, layout, and installing items as outlined.

Manhours exclude installing special fasteners and anchors, fittings, conduit, raceway, and scaffolding. See respective tables for these time frames.

For individual heights greater than 25 feet increase manhours 10% for each additional 5 feet or fraction thereof.

HANDY AND SECTIONAL SWITCH BOXES
Wall Construction
(Exposed Installation)

MANHOURS EACH

Item	Manhours for Height to			
	10'	15'	20'	25'
Handy Type				
Box	0.37	0.41	0.44	0.47
Box Extension	0.24	0.27	0.29	0.30
Box Cover	0.14	0.15	0.16	0.17
Sectional Switch Box				
Box with K.O.	0.37	0.41	0.43	0.46
Box with Clamps	0.28	0.31	0.33	0.35

Wood Frame Construction
(Concealed Installation)

MANHOURS EACH

Item	Manhours for Height to			
	10'	15'	20'	25'
Handy Type				
Box	0.35	0.38	0.41	0.43
Box Extension	0.23	0.25	0.27	0.29
Box Cover	0.13	0.14	0.15	0.16
Sectional Switch Box				
Box with K.O.	0.35	0.38	0.41	0.43
Box with Clamps	0.26	0.29	0.31	0.33

Manhours include checking out of job storage, handling, job hauling, layout, and installing items as outlined.

Manhours exclude installing special fasteners and anchors, fittings, conduit, raceway, and scaffolding. See respective tables for these time frames.

For individual heights greater than 25 feet increase manhours 10% for each additional 5 feet or fraction thereof.

S.C. PULL AND HINGE COVERED BOXES
Flat Concrete Slab Construction
(Concealed Installation)

MANHOURS EACH

Item	Manhours for Height to			
	10′	15′	20′	25′
S.C. Pull Boxes				
4″ x 4″ x 4″	1.11	1.22	1.31	1.38
4″ x 6″ x 4″	1.11	1.22	1.31	1.38
6″ x 6″ x 4″	1.11	1.22	1.31	1.38
6″ x 8″ x 4″	1.22	1.34	1.44	1.52
8″ x 8″ x 4″	1.22	1.34	1.44	1.52
8″ x 12″ x 4″	1.46	1.61	1.72	1.82
12″ x 12″ x 4″	1.71	1.88	2.01	2.13
12″ x 24″ x 4″	3.34	3.67	3.93	4.17
12″ x 12″ x 6″	2.42	2.66	2.85	3.02
12″ x 18″ x 6″	2.66	2.93	3.13	3.32
12″ x 24″ x 6″	3.34	3.67	3.93	4.17
18″ x 24″ x 6″	3.96	4.36	4.67	4.96
18″ x 30″ x 6″	4.51	4.96	5.31	5.63
24″ x 36″ x 6″	5.72	6.29	6.73	7.14
Hinge Covered Boxes				
6″ x 6″ x 4″	1.10	1.21	1.29	1.37
6″ x 8″ x 4″	1.22	1.34	1.44	1.52
8″ x 8″ x 4″	1.22	1.34	1.44	1.52
8″ x 12″ x 4″	1.46	1.61	1.72	1.82
12″ x 12″ x 4″	1.71	1.88	2.01	2.13
12″ x 18″ x 4″	2.66	2.93	3.13	3.32
12″ x 24″ x 4″	2.90	3.19	3.41	3.62
12″ x 12″ x 6″	2.42	2.66	2.85	3.02
12″ x 18″ x 6″	2.66	2.93	3.13	3.32
12″ x 24″ x 6″	3.12	3.43	3.67	3.89
18″ x 24″ x 6″	3.96	4.36	4.67	4.96

Manhours include checking out of job storage, handling, job hauling, layout, and installing items as outlined.

Manhours exclude installing special fasteners and anchors, fittings, conduit, raceway, and scaffolding. See respective tables for these time frames.

For individual heights greater than 25 feet increase manhours 10% for each additional 5 feet or fraction thereof.

S.C. PULL AND HINGE COVERED BOXES
Concrete Pan Form Construction
(Concealed Installation)

MANHOURS EACH

Item	Manhours for Height to			
	10'	15'	20'	25'
S.C. Pull Boxes				
4" x 4" x 4"	1.39	1.53	1.64	1.73
4" x 6" x 4"	1.39	1.53	1.64	1.73
6" x 6" x 4"	1.39	1.53	1.64	1.73
6" x 8" x 4"	1.53	1.68	1.80	1.91
8" x 8" x 4"	1.53	1.68	1.80	1.91
8" x 12" x 4"	1.83	2.01	2.15	2.28
12" x 12" x 4"	2.14	2.35	2.52	2.67
12" x 24" x 4"	4.18	4.60	4.92	5.21
12" x 12" x 6"	3.03	3.33	3.57	3.78
12" x 18" x 6"	3.33	3.66	3.92	4.15
12" x 24" x 6"	4.18	4.60	4.92	5.21
18" x 24" x 6"	4.95	5.45	5.83	6.18
18" x 30" x 6"	5.64	6.20	6.64	7.04
24" x 36" x 6"	7.15	7.87	8.42	8.34
Hinge Covered Boxes				
6" x 6" x 4"	1.38	1.52	1.62	1.72
6" x 8" x 4"	1.53	1.68	1.80	1.91
8" x 8" x 4"	1.53	1.68	1.80	1.91
8" x 12" x 4"	1.83	2.01	2.15	2.28
12" x 12" x 4"	2.14	2.35	2.52	2.67
12" x 18" x 4"	3.33	3.66	3.92	4.15
12" x 24" x 4"	3.63	3.99	4.27	4.53
12" x 12" x 6"	3.06	3.37	3.60	3.82
12" x 18" x 6"	3.33	3.66	3.92	4.15
12" x 24" x 6"	3.90	4.29	4.59	4.87
18" x 24" x 6"	4.95	5.45	5.83	6.18

Manhours include checking out of job storage, handling, job hauling, layout, and installing items as outlined.

Manhours exclude installing special fasteners and anchors, fittings, conduit, raceway, and scaffolding. See respective tables for these time frames.

For individual heights greater than 25 feet increase manhours 10% for each additional 5 feet or fraction thereof.

S.C. PULL AND HINGE COVERED BOXES
Concrete Wall Construction
(Concealed Installation)

MANHOURS EACH

Item	Manhours for Height to			
	10'	15'	20'	25'
S.C. Pull Boxes				
4" x 4" x 4"	1.15	1.27	1.35	1.43
4" x 6" x 4"	1.15	1.27	1.35	1.43
6" x 6" x 4"	1.15	1.27	1.35	1.43
6" x 8" x 4"	1.27	1.40	1.50	1.59
8" x 8" x 4"	1.27	1.40	1.50	1.59
8" x 12" x 4"	1.52	1.67	1.79	1.90
12" x 12" x 4"	1.78	1.96	2.10	2.22
12" x 24" x 4"	3.47	3.82	4.08	4.33
12" x 12" x 6"	2.52	2.77	2.97	3.14
12" x 18" x 6"	2.77	3.05	3.26	3.46
12" x 24" x 6"	3.47	3.82	4.08	4.33
18" x 24" x 6"	4.12	4.53	4.85	5.14
18" x 30" x 6"	4.69	5.16	5.52	5.85
24" x 36" x 6"	5.99	6.59	7.05	7 47
Hinge Covered Boxes				
6" x 6" x 4"	1.14	1.25	1.34	1.42
6" x 8" x 4"	1.27	1.40	1.50	1.59
8" x 8" x 4"	1.27	1.40	1.50	1.59
8" x 12" x 4"	1.52	1.67	1.79	1.90
12" x 12" x 4"	1.78	1.96	2.10	2.22
12" x 18" x 4"	2.77	3.05	3.26	3.46
12" x 24" x 4"	3.02	3.32	3.55	3.77
12" x 12" x 6"	2.52	3.82	4.08	4.33
12" x 18" x 6"	2.77	3.05	3.26	3.46
12" x 24" x 6"	3.24	3.56	3.81	4.04
18" x 24" x 6"	4.12	4.53	4.85	5.14

Manhours include checking out of job storage, handling, job hauling, layout, and installing items as outlined.

Manhours exclude installing special fasteners and anchors, fittings, conduit, raceway, and scaffolding. See respective tables for these time frames.

For individual heights greater than 25 feet increase manhours 10% for each additional 5 feet or fraction thereof.

S.C. PULL AND HINGE COVERED BOXES
Concrete Block, Brick, and Hollow Tile Walls
(Concealed Installation)

MANHOURS EACH

Item	Manhours for Height to			
	10'	15'	20'	25'
S.C. Pull Boxes				
4" x 4" x 4"	1.22	1.34	1.44	1.52
4" x 6" x 4"	1.22	1.34	1.44	1.52
6" x 6" x 4"	1.22	1.34	1.44	1.52
6" x 8" x 4"	1.34	1.47	1.58	1.67
8" x 8" x 4"	1.34	1.47	1.58	1.67
8" x 12" x 4"	1.61	1.77	1.89	2.01
12" x 12" x 4"	1.88	2.07	2.21	2.35
12" x 24" x 4"	3.67	4.04	4.32	4.58
12" x 12" x 6"	2.66	2.93	3.13	3.32
12" x 18" x 6"	2.93	3.22	3.45	3.66
12" x 24" x 6"	3.67	4.04	4.32	4.58
18" x 24" x 6"	4.36	4.80	5.13	5.44
18" x 30" x 6"	4.96	5.46	5.84	6.19
24" x 36" x 6"	6.29	6.92	7.40	7.85
Hinge Covered Boxes				
6" x 6" x 4"	1.21	1.33	1.42	1.51
6" x 8" x 4"	1.34	1.47	1.58	1.67
8" x 8" x 4"	1.34	1.47	1.58	1.67
8" x 12" x 4"	1.61	1.77	1.89	2.01
12" x 12" x 4"	1.88	2.07	2.21	2.35
12" x 18" x 4"	2.93	3.22	3.45	3.66
12" x 24" x 4"	3.19	3.51	3.75	3.98
12" x 12" x 6"	2.66	2.93	3.13	3.32
12" x 18" x 6"	2.93	3.22	3.45	3.66
12" x 24" x 6"	3.43	3.77	4.04	4.28
18" x 24" x 6"	4.36	4.80	5.13	5.44

Manhours include checking out of job storage, handling, job hauling, layout, and installing items as outlined.

Manhours exclude installing special fasteners and anchors, fittings, conduit, raceway, and scaffolding. See respective tables for these time frames.

For individual heights greater than 25 feet increase manhours 10% for each additional 5 feet or fraction thereof.

S.C. PULL AND HINGE COVERED BOXES
Wall and Ceiling Construction
(Exposed Installation)

MANHOURS EACH

Item	Manhours for Height to			
	10'	15'	20'	25'
S.C. Pull Boxes				
4" x 4" x 4"	1.00	1.10	1.18	1.25
4" x 6" x 4"	1.00	1.10	1.18	1.25
6" x 6" x 4"	1.00	1.10	1.18	1.25
6" x 8" x 4"	1.10	1.21	1.29	1.37
8" x 8" x 4"	1.10	1.21	1.29	1.37
8" x 12" x 4"	1.31	1.44	1.54	1.63
12" x 12" x 4"	1.54	1.69	1.81	1.92
12" x 24" x 4"	3.01	3.31	3.54	3.76
12" x 12" x 6"	2.18	2.40	2.57	2.72
12" x 18" x 6"	2.39	2.63	2.81	2.98
12" x 24" x 6"	3.01	3.31	3.54	3.76
18" x 24" x 6"	3.56	3.92	4.19	4.44
18" x 30" x 6"	4.06	4.47	4.78	5.07
24" x 36" x 6"	5.15	5.67	6.06	6.43
Hinge Covered Boxes				
6" x 6" x 4"	0.99	1.09	1.17	1.24
6" x 8" x 4"	1.10	1.21	1.29	1.37
8" x 8" x 4"	1.10	1.21	1.29	1.37
8" x 12" x 4"	1.31	1.44	1.54	1.63
12" x 12" x 4"	1.54	1.69	1.81	1.92
12" x 18" x 4"	2.39	2.63	2.81	2.98
12" x 24" x 4"	2.61	2.87	3.07	3.26
12" x 12" x 6"	2.18	2.40	2.57	2.72
12" x 18" x 6"	2.39	2.63	2.81	2.98
12" x 24" x 6"	2.81	3.09	3.31	3.51
18" x 24" x 6"	3.56	3.92	4.19	4.44

Manhours include checking out of job storage, handling, job hauling, layout, and installing items as outlined.

Manhours exclude installing special fasteners and anchors, fittings, conduit, raceway, and scaffolding. See respective tables for these time frames.

For individual heights greater than 25 feet increase manhours 10% for each additional 5 feet or fraction thereof.

S.C. PULL AND HINGE COVERED BOXES

Wood Frame Construction

(Concealed Installation)

MANHOURS EACH

Item	Manhours for Height to			
	10'	15'	20'	25'
S.C. Pull Boxes				
4" x 4" x 4"	0.94	1.03	1.11	1.17
4" x 6" x 4"	0.94	1.03	1.11	1.17
6" x 6" x 4"	0.94	1.03	1.11	1.17
6" x 8" x 4"	1.04	1.14	1.22	1.30
8" x 8" x 4"	1.04	1.14	1.22	1.30
8" x 12" x 4"	1.24	1.36	1.46	1.55
12" x 12" x 4"	1.45	1.60	1.71	1.81
12" x 24" x 4"	2.84	3.12	3.34	3.54
12" x 12" x 6"	2.06	2.27	2.42	2.57
12" x 18" x 6"	2.26	2.49	2.66	2.82
12" x 24" x 6"	2.84	3.12	3.34	3.54
18" x 24" x 6"	3.37	3.71	3.97	4.20
18" x 30" x 6"	3.83	4.21	4.51	4.78
24" x 36" x 6"	4.86	5.35	5.72	6.06
Hinge Covered Boxes				
6" x 6" x 4"	0.94	1.03	1.11	1.17
6" x 8" x 4"	1.04	1.14	1.22	1.30
8" x 8" x 4"	1.04	1.14	1.22	1.30
8" x 12" x 4"	1.25	1.38	1.47	1.56
12" x 12" x 4"	1.45	1.60	1.71	1.81
12" x 18" x 4"	2.26	2.49	2.66	2.82
12" x 24" x 4"	2.47	2.72	2.91	3.08
12" x 12" x 6"	2.06	2.27	2.42	2.57
12" x 18" x 6"	2.26	2.49	2.66	2.82
12" x 24" x 6"	2.65	2.92	3.12	3.31
18" x 24" x 6"	3.37	3.71	3.97	4.20

Manhours include checking out of job storage, handling, job hauling, layout, and installing items as outlined.

Manhours exclude installing special fasteners and anchors, fittings, conduit, raceway, and scaffolding. See respective tables for these time frames.

For individual heights greater than 25 feet increase manhours 10% for each additional 5 feet or fraction thereof.

S.C. PULL AND HINGE COVERED BOXES
Suspended Ceiling Construction
(Concealed Installation)

MANHOURS EACH

Item	Manhours for Height to			
	10'	15'	20'	25'
S.C. Pull Boxes				
4" x 4" x 4"	1.15	1.27	1.35	1.43
4" x 6" x 4"	1.15	1.27	1.35	1.43
6" x 6" x 4"	1.15	1.27	1.35	1.43
6" x 8" x 4"	1.27	1.40	1.50	1.59
8" x 8" x 4"	1.27	1.40	1.50	1.59
8" x 12" x 4"	1.52	1.67	1.79	1.90
12" x 12" x 4"	1.78	1.96	2.10	2.22
12" x 24" x 4"	3.47	3.82	4.08	4.33
12" x 12" x 6"	2.52	2.77	2.97	3.14
12" x 18" x 6"	2.77	3.05	3.26	3.46
12" x 24" x 6"	3.47	3.82	4.08	4.33
18" x 24" x 6"	4.12	4.53	4.85	5.14
18" x 30" x 6"	4.69	5.16	5.52	5.85
24" x 36" x 6"	5.99	6.59	7.05	7.47
Hinge Covered Boxes				
6" x 6" x 4"	1.14	1.25	1.34	1.42
6" x 8" x 4"	1.27	1.40	1.50	1.59
8" x 8" x 4"	1.27	1.40	1.50	1.59
8" x 12" x 4"	1.52	1.67	1.79	1.90
12" x 12" x 4"	1.78	1.96	2.10	2.22
12" x 18" x 4"	2.77	3.05	3.26	3.46
12" x 24" x 4"	3.02	3.32	3.66	3.77
12" x 12" x 6"	2.52	3.82	4.08	4.33
12" x 18" x 6"	2.77	3.05	3.26	3.46
12" x 24" x 6"	3.24	3.56	3.81	4.04
18" x 24" x 6"	4.12	4.53	4.85	5.14

Manhours include checking out of job storage, handling, job hauling, layout, and installing items as outlined.

Manhours exclude installing special fasteners and anchors, fittings, conduit, raceway, and scaffolding. See respective tables for these time frames.

For individual heights greater than 25 feet increase manhours 10% for each additional 5 feet or fraction thereof.

RIGID STEEL OR ALUMINUM
Oval Conduit Bodies

MANHOURS EACH

Types	Size (Inches)	Manhours for Height to							
		Threaded				Threadless			
		10′	15′	20′	25′	10′	15′	20′	25′
E	1/2	.48	.53	.56	.60	.38	.42	.45	.47
	3/4	.61	.67	.72	.76	.49	.54	.58	.61
	1	.72	.79	.85	.90	.58	.64	.68	.72
	1-1/4	1.03	1.13	1.21	1.29	.82	.90	.96	1.02
	1-1/2	1.35	1.49	1.59	1.68	1.08	1.19	1.27	1.35
	2	2.04	2.24	2.40	2.55	1.63	1.79	1.92	2.30
C, LB, LL & LR	1/2	.76	.84	.89	.95	.61	.67	.72	.76
	3/4	.91	1.00	1.07	1.14	.73	.80	.86	.91
	1	1.16	1.28	1.37	1.45	.93	1.02	1.09	1.16
	1-1/4	1.79	1.97	2.11	2.23	1.43	1.57	1.68	1.78
	1-1/2	2.20	2.42	2.59	2.74	1.76	1.94	2.07	2.20
	2	2.75	3.03	3.24	3.43	2.20	2.42	2.59	2.74
	2-1/2	3.92	4.31	4.61	4.89	3.14	3.45	3.70	3.92
	3	5.57	6.13	6.56	6.95	3.46	4.91	5.25	5.56
	3-1/2	7.15	7.87	8.42	8.92	5.72	6.29	6.73	7.14
	4	8.25	9.08	9.71	10.29	6.60	7.26	7.77	8.23
LRL	1/2	.87	.96	1.02	1.09	.70	.77	.82	.87
	3/4	1.02	1.12	1.20	1.27	.82	.90	.97	1.02
	1	1.28	1.41	1.51	1.60	1.02	1.12	1.20	1.27
	1-1/4	1.79	1.97	2.11	2.23	1.43	1.57	1.68	1.78
	1-1/2	2.20	2.42	2.59	2.74	1.76	1.94	2.07	2.20
	2	2.75	3.03	3.24	3.43	2.20	2.42	2.59	2.74
	2-1/2	4.47	4.92	5.26	5.58	3.58	3.94	4.21	4.47
	3	5.57	6.13	6.56	6.95	4.46	4.91	5.25	5.56
	3-1/2	7.15	7.87	8.42	8.92	5.72	6.29	6.73	7.14
	4	8.25	9.08	9.71	10.29	6.60	7.26	7.77	8.23

Manhours include checking body out of job storage, handling, job hauling, and complete installation including cover plates and gaskets, and cover duplex receptacle or cover lamp holder where required. Manhours exclude installation of other fittings, conduit, and scaffolding. See respective tables for these time frames.

RIGID STEEL OR ALUMINUM
Oval Conduit Bodies

MANHOURS EACH

Types	Size (Inches)	Manhours for Height to							
		Threaded				Threadless			
		10'	15'	20'	25'	10'	15'	20'	25'
T & TB	1/2	.99	1.09	1.17	1.24	.79	.87	.93	.99
	3/4	1.19	1.31	1.40	1.48	.95	1.05	1.12	1.19
	1	1.46	1.61	1.72	1.82	1.17	1.29	1.38	1.46
	1-1/4	2.54	2.79	2.99	3.17	2.03	2.23	2.39	2.53
	1-1/2	2.75	3.03	3.24	3.43	2.20	2.42	2.59	2.74
	2	3.44	3.78	4.05	4.29	2.75	3.03	3.24	3.43
	2-1/2	5.29	5.82	6.23	6.60	4.23	4.65	4.98	5.28
	3	6.94	7.63	8.17	8.66	5.55	6.11	6.53	6.92
	3-1/2	8.66	9.53	10.19	10.80	6.93	7.62	8.16	8.65
	4	6.96	10.66	11.41	12.09	7.75	8.53	9.12	9.67
X	1/2	1.41	1.55	1.66	1.76	1.13	1.24	1.33	1.41
	3/4	1.65	1.82	1.94	2.06	1.32	1.45	1.55	1.65
	1	1.93	2.12	2.27	2.41	1.54	1.69	1.81	1.92
	1-1/4	3.09	3.40	3.64	3.86	2.47	2.72	2.91	3.08
	1-1/2	3.44	3.78	4.05	4.29	2.75	3.03	3.24	3.43
	2	3.99	4.39	4.70	4.98	3.19	3.51	3.75	3.98
	2-1/2	6.12	6.73	7.20	7.64	4.90	5.39	5.77	6.11
	3	8.04	8.84	9.46	10.03	6.43	7.07	7.57	8.02
	3-1/2	9.76	10.74	11.40	12.18	7.81	8.59	9.19	9.74
	4	11.07	12.18	13.03	13.81	8.86	9.75	10.43	11.05

Manhours include checking body out of job storage, handling, job hauling, and complete installation including cover plates and gaskets where required.

Manhours exclude installation of other fittings, conduit, and scaffolding. See respective tables for these time frames.

ELECTRIC METALLIC TUBING

Oval Conduit Bodies

MANHOURS EACH

Types	Size (Inches)	Manhours for Height to							
		Set Screw				Threadless			
		10'	15'	20'	25'	10'	15'	20'	25'
C-LB-LL-LR	1/2	.53	.58	.62	.66	.42	.46	.49	.52
	3/4	.67	.74	.79	.84	.54	.59	.64	.67
	1	.87	.96	1.02	1.09	.70	.77	.82	.87
	1-1/4	1.10	1.21	1.29	1.37	.88	.97	1.04	1.10
	1-1/2	1.46	1.61	1.72	1.82	1.17	1.29	1.38	1.46
	2	1.87	2.06	2.20	2.33	1.50	1.65	1.77	1.87
T	1/2	.74	.81	.87	.92	.59	.65	.69	.74
	3/4	.91	1.00	1.07	1.14	.73	.80	.86	.91
	1	1.16	1.28	1.37	1.45	.93	1.02	1.09	1.16
	1-1/4	1.46	1.61	1.72	1.82	1.17	1.29	1.38	1.46
	1-1/2	1.68	1.85	1.98	2.10	1.34	1.47	1.58	1.67
	2	2.16	2.38	2.54	2.69	1.73	1.90	2.04	2.16
X	1/2	1.14	1.25	1.34	1.42	.91	1.00	1.07	1.14
	3/4	1.33	1.46	1.57	1.66	1.06	1.17	1.25	1.32
	1	1.56	1.72	1.84	1.95	1.25	1.38	1.47	1.56
E	1-1/4	.74	.81	.87	.92	.59	.65	.69	.74
	1-1/2	.94	1.03	1.11	1.17	.75	.83	.88	.94
	2	1.58	1.74	1.86	1.97	1.26	1.39	1.48	1.57
Entrance Caps	1-1/4	.74	.81	.87	.92	.59	.65	.69	.74
	1-1/2	.94	1.03	1.11	1.17	.75	.83	.88	.94
	2	1.58	1.74	1.86	1.97	1.26	1.39	1.48	1.57

Manhours include checking body out of job storage, handling, job hauling, and complete installation including plates and gaskets where required.

Manhours exclude installation of other fittings, electric metallic tubing, and scaffolding. See respective tables for these time frames.

PLASTIC CONDUIT
Oval Conduit Bodies

MANHOURS EACH

Types	Size (Inches)	Cement Joint For Height to			
		10′	15′	20′	25′
C-LB-LL-LR	1/2	.54	.59	.64	.67
	3/4	.67	.74	.79	.84
	1	.87	.96	1.02	1.09
	1-1/4	1.25	1.38	1.47	1.56
	1-1/2	1.52	1.67	1.79	1.90
	2	2.20	2.42	2.59	2.74
T	1/2	.74	.81	.87	.92
	3/4	.88	.97	1.04	1.10
	1	1.16	1.28	1.37	1.45
	1-1/4	1.52	1.67	1.79	1.90
	1-1/2	1.79	1.97	2.11	2.23
	2	2.54	2.79	2.99	3.17
X	1/2	1.18	1.30	1.39	1.47
	3/4	1.40	1.54	1.65	1.75
	1	1.63	1.79	1.92	2.03
	1-1/4	2.56	2.82	3.01	3.19
	1-1/2	2.86	3.15	3.37	3.57
	2	3.47	3.82	4.08	4.33

Manhours include checking body out of job storage, handling, job hauling, and complete installation including plates and gaskets where required.

Manhours exclude installation of other fittings, plastic conduit, and scaffolding. See respective tables for these time frames.

FS TYPE THREADED CAST DEVICE BOXES
Flat Concrete Slab Construction
(Concealed Installation)

MANHOURS EACH

Type	Number of Gangs	Number of Hubs	Hub Size (Inches)	Manhours for Height to			
				10'	15'	20'	25'
FS	1	1	1/2	.81	.89	.95	1.01
FS	1	1	3/4	.88	.97	1.04	1.10
FS	1	1	1	1.05	1.16	1.24	1.31
FS	2	1	1/2	.81	.89	.95	1.01
FS	2	1	3/4	.88	.97	1.04	1.10
FS	2	1	1	1.02	1.12	1.20	1.27
FS	3	1	1/2	.92	1.01	1.08	1.15
FS	3	1	3/4	.99	1.09	1.17	1.24
FS	3	1	1	1.13	1.24	1.33	1.41
FS	4	1	1/2	1.07	1.18	1.26	1.33
FS	4	1	3/4	1.19	1.31	1.40	1.48
FS	4	1	1	1.35	1.49	1.59	1.68
FSC	1	2	1/2	.97	1.07	1.14	1.21
FSC	1	2	3/4	1.13	1.24	1.33	1.41
FSC	1	2	1	1.32	1.45	1.55	1.65
FSC	2	2	1/2	.97	1.07	1.14	1.21
FSC	2	2	3/4	1.13	1.24	1.33	1.41
FSC	2	2	1	1.32	1.45	1.55	1.65
FSC	3	2	1/2	1.08	1.19	1.27	1.35
FSC	3	2	3/4	1.24	1.36	1.46	1.55
FSC	3	2	1	1.43	1.57	1.68	1.78
FSD	2	2	1/2	.97	1.07	1.14	1.21
FSD	2	2	3/4	1.13	1.24	1.33	1.41
FSD	2	2	1	1.32	1.45	1.55	1.65
FSD	3	4	1/2	1.45	1.60	1.71	1.81
FSD	3	4	3/4	1.61	1.77	1.89	2.01
FSD	3	4	1	1.87	2.06	2.20	2.33
FSD	4	5	1/2	2.15	2.37	2.53	2.68
FSD	4	5	3/4	2.42	2.66	2.85	3.02
FSD	4	5	1	2.97	3.27	3.50	3.71

Manhours include checking out of job storage, handling, job hauling, layout, and installing of items as outlined.

Manhours exclude installation of special fasteners and anchors, fittings, conduit, and scaffolding. See respective tables for these time frames.

For heights above 25 feet increase manhours 10% for each additional 5 feet or fraction thereof.

FS TYPE THREADED CAST DEVICE BOXES
Flat Concrete Slab Construction
(Concealed Installation)

MANHOURS EACH

Type	Number of Gangs	Number of Hubs	Hub Size (Inches)	Manhours for Height to			
				10′	15′	20′	25′
FSL	1	2	1/2	.97	1.07	1.14	1.21
FSL	1	2	3/4	1.13	1.24	1.33	1.41
FSL	1	2	1	1.32	1.45	1.55	1.65
FSR	1	2	1/2	.97	1.07	1.14	1.21
FSR	1	2	3/4	1.13	1.24	1.33	1.41
FSR	1	2	1	1.32	1.45	1.55	1.65
FSS	1	2	1/2	.97	1.07	1.14	1.21
FSS	1	2	3/4	1.13	1.24	1.33	1.41
FSS	1	2	1	1.32	1.45	1.55	1.65
FSS	2	3	1/2	1.11	1.22	1.31	1.38
FSS	2	3	3/4	1.32	1.45	1.55	1.65
FSS	2	3	1	1.51	1.66	1.78	1.88
FSCC	1	3	1/2	1.11	1.22	1.31	1.38
FSCC	1	3	3/4	1.32	1.45	1.55	1.65
FSCC	1	3	1	1.51	1.66	1.78	1.88
FST	1	3	1/2	1.11	1.22	1.31	1.38
FST	1	3	3/4	1.32	1.45	1.55	1.65
FST	1	3	1	1.51	1.66	1.78	1.88
FSX	1	4	1/2	1.32	1.45	1.55	1.65
FSX	1	4	3/4	1.60	1.76	1.88	2.00
FSX	1	4	1	1.87	2.06	2.20	2.33
FSCD	1	4	1/2	1.32	1.45	1.55	1.65
FSCD	1	4	3/4	1.60	1.76	1.88	2.00
FSCD	1	4	1	1.87	2.06	2.20	2.33

Manhours include checking out of job storage, handling, job hauling, layout, and installing of items as outlined.

Manhours exclude installation of special fasteners and anchors, fittings, conduit, and scaffolding. See respective tables for these time frames.

For heights above 25 feet increase manhours 10% for each additional 5 feet or fraction thereof.

FS TYPE THREADED CAST DEVICE BOXES
Concrete Pan Form Construction
(Concealed Installation)

MANHOURS EACH

Type	Number of Gangs	Number of Hubs	Hub Size (Inches)	Manhours for Height to			
				10'	15'	20'	25'
FS	1	1	1/2	.93	1.02	1.09	1.16
FS	1	1	3/4	1.01	1.11	1.19	1.26
FS	1	1	1	1.21	1.33	1.42	1.51
FS	2	1	1/2	.93	1.02	1.09	1.16
FS	2	1	3/4	1.01	1.11	1.19	1.26
FS	2	1	1	1.17	1.29	1.38	1.46
FS	3	1	1/2	1.06	1.17	1.25	1.32
FS	3	1	3/4	1.14	1.25	1.34	1.42
FS	3	1	1	1.30	1.43	1.53	1.62
FS	4	1	1/2	1.23	1.35	1.45	1.53
FS	4	1	3/4	1.37	1.51	1.61	1.71
FS	4	1	1	1.55	1.71	1.82	1.93
FSC	1	2	1/2	1.12	1.23	1.32	1.40
FSC	1	2	3/4	1.30	1.43	1.53	1.62
FSC	1	2	1	1.52	1.67	1.79	1.90
FSC	2	2	1/2	1.12	1.23	1.32	1.40
FSC	2	2	3/4	1.30	1.43	1.53	1.62
FSC	2	2	1	1.52	1.67	1.79	1.90
FSC	3	2	1/2	1.24	1.36	1.46	1.55
FSC	3	2	3/4	1.43	1.57	1.68	1.78
FSC	3	2	1	1.64	1.80	1.93	2.05
FSD	2	2	1/2	1.12	1.23	1.32	1.40
FSD	2	2	3/4	1.30	1.43	1.53	1.62
FSD	2	2	1	1.52	1.67	1.79	1.90
FSD	3	4	1/2	1.67	1.84	1.97	2.08
FSD	3	4	3/4	1.85	2.04	2.18	2.31
FSD	3	4	1	2.15	2.37	2.53	2.68
FSD	4	5	1/2	2.47	2.72	2.91	3.08
FSD	4	5	3/4	2.78	3.06	3.27	3.47
FSD	4	5	1	3.42	3.76	4.03	4.27

Manhours include checking out of job storage, handling, job hauling, layout, and installing of items as outlined.

Manhours exclude installation of special fasteners and anchors, fittings, conduit, and scaffolding. See respective tables for these time frames.

For heights above 25 feet increase manhours 10% for each additional 5 feet or fraction thereof.

FS TYPE THREADED CAST DEVICE BOXES
Concrete Pan Form Construction
(Concealed Installation)

MANHOURS EACH

Type	Number of Gangs	Number of Hubs	Hub Size (Inches)	Manhours for Height to			
				10'	15'	20'	25'
FSL	1	2	1/2	1.12	1.23	1.32	1.40
FSL	1	2	3/4	1.30	1.43	1.53	1.62
FSL	1	2	1	1.52	1.67	1.79	1.90
FSR	1	2	1/2	1.12	1.23	1.32	1.40
FSR	1	2	3/4	1.30	1.43	1.53	1.62
FSR	1	2	1	1.52	1.67	1.79	1.90
FSS	1	2	1/2	1.12	1.23	1.32	1.40
FSS	1	2	3/4	1.30	1.43	1.53	1.62
FSS	1	2	1	1.52	1.67	1.79	1.90
FSS	2	3	1/2	1.28	1.41	1.51	1.60
FSS	2	3	3/4	1.52	1.67	1.79	1.90
FSS	2	3	1	1.74	1.91	2.05	2.17
FSCC	1	3	1/2	1.28	1.41	1.51	1.60
FSCC	1	3	3/4	1.52	1.67	1.79	1.90
FSCC	1	3	1	1.74	1.91	2.05	2.17
FST	1	3	1/2	1.28	1.41	1.51	1.60
FST	1	3	3/4	1.52	1.67	1.79	1.90
FST	1	3	1	1.74	1.91	2.05	2.17
FSX	1	4	1/2	1.52	1.67	1.79	1.90
FSX	1	4	3/4	1.84	2.02	2.17	2.30
FSX	1	4	1	2.15	2.37	2.53	2.68
FSCD	1	4	1/2	1.52	1.67	1.79	1.90
FSCD	1	4	3/4	1.84	2.02	2.17	2.30
FSCD	1	4	1	2.15	2.37	2.53	2.68

Manhours include checking out of job storage, handling, job hauling, layout, and installing of items as outlined.

Manhours exclude installation of special fasteners and anchors, fittings, conduit, and scaffolding. See respective tables for these time frames.

For heights above 25 feet increase manhours 10% for each additional 5 feet or fraction thereof.

FS TYPE THREADED CAST DEVICE BOXES
Concrete Wall Construction
(Concealed Installation)

MANHOURS EACH

Type	Number of Gangs	Number of Hubs	Hub Size (Inches)	Manhours for Height to			
				10′	15′	20′	25′
FS	1	1	1/2	.81	.89	.95	1.01
FS	1	1	3/4	.88	.97	1.04	1.10
FS	1	1	1	1.05	1.16	1.24	1.31
FS	2	1	1/2	.81	.89	.95	1.01
FS	2	1	3/4	.88	.97	1.04	1.10
FS	2	1	1	1.02	1.12	1.20	1.27
FS	3	1	1/2	.92	1.01	1.08	1.15
FS	3	1	3/4	.99	1.09	1.17	1.24
FS	3	1	1	1.13	1.24	1.33	1.41
FS	4	1	1/2	1.07	1.18	1.26	1.33
FS	4	1	3/4	1.19	1.31	1.40	1.48
FS	4	1	1	1.35	1.49	1.59	1.68
FSC	1	2	1/2	.97	1.07	1.14	1.21
FSC	1	2	3/4	1.13	1.24	1.33	1.41
FSC	1	2	1	1.32	1.45	1.55	1.65
FSC	2	2	1/2	.97	1.07	1.14	1.21
FSC	2	2	3/4	1.13	1.24	1.33	1.41
FSC	2	2	1	1.32	1.45	1.55	1.65
FSC	3	2	1/2	1.08	1.19	1.27	1.35
FSC	3	2	3/4	1.24	1.36	1.46	1.55
FSC	3	2	1	1.43	1.57	1.68	1.78
FSD	2	2	1/2	.97	1.07	1.14	1.21
FSD	2	2	3/4	1.13	1.24	1.33	1.41
FSD	2	2	1	1.32	1.45	1.55	1.65
FSD	3	4	1/2	1.45	1.60	1.71	1.81
FSD	3	4	3/4	1.61	1.77	1.89	2.01
FSD	3	4	1	1.87	2.06	2.20	2.33
FSD	4	5	1/2	2.15	2.37	2.53	2.68
FSD	4	5	3/4	2.42	2.66	2.85	3.02
FSD	4	5	1	2.97	3.27	3.50	3.71

Manhours include checking out of job storage, handling, job hauling, layout, and installing of items as outlined.

Manhours exclude installation of special fasteners and anchors, fittings, conduit, and scaffolding. See respective tables for these time frames.

For heights above 25 feet increase manhours 10% for each additional 5 feet or fraction thereof.

FS TYPE THREADED CAST DEVICE BOXES
Concrete Wall Construction
(Concealed Installation)

MANHOURS EACH

Type	Number of Gangs	Number of Hubs	Hub Size (Inches)	Manhours for Height to			
				10'	15'	20'	25'
FSL	1	2	1/2	.97	1.07	1.14	1.21
FSL	1	2	3/4	1.13	1.24	1.33	1.41
FSL	1	2	1	1.32	1.45	1.55	1.65
FSR	1	2	1/2	.97	1.07	1.14	1.21
FSR	1	2	3/4	1.13	1.24	1.33	1.41
FSR	1	2	1	1.32	1.45	1.55	1.65
FSS	1	2	1/2	.97	1.07	1.14	1.21
FSS	1	2	3/4	1.13	1.24	1.33	1.41
FSS	1	2	1	1.32	1.45	1.55	1.65
FSS	2	3	1/2	1.11	1.22	1.31	1.38
FSS	2	3	3/4	1.32	1.45	1.55	1.65
FSS	2	3	1	1.51	1.66	1.78	1.88
FSCC	1	3	1/2	1.11	1.22	1.31	1.38
FSCC	1	3	3/4	1.32	1.45	1.55	1.65
FSCC	1	3	1	1.51	1.66	1.78	1.88
FST	1	3	1/2	1.11	1.22	1.31	1.38
FST	1	3	3/4	1.32	1.45	1.55	1.65
FST	1	3	1	1.51	1.66	1.78	1.88
FSX	1	4	1/2	1.32	1.45	1.55	1.65
FSX	1	4	3/4	1.60	1.76	1.88	2.00
FSX	1	4	1	1.87	2.06	2.20	2.33
FSCD	1	4	1/2	1.32	1.45	1.55	1.65
FSCD	1	4	3/4	1.60	1.76	1.88	2.00
FSCD	1	4	1	1.87	2.06	2.20	2.33

Manhours include checking out of job storage, handling, job hauling, layout, and installing of items as outlined.

Manhours exclude installation of special fasteners and anchors, fittings, conduit, and scaffolding. See respective tables for these time frames.

For heights above 25 feet increase manhours 10% for each additional 5 feet or fraction thereof.

FS TYPE THREADED CAST DEVICE BOXES
Concrete Block, Brick, and Hollow Tile Walls
(Concealed Installation)

MANHOURS EACH

Type	Number of Gangs	Number of Hubs	Hub Size (Inches)	Manhours for Height to			
				10'	15'	20'	25'
FS	1	1	1/2	.85	.94	1.00	1.06
FS	1	1	3/4	.92	1.01	1.08	1.15
FS	1	1	1	1.10	1.21	1.29	1.37
FS	2	1	1/2	.85	.94	1.00	1.06
FS	2	1	3/4	.92	1.01	1.08	1.15
FS	2	1	1	1.07	1.18	1.26	1.33
FS	3	1	1/2	.97	1.07	1.14	1.21
FS	3	1	3/4	1.04	1.14	1.22	1.30
FS	3	1	1	1.19	1.31	1.40	1.48
FS	4	1	1/2	1.12	1.23	1.32	1.40
FS	4	1	3/4	1.25	1.38	1.47	1.56
FS	4	1	1	1.42	1.56	1.67	1.77
FSC	1	2	1/2	1.02	1.12	1.20	1.27
FSC	1	2	3/4	1.19	1.31	1.40	1.48
FSC	1	2	1	1.39	1.53	1.64	1.73
FSC	2	2	1/2	1.02	1.12	1.20	1.27
FSC	2	2	3/4	1.19	1.31	1.40	1.48
FSC	2	2	1	1.39	1.53	1.64	1.73
FSC	3	2	1/2	1.13	1.24	1.33	1.41
FSC	3	2	3/4	1.30	1.43	1.53	1.62
FSC	3	2	1	1.50	1.65	1.77	1.87
FSD	2	2	1/2	1.02	1.12	1.20	1.27
FSD	2	2	3/4	1.19	1.31	1.40	1.48
FSD	2	2	1	1.39	1.53	1.64	1.73
FSD	3	4	1/2	1.52	1.67	1.79	1.90
FSD	3	4	3/4	1.69	1.86	1.99	2.11
FSD	3	4	1	1.96	2.16	2.31	2.45
FSD	4	5	1/2	2.26	2.49	2.66	2.82
FSD	4	5	3/4	2.54	2.79	2.99	3.17
FSD	4	5	1	3.12	3.43	3.67	3.89

Manhours include checking out of job storage, handling, job hauling, layout, and installing of items as outlined.

Manhours exclude installation of special fasteners and anchors, fittings, conduit, and scaffolding. See respective tables for these time frames.

For heights above 25 feet increase manhours 10% for each additional 5 feet or fraction thereof.

FS TYPE THREADED CAST DEVICE BOXES
Concrete Block, Brick, and Hollow Tile Walls
(Concealed Installation)

MANHOURS EACH

Type	Number of Gangs	Number of Hubs	Hub Size (Inches)	Manhours for Height to			
				10'	15'	20'	25'
FSL	1	2	1/2	1.02	1.12	1.20	1.27
FSL	1	2	3/4	1.19	1.31	1.40	1.48
FSL	1	2	1	1.39	1.53	1.64	1.73
FSR	1	2	1/2	1.02	1.12	1.20	1.27
FSR	1	2	3/4	1.19	1.31	1.40	1.48
FSR	1	2	1	1.39	1.53	1.64	1.73
FSS	1	2	1/2	1.02	1.12	1.20	1.27
FSS	1	2	3/4	1.19	1.31	1.40	1.48
FSS	1	2	1	1.39	1.53	1.64	1.73
FSS	2	3	1/2	1.17	1.29	1.38	1.46
FSS	2	3	3/4	1.39	1.53	1.64	1.73
FSS	2	3	1	1.59	1.75	1.87	1.98
FSCC	1	3	1/2	1.17	1.29	1.38	1.46
FSCC	1	3	3/4	1.39	1.53	1.64	1.73
FSCC	1	3	1	1.59	1.75	1.87	1.98
FST	1	3	1/2	1.17	1.29	1.38	1.46
FST	1	3	3/4	1.39	1.53	1.64	1.73
FST	1	3	1	1.59	1.75	1.87	1.98
FSX	1	4	1/2	1.39	1.53	1.64	1.73
FSX	1	4	3/4	1.68	1.85	1.98	2.10
FSX	1	4	1	1.96	2.16	2.31	2.45
FSCD	1	4	1/2	1.39	1.53	1.64	1.73
FSCD	1	4	3/4	1.68	1.85	1.98	2.10
FSCD	1	4	1	1.96	2.16	2.31	2.45

Manhours include checking out of job storage, handling, job hauling, layout, and installing of items as outlined.

Manhours exclude installation of special fasteners and anchors, fittings, conduit, and scaffolding. See respective tables for these time frames.

For heights above 25 feet increase manhours 10% for each additional 5 feet or fraction thereof.

FS TYPE THREADED CAST DEVICE BOXES
Wall and Ceiling Construction
(Exposed Installation)

MANHOURS EACH

Type	Number of Gangs	Number of Hubs	Hub Size (Inches)	Manhours for Height to			
				10'	15'	20'	25'
FS	1	1	1/2	.77	.85	.91	.96
FS	1	1	3/4	.84	.92	.99	1.05
FS	1	1	1	1.00	1.10	1.18	1.25
FS	2	1	1/2	.77	.85	.91	.96
FS	2	1	3/4	.84	.92	.99	1.05
FS	2	1	1	.97	1.07	1.15	1.21
FS	3	1	1/2	.87	.96	1.02	1.09
FS	3	1	3/4	.94	1.03	1.11	1.17
FS	3	1	1	1.07	1.18	1.26	1.33
FS	4	1	1/2	1.02	1.12	1.20	1.27
FS	4	1	3/4	1.13	1.24	1.33	1.41
FS	4	1	1	1.28	1.41	1.51	1.60
FSC	1	2	1/2	.92	1.01	1.08	1.15
FSC	1	2	3/4	1.07	1.18	1.26	1.33
FSC	1	2	1	1.25	1.38	1.47	1.60
FSC	2	2	1/2	.92	1.01	1.08	1.15
FSC	2	2	3/4	1.07	1.18	1.26	1.33
FSC	2	2	1	1.25	1.38	1.47	1.60
FSC	3	2	1/2	1.03	1.13	1.21	1.29
FSC	3	2	3/4	1.18	1.30	1.39	1.47
FSC	3	2	1	1.36	1.50	1.60	1.70
FSD	2	2	1/2	.92	1.01	1.08	1.15
FSD	2	2	3/4	1.07	1.18	1.26	1.33
FSD	2	2	1	1.25	1.38	1.47	1.60
FSD	3	4	1/2	1.38	1.52	1.62	1.72
FSD	3	4	3/4	1.53	1.68	1.80	1.91
FSD	3	4	1	1.78	1.96	2.10	2.22
FSD	4	5	1/2	2.04	2.24	2.40	2.55
FSD	4	5	3/4	2.30	2.53	2.71	2.87
FSD	4	5	1	2.82	3.10	3.32	3.52

Manhours include checking out of job storage, handling, job hauling, layout, and installing of items as outlined.

Manhours exclude installation of special fasteners and anchors, fittings, conduit, and scaffolding. See respective tables for these time frames.

For heights above 25 feet increase manhours 10% for each additional 5 feet or fraction thereof.

FS TYPE THREADED CAST DEVICE BOXES
Wall and Ceiling Construction
(Concealed Installation)

MANHOURS EACH

Type	Number of Gangs	Number of Hubs	Hub Size (Inches)	Manhours for Height to			
				10'	15'	20'	25'
FSL	1	2	1/2	.92	1.01	1.08	1.15
FSL	1	2	3/4	1.07	1.18	1.26	1.33
FSL	1	2	1	1.25	1.38	1.47	1.60
FSR	1	2	1/2	.92	1.01	1.08	1.15
FSR	1	2	3/4	1.07	1.18	1.26	1.33
FSR	1	2	1	1.25	1.38	1.47	1.60
FSS	1	2	1/2	.92	1.01	1.08	1.15
FSS	1	2	3/4	1.07	1.18	1.26	1.33
FSS	1	2	1	1.25	1.38	1.47	1.60
FSS	2	3	1/2	1.05	1.16	1.24	1.31
FSS	2	3	3/4	1.25	1.38	1.47	1.60
FSS	2	3	1	1.43	1.57	1.68	1.78
FSCC	1	3	1/2	1.05	1.16	1.24	1.31
FSCC	1	3	3/4	1.25	1.38	1.47	1.60
FSCC	1	3	1	1.43	1.57	1.68	1.78
FST	1	3	1/2	1.05	1.16	1.24	1.31
FST	1	3	3/4	1.25	1.38	1 47	1.60
FST	1	3	1	1.43	1.57	1.68	1.78
FSX	1	4	1/2	1.25	1.38	1.47	1.60
FSX	1	4	3/4	1.52	1.67	1.79	1.90
FSX	1	4	1	1.78	1.96	2.10	2.22
FSCD	1	4	1/2	1.25	1.38	1.47	1.60
FSCD	1	4	3/4	1.52	1.67	1.79	1.90
FSCD	1	4	1	1.78	1.96	2.10	2.22

Manhours include checking out of job storage, handling, job hauling, layout, and installing of items as outlined.

Manhours exclude installation of special fasteners and anchors, fittings, conduit, and scaffolding. See respective tables for these time frames.

For heights above 25 feet increase manhours 10% for each additional 5 feet or fraction thereof.

FS TYPE THREADED CAST DEVICE BOXES

Wood Frame Construction

(Concealed Installation)

MANHOURS EACH

Type	Number of Gangs	Number of Hubs	Hub Size (Inches)	Manhours for Height to			
				10'	15'	20'	25'
FS	1	1	1/2	.73	.80	.86	.91
FS	1	1	3/4	.79	.87	.93	.99
FS	1	1	1	.95	1.05	1.12	1.19
FS	2	1	1/2	.73	.80	.86	.91
FS	2	1	3/4	.79	.87	.93	.99
FS	2	1	1	.92	1.01	1.08	1.15
FS	3	1	1/2	.83	.91	.98	1.04
FS	3	1	3/4	.89	.98	1.05	1.11
FS	3	1	1	1.02	1.12	1.20	1.27
FS	4	1	1/2	.96	1.06	1.13	1.20
FS	4	1	3/4	1.07	1.18	1.26	1.33
FS	4	1	1	1.22	1.34	1.45	1.52
FSC	1	2	1/2	.87	.96	1.02	1.09
FSC	1	2	3/4	1.02	1.12	1.20	1.27
FSC	1	2	1	1.19	1.31	1.39	1.47
FSC	2	2	1/2	.87	.96	1.02	1.09
FSC	2	2	3/4	1.02	1.12	1.20	1.27
FSC	2	2	1	1.19	1.31	1.39	1.47
FSC	3	2	1/2	.97	1.07	1.14	1.21
FSC	3	2	3/4	1.12	1.23	1.32	1.40
FSC	3	2	1	1.29	1.42	1.52	1.61
FSD	2	2	1/2	.87	.96	1.02	1.09
FSD	2	2	3/4	1.02	1.12	1.20	1.27
FSD	2	2	1	1.19	1.31	1.39	1.47
FSD	3	4	1/2	1.31	1.44	1.54	1.63
FSD	3	4	3/4	1.45	1.60	1.71	1.81
FSD	3	4	1	1.68	1.85	1.98	2.10
FSD	4	5	1/2	1.94	2.13	2.28	2.42
FSD	4	5	3/4	2.18	2.40	2.57	2.72
FSD	4	5	1	2.67	2.94	3.14	3.33

Manhours include checking out of job storage, handling, job hauling, layout, and installing of items as outlined.

Manhours exclude installation of special fasteners and anchors, fittings, conduit, and scaffolding. See respective tables for these time frames.

For heights above 25 feet increase manhours 10% for each additional 5 feet or fraction thereof.

FS TYPE THREADED CAST DEVICE BOXES
Wood Frame Construction
(Concealed Installation)

MANHOURS EACH

Type	Number of Gangs	Number of Hubs	Hub Size (Inches)	Manhours for Height to			
				10'	15'	20'	25'
FSL	1	2	1/2	.87	.96	1.02	1.09
FSL	1	2	3/4	1.02	1.12	1.20	1.27
FSL	1	2	1	1.19	1.31	1.39	1.47
FSR	1	2	1/2	.87	.96	1.02	1.09
FSR	1	2	3/4	1.02	1.12	1.20	1.27
FSR	1	2	1	1.19	1.31	1.39	1.47
FSS	1	2	1/2	.87	.96	1.02	1.09
FSS	1	2	3/4	1.02	1.12	1.20	1.27
FSS	1	2	1	1.19	1.31	1.38	1.47
FSS	2	3	1/2	1.00	1.10	1.18	1.25
FSS	2	3	3/4	1.19	1.31	1.38	1.47
FSS	2	3	1	1.36	1.50	1.60	1.70
FSCC	1	3	1/2	1.00	1.10	1.18	1.25
FSCC	1	3	3/4	1.19	1.31	1.38	1.47
FSCC	1	3	1	1.36	1.50	1.60	1.70
FST	1	3	1/2	1.00	1.10	1.18	1.25
FST	1	3	3/4	1.19	1.31	1.38	1.47
FST	1	3	1	1.36	1.50	1.60	1.70
FSX	1	4	1/2	1.19	1.31	1.38	1.47
FSX	1	4	3/4	1.44	1.58	1.69	1.80
FSX	1	4	1	1.68	1.85	1.98	2.10
FSCD	1	4	1/2	1.19	1.31	1.38	1.47
FSCD	1	4	3/4	1.44	1.58	1.69	1.80
FSCD	1	4	1	1.68	1.85	1.98	2.10

Manhours include checking out of job storage, handling, job hauling, layout, and installing of items as outlined.

Manhours exclude installation of special fasteners and anchors, fittings, conduit, and scaffolding. See respective tables for these time frames.

For heights above 25 feet increase manhours 10% for each additional 5 feet or fraction thereof.

FS TYPE THREADED CAST DEVICE BOXES
Suspended Ceiling Construction
(Concealed Installation)

MANHOURS EACH

Type	Number of Gangs	Number of Hubs	Hub Size (Inches)	Manhours for Height to			
				10'	15'	20'	25'
FS	1	1	1/2	.81	.89	.95	1.01
FS	1	1	3/4	.88	.97	1.04	1.10
FS	1	1	1	1.05	1.16	1.24	1.31
FS	2	1	1/2	.81	.89	.95	1.01
FS	2	1	3/4	.88	.97	1.04	1.10
FS	2	1	1	1.02	1.12	1.20	1.27
FS	3	1	1/2	.92	1.01	1.08	1.15
FS	3	1	3/4	.99	1.09	1.17	1.24
FS	3	1	1	1.13	1.24	1.33	1.41
FS	4	1	1/2	1.07	1.18	1.26	1.33
FS	4	1	3/4	1.19	1.31	1.40	1.48
FS	4	1	1	1.35	1.49	1.59	1.68
FSC	1	2	1/2	.97	1.07	1.14	1.21
FSC	1	2	3/4	1.13	1.24	1.33	1.41
FSC	1	2	1	1.32	1.45	1.55	1.65
FSC	2	2	1/2	.97	1.07	1.14	1.21
FSC	2	2	3/4	1.13	1.24	1.33	1.41
FSC	2	2	1	1.32	1.45	1.55	1.65
FSC	3	2	1/2	1.08	1.19	1.27	1.35
FSC	3	2	3/4	1.24	1.36	1.46	1.55
FSC	3	2	1	1.43	1.57	1.68	1.78
FSD	2	2	1/2	.97	1.07	1.14	1.21
FSD	2	2	3/4	1.13	1.24	1.33	1.41
FSD	2	2	1	1.32	1.45	1.55	1.65
FSD	3	4	1/2	1.45	1.60	1.71	1.81
FSD	3	4	3/4	1.61	1.77	1.89	2.01
FSD	3	4	1	1.87	2.06	2.20	2.33
FSD	4	5	1/2	2.15	2.37	2.53	2.68
FSD	4	5	3/4	2.42	2.66	2.85	3.02
FSD	4	5	1	2.97	3.27	3.50	3.71

Manhours include checking out of job storage, handling, job hauling, layout, and installing of items as outlined.

Manhours exclude installation of special fasteners and anchors, fittings, conduit, and scaffolding. See respective tables for these time frames.

For heights above 25 feet increase manhours 10% for each additional 5 feet or fraction thereof.

FS TYPE THREADED CAST DEVICE BOXES
Suspended Ceiling Construction
(Concealed Installation)

MANHOURS EACH

Type	Number of Gangs	Number of Hubs	Hub Size (Inches)	Manhours for Height to			
				10'	15'	20'	25'
FSL	1	2	1/2	.97	1.07	1.14	1.21
FSL	1	2	3/4	1.13	1.24	1.33	1.41
FSL	1	2	1	1.32	1.45	1.55	1.65
FSR	1	2	1/2	.97	1.07	1.14	1.21
FSR	1	2	3/4	1.13	1.24	1.33	1.41
FSR	1	2	1	1.32	1.45	1.55	1.65
FSS	1	2	1/2	.97	1.07	1.14	1.21
FSS	1	2	3/4	1.13	1.24	1.33	1.41
FSS	1	2	1	1.32	1.45	1.55	1.65
FSS	2	3	1/2	1.11	1.22	1.31	1.38
FSS	2	3	3/4	1.32	1.45	1.55	1.65
FSS	2	3	1	1.51	1.66	1.78	1.88
FSCC	1	3	1/2	1.11	1.22	1.31	1.38
FSCC	1	3	3/4	1.32	1.45	1.55	1.65
FSCC	1	3	1	1.51	1.66	1.78	1.88
FST	1	3	1/2	1.11	1.22	1.31	1.38
FST	1	3	3/4	1.32	1.45	1.55	1.65
FST	1	3	1	1.51	1.66	1.78	1.88
FSX	1	4	1/2	1.32	1.45	1.55	1.65
FSX	1	4	3/4	1.60	1.76	1.88	2.00
FSX	1	4	1	1.87	2.06	2.20	2.33
FSCD	1	4	1/2	1.32	1.45	1.55	1.65
FSCD	1	4	3/4	1.60	1.76	1.88	2.00
FSCD	1	4	1	1.87	2.06	2.20	2.33

Manhours include checking out of job storage, handling, job hauling, layout, and installing of items as outlined.

Manhours exclude installation of special fasteners and anchors, fittings, conduit, and scaffolding. See respective tables for these time frames.

For heights above 25 feet increase manhours 10% for each additional 5 feet or fraction thereof.

FS TYPE THREADLESS CAST DEVICE BOXES

Flat Concrete Slab Construction (Concealed Installation)

MANHOURS EACH

Type	Number of Gangs	Number of Hubs	Hub Size (Inches)	Manhours for Height to			
				10'	15'	20'	25'
FS	1	1	1/2	.70	.77	.82	.87
FS	1	1	3/4	.77	.85	.91	.96
FS	1	1	1	.91	1.00	1.07	1.14
FSC	1	2	1/2	.86	.95	1.01	1.07
FSC	1	2	3/4	1.02	1.12	1.20	1.27
FSC	1	2	1	1.21	1.33	1.42	1.51
FSS	1	2	1/2	.86	.95	1.01	1.07
FSS	1	2	3/4	1.02	1.12	1.20	1.27
FSS	1	2	1	1.21	1.33	1.42	1.51
For Electric Metallic Tubing							
FS	1	1	1/2	.56	.62	.66	.70
FS	1	1	3/4	.68	.75	.80	.85
FS	1	1	1	.81	.89	.95	1.01
FSC	1	2	1/2	.73	.80	.86	.91
FSC	1	2	3/4	.84	.92	.99	1.05
FSC	1	2	1	1.02	1.12	1.20	1.27

Concrete Pan Form Construction (Concealed Installation)

MANHOURS EACH

Type	Number of Gangs	Number of Hubs	Hub Size (Inches)	Manhours for Height to			
				10'	15'	20'	25'
FS	1	1	1/2	.81	.89	.95	1.01
FS	1	1	3/4	.89	.98	1.05	1.11
FS	1	1	1	1.05	1.16	1.24	1.31
FSC	1	2	1/2	.99	1.09	1.17	1.24
FSC	1	2	3/4	1.17	1.29	1.38	1.46
FSC	1	2	1	1.39	1.53	1.64	1.73
FSS	1	2	1/2	.99	1.09	1.17	1.24
FSS	1	2	3/4	1.17	1.29	1.38	1.46
FSS	1	2	1	1.39	1.53	1.64	1.73
For Electric Metallic Tubing							
FS	1	1	1/2	.64	.70	.75	.80
FS	1	1	3/4	.78	.86	.91	.97
FS	1	1	1	.93	1.02	1.09	1.16
FSC	1	2	1/2	.84	.92	.99	1.05
FSC	1	2	3/4	.92	1.07	1.14	1.21
FSC	1	2	1	1.17	1.29	1.38	1.46

Manhours include checking out of job storage, handling, job hauling, layout, and installting of rigid conduit or electric metallic tubing devices as outlined.

Manhours exclude installation of special fasteners and anchors, fittings, conduit, and scaffolding. See respective tables for these time frames.

For heights above 25 feet increase manhours 10% for each additional 5 feet or fraction thereof.

FS TYPE THREADLESS CAST DEVICE BOXES

Concrete Wall Construction

(Concealed Installation)

MANHOURS EACH

Type	Number of Gangs	Number of Hubs	Hub Size (Inches)	Manhours for Height to			
				10′	15′	20′	25′
FS	1	1	1/2	.70	.77	.82	.87
FS	1	1	3/4	.77	.85	.91	.96
FS	1	1	1	.91	1.00	1.07	1.14
FSC	1	2	1/2	.86	.95	1.01	1.07
FSC	1	2	3/4	1.02	1.12	1.20	1.27
FSC	1	2	1	1.21	1.33	1.42	1.51
FSS	1	2	1/2	.86	.95	1.01	1.07
FSS	1	2	3/4	1.02	1.12	1.20	1.27
FSS	1	2	1	1.21	1.33	1.42	1.51
For Electric Metallic Tubing							
FS	1	1	1/2	.56	.62	.66	.70
FS	1	1	3/4	.68	.75	.80	.85
FS	1	1	1	.81	.89	.95	1.01
FSC	1	2	1/2	.73	.80	.86	.91
FSC	1	2	3/4	.84	.92	.99	1.05
FSC	1	2	1	1.02	1.12	1.20	1.27

Concrete Block, Brick, and Hollow Tile Walls

(Concealed Installation)

MANHOURS EACH

Type	Number of Gangs	Number of Hubs	Hub Size (Inches)	Manhours for Height to			
				10′	15′	20′	25′
FS	1	1	1/2	.74	.81	.87	.92
FS	1	1	3/4	.81	.89	.95	1.01
FS	1	1	1	.96	1.06	1.13	1.20
FSC	1	2	1/2	.90	.99	1.06	1.12
FSC	1	2	3/4	1.07	1.18	1.26	1.33
FSC	1	2	1	1.27	1.40	1.49	1.58
FSS	1	2	1/2	.90	.99	1.06	1.12
FSS	1	2	3/4	1.07	1.18	1.26	1.33
FSS	1	2	1	1.27	1.40	1.49	1.58
For Electric Metallic Tubing							
FS	1	1	1/2	.59	.65	.69	.74
FS	1	1	3/4	.71	.78	.84	.89
FS	1	1	1	.85	.94	1.00	1.06
FSC	1	2	1/2	.77	.85	.91	.96
FSC	1	2	3/4	.88	.97	1.04	1.10
FSC	1	2	1	1.07	1.18	1.26	1.33

Manhours include checking out of job storage, handling, job hauling, layout, and installing of rigid conduit or electric metallic tubing devices as outlined.

Manhours exclude installation of special fasteners and anchors, fittings, conduit, and scaffolding. See respective tables for these time frames.

For heights above 25 feet increase manhours 10% for each additional 5 feet or fraction thereof.

FS TYPE THREADLESS CAST DEVICE BOXES

Wall and Ceiling Construction

(Exposed Installation)

MANHOURS EACH

Type	Number of Gangs	Number of Hubs	Hub Size (Inches)	Manhours for Height to			
				10'	15'	20'	25'
FS	1	1	1/2	.67	.74	.79	.84
FS	1	1	3/4	.73	.80	.86	.91
FS	1	1	1	.86	.95	1.01	1.07
FSC	1	2	1/2	.82	.90	.97	1.02
FSC	1	2	3/4	.97	1.07	1.14	1.21
FSC	1	2	1	1.15	1.27	1.35	1.43
FSS	1	2	1/2	.82	.90	.97	1.02
FSS	1	2	3/4	.97	1.07	1.14	1.21
FSS	1	2	1	1.15	1.27	1.35	1.43
For Electric Metallic Tubing							
FS	1	1	1/2	.53	.58	.62	.66
FS	1	1	3/4	.65	.72	.77	.81
FS	1	1	1	.77	.85	.91	.96
FSC	1	2	1/2	.69	.76	.81	.86
FSC	1	2	3/4	.80	.88	.94	1.00
FSC	1	2	1	.97	1.07	1.14	1.21

Wood Frame Construction (Concealed Installation)

MANHOURS EACH

Type	Number of Gangs	Number of Hubs	Hub Size (Inches)	Manhours for Height to			
				10'	15'	20'	25'
FS	1	1	1/2	.63	.69	.74	.79
FS	1	1	3/4	.69	.76	.81	.86
FS	1	1	1	.82	.90	.97	1.02
FSC	1	2	1/2	.77	.85	.91	.96
FSC	1	2	3/4	.92	1.01	1.08	1.19
FSC	1	2	1	1.09	1.20	1.28	1.36
FSS	1	2	1/2	.77	.85	.91	.96
FSS	1	2	3/4	.92	1.01	1.08	1.19
FSS	1	2	1	1.09	1.20	1.28	1.36
For Electric Metallic Tubing							
FS	1	1	1/2	.50	.55	.59	.62
FS	1	1	3/4	.61	.67	.72	.76
FS	1	1	1	.73	.80	.86	.91
FSC	1	2	1/2	.66	.73	.78	.82
FSC	1	2	3/4	.76	.84	.89	.95
FSC	1	2	1	.92	10.1	1.08	1.19

Manhours include checking out of job storage, handling, job hauling, layout, and installing of rigid conduit or electric metallic tubing devices as outlined.

Manhours exclude installation of special fasteners and anchors, fittings, conduit, and scaffolding. See respective tables for these time frames.

For heights above 25 feet increase manhours 10% for each additional 5 feet or fraction thereof.

FS TYPE THREADLESS CAST DEVICE BOXES
Suspended Ceiling Construction
(Concealed Installation)

MANHOURS EACH

Type	Number of Gangs	Number of Hubs	Hub Size (Inches)	Manhours for Height to			
				10'	15'	20'	25'
FS	1	1	1/2	.70	.77	.82	.87
FS	1	1	3/4	.77	.85	.91	.96
FS	1	1	1	.91	1.00	1.07	1.14
FSC	1	2	1/2	.86	.95	1.01	1.07
FSC	1	2	3/4	1.02	1.12	1.20	1.27
FSC	1	2	1	1.21	1.33	1.42	1.51
FSS	1	2	1/2	.86	.95	1.01	1.07
FSS	1	2	3/4	1.02	1.12	1.20	1.27
FSS	1	2	1	1.21	1.33	1.42	1.51
For Electric Metallic Tubing							
FS	1	1	1/2	.56	.62	.66	.70
FS	1	1	3/4	.68	.75	.80	.85
FS	1	1	1	.81	.89	.95	1.01
FSC	1	2	1/2	.73	.90	.86	.91
FSC	1	2	3/4	.84	.92	.99	1.05
FSC	1	2	1	1.02	1.12	1.20	1.27

Manhours include checking out of job storage, handling, job hauling, layout, and installing of rigid conduit or electric metallic tubing devices as outlined.

Manhours exclude installation of special fasteners and anchors, fittings, conduit, and scaffolding. See respective tables for these time frames.

For heights above 25 feet increase manhours 10% for each additional 5 feet or fraction thereof.

FS TYPE DEVICE BOX COVERS

MANHOURS EACH

Item Description	Manhours for Height to							
	Horizontal Runs				Vertical Runs			
	10′	15′	20′	25′	10′	15′	20′	25′
Switch Cover — 1-Gang	.10	.11	.12	.13	.11	.12	.13	.14
Switch Cover — 2-Gang	.10	.11	.12	.13	.11	.12	.13	.14
Switch Cover — 3-Gang	.13	.14	.15	.16	.14	.15	.16	.17
Switch Cover — 4-Gang	.13	.14	.15	.16	.14	.15	.16	.17
Switch Cover with Guard — 1-Gang	.22	.24	.26	.27	.23	.25	.27	.29
Switch Cover with Guard — 2-Gang	.22	.24	.26	.27	.23	.25	.27	.29
Switch Cover-Vaptit. Rod — 1-Gang	.25	.28	.29	.31	.26	.29	.31	.32
Switch Cover-Vaptit. Rod — 2-Gang	.25	.28	.29	.31	.26	.29	.31	.32
Switch Cover-Vaptit. Rod — 3-Gang	.37	.41	.44	.46	.39	.43	.46	.49
Switch Cover-Vaptit. Rod — 4-Gang	.37	.41	.44	.46	.39	.43	.46	.49
Switch Cover-Vaptit. RKR — 1-Gang	.25	.28	.29	.31	.26	.29	.31	.32
Switch Cover-Vaptit. RKR — 2-Gang	.25	.28	.29	.31	.26	.29	.31	.32
Single Receptacle Cover — 1-Gang	.10	.11	.12	.13	.11	.12	.13	.14
Single Receptacle Cover — 2-Gang	.10	.11	.12	.13	.11	.12	.13	.14
Single Receptacle Hinge Strap — 1-Gang	.10	.11	.12	.13	.11	.12	.13	.14
Single Receptacle Hinge Cast — 1-Gang	.10	.11	.12	.13	.11	.12	.13	.14
Duplex Receptacle Cover — 1-Gang	.10	.11	.12	.13	.11	.12	.13	.14
Duplex Receptacle Cover — 2-Gang	.10	.11	.12	.13	.11	.12	.13	.14
Duplex Spring Receptacle Cover — 1-Gang	.10	.11	.12	.13	.11	.12	.13	.14
1 Switch 1 Receptacle Cover — 2-Gang	.10	.11	.12	.13	.11	.12	.13	.14
1 Switch 1 Duplex Recept. Cover — 2-Gang	.10	.11	.12	.13	.11	.12	.13	.14
Despard Cover — 1-Gang	.10	.11	.12	.13	.11	.12	.13	.14
Despard Cover — 2-Gang	.10	.11	.12	.13	.11	.12	.13	.14
Stamped Pilot Cover — 1-Gang	.10	.11	.12	.13	.11	.12	.13	.14
Cast Pilot Cover — 1-Gang	.10	.11	.12	.13	.11	.12	.13	.14
Blank Cover — 1-Gang	.10	.11	.12	.13	.11	.12	.13	.14
Blank Cover — 2-Gang	.10	.11	.12	.13	.11	.12	.13	.14
Blank Cover — 3-Gang	.13	.14	.15	.16	.14	.15	.16	.17
Blank Cover — 4-Gang	.13	.14	.15	.16	.14	.15	.16	.17

MOISTURE PROOF THREADED DEVICE BOXES AND COVERS

MANHOURS EACH

Item Description	Manhours for Height to							
	Horizontal Runs				Vertical runs			
	10′	15′	20′	25′	10′	15′	20′	25′
Bell Boxes								
1-Gang — 3-1/2″	.81	.89	.95	1.01	.85	.94	1.00	1.06
1-Gang — 3-3/4″	.88	.97	1.04	1.10	.92	1.01	1.08	1.15
1-Gang — 4-1/2″	.97	1.07	1.14	1.21	1.02	1.12	1.20	1.27
1-Gang — 4-3/4″	1.13	1.24	1.33	1.41	1.19	1.31	1.40	1.48
1-Gang — 5-1/2″	1.20	1.32	1.41	1.50	1.26	1.39	1.48	1.57
1-Gang — 5-3/4″	1.32	1.45	1.55	1.65	1.39	1.53	1.64	1.73
Universal Boxes								
1-Gang — 3-1/2″	.81	.89	.95	1.01	.85	.94	1.00	1.06
1-Gang — 3-3/4″	.88	.97	1.04	1.10	.92	1.01	1.08	1.15
1-Gang — 4-1/2″	.97	1.07	1.14	1.21	1.02	1.12	1.20	1.27
1-Gang — 4-3/4″	1.13	1.24	1.33	1.41	1.19	1.31	1.40	1.48
1-Gang — 5-1/2″	1.20	1.32	1.41	1.50	1.26	1.39	1.48	1.57
1-Gang — 5-3/4″	1.32	1,45	1.55	1.65	1.39	1.53	1.64	1.73
2-Gang — 3-1/2″	.97	1.07	1.14	1.21	1.02	1.12	1.20	1.27
2-Gang — 3-3/4″	1.13	1.24	1.33	1.41	1.19	1.31	1.40	1.48
2-Gang — 4-1/2″	1.20	1.32	1.41	1.50	1.26	1.39	1.48	1.57
2-Gang — 4-3/4″	1.32	1.45	1.55	1.65	1.39	1.53	1.64	1.73
2-Gang — 5-1/2″	1.45	1.60	1.71	1.81	1.52	1.67	1.79	1.90
2-Gang — 5-3/4″	1.61	1.77	1.89	2.01	1.69	1.86	1.99	2.11
Covers								
1-Gang Duplex Device Vert. — 1-Lid	.13	.14	.15	.16	.14	.15	.16	.17
1-Gang Duplex Device Vert. — Lock	.13	.14	.15	.16	.14	.15	.16	.17
1-Gang Duplex Device Horz. — 1-Lid	.13	.14	.15	.16	.14	.15	.16	.17
1-Gang Duplex Device Horz. — 2-Lid	.13	.14	.15	.16	.14	.15	.16	.17
1-Gang Duplex Device Horz. — Lock	.13	.14	.15	.16	.14	.15	.16	.17
1-Gang Duplex Comb. Horz. — 2-Lid	.13	.14	.15	.16	.14	.15	.16	.17
1-Gang Signal Device Vert. — 1-Lid	.13	.14	.15	.16	.14	.15	.16	.17
1-Gang Signal Device Horz. — 1-Lid	.13	.14	.15	.16	.14	.15	.16	.17
1-Gang Switch Lever	.25	.28	.29	.31	.26	.29	.31	.32
2-Gang Switch Lever	.25	.28	.29	.31	.26	.29	.31	.32
1-Gang Blank	.10	.11	.12	.13	.11	.12	.13	.14
2-Gang Blank	.10	.11	.12	.13	.11	.12	.13	.14

Manhours include checking out of job storage, handling, job hauling, layout, where required, and installing items as outlined.

Manhours exclude installation of conduit, wire, and scaffolding. See respective tables for these time frames.

For heights above 25 feet increase manhours 10% for each additional 5 feet or fraction thereof.

VAPOLET AND SEH TYPE THREADED CAST OUTLET AND JUNCTION BOXES

MANHOURS EACH

Item Description	Manhours for Height to							
	Horizontal Runs				Vertical Runs			
	10′	15′	20′	25′	10′	15′	20′	25′
Round Vapolets and Covers								
4-1/2″ Round 2 − 1/2″ Taps	1.21	1.33	1.42	1.51	1.27	1.40	1.49	1.58
4-1/2″ Round 2 − 3/4″ Taps	1.32	1.45	1.55	1.65	1.39	1.53	1.64	1.73
4-1/2″ Round 2 − 1″ Taps	1.46	1.61	1.72	1.82	1.53	1.68	1.80	1.91
4-1/2″ Round 4 − 1/2″ Taps	1.62	1.78	1.91	2.02	1.70	1.87	2.00	2.12
4-1/2″ Round 4 − 3/4″ Taps	1.79	1.97	2.11	2.23	1.88	2.07	2.21	2.35
4-1/2″ Round 4 − 1″ Taps	1.85	2.04	2.18	2.31	1.94	2.13	2.28	2.42
Round Hub Boxes and Covers								
4″ Round Box 1 − 1/2″ Hub	1.03	1.13	1.21	1.29	1.08	1.19	1.27	1.35
4″ Round Box 1 − 3/4″ Hub	1.10	1.21	1.29	1.37	1.16	1.28	1.37	1.45
4″ Round Box 1 −1″ Hub	1.20	1.32	1.41	1.50	1.26	1.39	1.48	1.57
4″ Round Box 2 − 1/2″ Hub	1.21	1.33	1.42	1.51	1.27	1.40	1.49	1.58
4″ Round Box 2 − 3/4″ Hub	1.32	1.45	1.55	1.65	1.39	1.53	1.64	1.73
4″ Round Box 2 − 1″ Hub	1.46	1.61	1.72	1.82	1.53	1.68	1.80	1.91
4″ Round Box 3 − 1/2″ Hub	1.46	1.61	1.72	1.82	1.53	1.68	1.80	.191
4″ Round Box 3 − 3/4″ Hub	1.66	1.83	1.95	2.07	1.74	1.91	2.05	2.17
4″ Round Box 3 − 1″ Hub	1.87	2.06	2.20	2.33	1.96	2.16	2.31	2.45
4″ Round Box 4 − 1/2″ Hub	1.62	1.78	1.91	2.02	1.70	1.87	2.00	2.12
4″ Round Box 4 − 3/4″ Hub	1.82	2.00	2.14	2.27	1.91	2.10	2.25	2.38
4″ Round Box 4 − 1″ Hub	2.02	2.22	2.38	2.52	2.12	2.33	2.50	2.64
4″ Hub Box Cover − Blank	.18	.20	.21	.22	.19	.21	.22	.24
4″ Hub Box Cover − 1/2″ Hub	.33	.36	.39	.41	.35	.39	.41	.44
4″ Hub Box Cover − 3/4″ Hub	.61	.67	.72	.76	.64	.70	.75	.80
Junction Boxes and Appurtenants								
3″ WP Junction Box − 3/4″ Tap	1.32	1.45	1.55	1.65	1.39	1.53	1.64	1.73
4″ WP Junction Box − 1″ Tap	1.46	1.61	1.72	1.82	1.53	1.68	1.80	1.91
4″ Extension Collar − 3/4″ Tap	.91	1.00	1.07	1.14	.96	1.06	1.13	1.20
4″ Adapter Plate	.33	.36	.39	.41	.35	.39	.41	.44
4″ Pendant Cover − 1/2″ Tap	.33	.36	.39	.41	.35	.39	.41	.44
4″ Pendant Cover − 3/4″ Tap	.48	.53	.56	.60	.50	.55	.59	.62
4″ WP Ball Aligner − 1/2″ Tap	.38	.42	.45	.47	.49	.54	.58	.61
4″ WP Ball Aligner − 3/4″ Tap	.54	.59	.64	.67	.70	.77	.82	.87

Manhours include checking out of job storage, handling, job hauling, layout, and installing of items as outlined.

Manhours exclude installation of conduit, wire, and scaffolding. See respective tables for these time frames.

For heights above 25 feet increase manhours 10% for each additional 5 feet or fraction thereof.

SINGLE AND DUPLEX CAST IRON AND STEEL K.O. TYPE FLOOR BOXES

MANHOURS EACH

Item Description	Size (Inches)	Gr. Floor Manhours
Utility Receptacle Box – Ungr.	2-1/2	
Utility Receptacle Box – Gr.	2-1/2	1.10
Cast Iron Round Nonadjustable – No Rec.	3-1/2	1.60
Cast Iron Round Nonadjustable – Ungr. Rec.	3-1/2	1.87
Cast Iron Round Nonadjustable – Gr. Rec.	3-1/2	1.87
Cast Iron Round Nonadjustable – No Rec.	4-1/4	1.60
Cast Iron Round Nonadjustable – Ungr. Rec.	4-1/4	1.87
Cast Iron Round Nonadjustable – Gr. Rec.	4-1/4	1.87
Duplex Cast Iron Nonadjustable – No Rec.	–	1.87
Duplex Cast Iron Nonadjustable – Ungr. Rec.	–	2.42
Duplex Cast Iron Nonadjustable – Gr. Rec.	–	2.42
Cast Iron Round Adjustable – No Rec.	4-1/4	1.87
Cast Iron Round Adjustable – Ungr. Rec.	4-1/4	2.15
Cast Iron Round Adjustable – Gr. Rec.	4-1/4	2.15
Cast Iron Round Adjustable – No Rec.	5	1.87
Cast Iron Round Adjustable – Ungr. Rec.	5	2.15
Cast Iron Round Adjustable – Gr. Rec.	5	2.15
Octagon K.O. Adjustable – No. Rec.	3-1/4	1.32
Octagon K.O. Adjustable – Ungr. Rec.	3-1/4	1.60
Octagon K.O. Adjustable – Gr. Rec.	3-1/4	1.60
Octagon K.O. Adjustable – No. Rec.	4	1.32
Octagon K.O. Adjustable – Ungr. Rec.	4	1.54
Octagon K.O. Adjustable – Gr. Rec.	4	1.54
Square K.O. Adjustable – No Rec.	4-11/16	1.32
Square K.O. Adjustable – Ungr. Rec.	4-11/16	1.60
Square K.O. Adjustable – Gr. Rec.	4-11/16	1.60

Manhours include checking out of job storage, handling, job hauling, layout, installing, and raceway termination labor for items as outlined.

Manhours exclude installation of raceway, wire, box support or fasteners, nozzles or service outlets, and scaffolding. See respective tables for these time frames.

If boxes are to be installed above ground floor, increase manhours 3% for each 5 feet or fraction thereof.

CAST IRON JUNCTION AND GANG FLOOR BOXES

MANHOURS EACH

Item Description	Manhours for Height to				
	Gr. Fl.	10'	15'	20'	25'
Adjustable Cast Iron Floor Junction Boxes					
6-1/4" x 8-1/4" x 4-3/8" Deep	1.87	1.96	2.16	2.31	2.45
6-1/4" x 8-1/4" x 6" Deep	2.15	2.26	2.48	2.66	2.82
10" x 12" x 4-1/16" Deep	2.42	2.54	2.80	2.99	3.17
10" x 12" x 4-13/16" Deep	2.70	2.84	3.12	3.34	3.54
Nonadjustable Cast Iron Floor Junction Boxes					
6" x 6" x 4" Deep	1.60	1.68	1.85	1.98	2.10
6" x 8" x 4" Deep	1.60	1.68	1.85	1.98	2.10
6" x 12" x 4" Deep	1.82	1.91	2.10	2.25	2.38
6" x 12" x 6" Deep	2.20	2.31	2.54	2.72	2.88
8" x 8" x 4" Deep	2.86	3.00	3.30	3.53	3.75
8" x 8" x 6" Deep	2.86	3.00	3.30	3.53	3.75
8" x 8" x 8" Deep	4.68	4.91	5.41	5.78	6.13
8" x 12" x 6" Deep	3.96	4.16	4.57	4.89	5.19
12" x 12" x 4" Deep	4.51	4.74	5.21	5.57	5.91
12" x 12" x 6" Deep	4.51	4.74	5.21	5.57	5.91
12" x 12" x 8" Deep	5.78	6.07	6.68	7.14	7.57
Adjustable Gang Floor Boxes					
1-Gang	1.82	1.91	2.10	2.25	2.38
2-Gang	2.42	2.54	2.80	2.99	3.17
3-Gang	3.74	3.93	4.32	4.62	4.90
4-Gang	4.73	4.97	5.46	5.84	6.20
5-Gang	5.83	6.12	6.73	7.20	7.64
6-Gang	6.93	7.28	8.00	8.56	9.08

Manhours include checking out of job storage, handling, job hauling, laying out, and installing of boxes as outlined including placement of gaskets and cover plates where required.

Manhours exclude on job drilling and tapping and installation of raceway, wire, box supports or fasteners, nozzle or service outlets, and scaffolding. See respective tables for these time frames.

For heights above 25 feet increase manhours 10% for each additional 5 feet or fraction thereof.

GALVANIZED RIGID STEEL CONDUIT
Underground Construction
(Concealed Installation)

MANHOURS PER HUNDRED LINEAR FEET

Size (Inches)	Manhours Underground
1/2	5.74
3/4	7.03
1	7.03
1-1/4	7.03
1-1/2	8.61
2	11.48
2-1/2	16.13
3	21.19
3-1/2	22.47
4	30.49
5	41.68
6	53.86

Flat Concrete Slab or Concrete Pan Form
(Concealed Installation)

MANHOURS PER HUNDRED LINEAR FEET

Size (Inches)	Manhours for Height to			
	10'	15'	20'	25'
1/2	6.38	7.02	7.51	7.96
3/4	7.81	8.59	9.19	9.74
1	8.00	8.80	9.42	9.98
1-1/4	8.03	8.83	9.45	10.01
1-1/2	9.57	10.53	11.26	11.94
2	12.76	14.04	15.02	15.92
2-1/2	17.82	19.60	20.97	22.23
3	23.54	25.89	27.71	29.37
3-1/2	24.97	27.47	29.39	31.15
4	33.88	37.27	39.88	42.27
5	46.31	50.94	54.51	57.78
6	59.84	65.82	70.43	74.66

Manhours include checking out of job storage, handling, job hauling, laying out, and installing conduit as outlined.

Manhours exclude excavation, placing of concrete, backfill, fabrication of offset bends, installation of fittings and wire. See respective tables for these time frames.

If parallel runs are to be installed at the same time apply the following percentages of the manhours for each additional run:

For Second Parallel Run—97%
For Third Parallel Run—94%
For Fourth Parallel Run—90%
For Fifth Parallel Run—85%
For Each Additional Run—82%

GALVANIZED RIGID STEEL CONDUIT

Concrete Walls

(Concealed Installation)

MANHOURS PER HUNDRED LINEAR FEET

Size	Manhours for Height to			
(Inches)	10'	15'	20'	25'
1/2	6.64	7.30	7.82	8.29
3/4	8.12	8.93	9.56	10.13
1	8.32	9.15	9.79	10.38
1-1/4	8.35	9.19	9.83	10.42
1-1/2	9.95	10.95	11.71	12.41
2	13.27	14.60	15.62	16.56
2-1/2	18.53	20.38	21.81	23.12
3	24.48	26.93	28.81	30.54
3-1/2	25.97	28.57	30.57	32.40
4	35.24	38.76	41.48	43.97
5	48.16	52.98	56.68	60.09
6	62.23	68.45	73.24	77.64

Concrete Block, Brick, and Hollow Tile Walls

(Concealed Installation)

MANHOURS PER HUNDRED LINEAR FEET

Size	Manhours for Height to			
(Inches)	10'	15'	20'	25'
1/2	7.02	7.72	8.26	8.76
3/4	8.59	9.45	10.11	10.72
1	8.80	9.68	10.36	10.98
1-1/4	8.83	9.71	10.39	11.02
1-1/2	10.53	11.58	12.39	13.14
2	14.04	15.44	16.53	17.52
2-1/2	19.60	21.56	23.07	24.45
3	24.79	27.27	29.18	30.93
3-1/2	27.47	30.22	32.33	34.27
4	37.27	41.00	43.87	46.50
5	50.94	56.03	59.96	63.55
6	65.82	72.40	77.47	82.12

Manhours include checking out of job storage, handling, job hauling, layout, and installing conduit as outlined.

Manhours exclude fabrication of offset bends, installation of fittings and wire, and placement of scaffolding. See respective tables for these time frames.

For heights above 25 feet increase manhours 10% for each additional 5 feet or fraction thereof.

If parallel runs are to be installed at the same time apply the following percentages of the manhours for each additional run:

For Second Parallel Run—97%
For Third Parallel Run—94%
For Fourth Parallel Run—90%
For Fifth Parallel Run—85%
For Each Additional Run—82%

GALVANIZED RIGID STEEL CONDUIT
Walls and Ceilings
(Exposed Installation)

MANHOURS PER HUNDRED LINEAR FEET

Size (Inches)	Manhours for Height to			
	10'	15'	20'	25'
1/2	5.74	6.31	6.76	7.16
3/4	7.03	7.73	8.27	8.77
1	7.20	7.92	8.47	8.98
1-1/4	7.23	7.95	8.51	9.02
1-1/2	8.61	9.47	10.13	10.74
2	11.48	12.63	13.51	14.32
2-1/2	16.04	17.64	18.88	20.01
3	21.19	23.31	24.94	26.44
3-1/2	22.47	24.72	26.45	28.03
4	30.49	33.54	35.89	38.04
5	41.68	45.85	49.06	52.00
6	53.86	59.25	63.39	67.20

Wood Frame Construction
(Concealed Installation)

MANHOURS PER HUNDRED LINEAR FEET

Size (Inches)	Manhours for Height to			
	10'	15'	20'	25'
1/2	5.42	5.96	6.38	6.76
3/4	6.64	7.30	7.82	8.28
1	6.80	7.48	8.00	8.48
1-1/4	6.83	7.51	8.04	8.52
1-1/2	8.13	8.94	9.57	10.14
2	10.85	11.94	12.77	13.54
2-1/2	15.15	16.67	17.83	18.90
3	20.00	22.00	23.54	24.95
3-1/2	21.22	23.34	24.98	26.47
4	28.80	31.68	33.90	35.93
5	39.36	43.30	46.33	49.11
6	50.86	55.95	59.86	63.45

Manhours include checking out of job storage, handling, job hauling, layout, and installing conduit as outlined.

Manhours exclude fabrication of offset bends, installation of fittings and wire, and placement of scaffolding. See respective tables for these time frames.

For heights above 25 feet increase manhours 10% for each additional 5 feet or fraction thereof.

If parallel runs are to be installed at the same time apply the following percentages of the manhours for each additional run:

For Second Parallel Run—97%
For Third Parallel Run—94%
For Fourth Parallel Run—90%
For Fifth Parallel Run—85%
For Each Additional Run—82%

GALVANIZED RIGID STEEL CONDUIT
Suspended Ceiling Construction
(Concealed Installation)

MANHOURS PER HUNDRED LINEAR FEET

Size (Inches)	Manhours for Height to			
	10'	15'	20'	25'
1/2	6.64	7.30	7.82	8.29
3/4	8.12	8.93	9.56	10.13
1	8.32	9.15	9.79	10.38
1-1/4	8.35	9.19	9.83	10.42
1-1/2	9.95	10.95	11.71	12.41
2	13.27	14.60	15.62	16.56
2-1/2	18.53	20.38	21.81	23.12
3	24.48	26.93	28.81	30.54
3-1/2	25.97	28.57	30.57	32.40
4	35.24	38.76	41.48	43.97
5	48.16	52.98	56.68	60.09
6	62.23	68.45	73.24	77.64

Equipment Connections
(Exposed Installation)

MANHOURS PER HUNDRED LINEAR FEET

Size (Inches)	Manhours for Height to			
	10'	15'	20'	25'
1/2	5.74	6.31	6.76	7.16
3/4	7.03	7.73	8.27	8.77
1	7.20	7.92	8.47	8.98
1-1/4	7.23	7.95	8.51	9.02
1-1/2	8.61	9.47	10.13	10.74
2	11.84	12.63	13.51	14.32
2-1/2	16.04	17.64	18.88	20.01
3	21.19	23.31	24.94	26.44
3-1/2	22.47	24.72	26.45	28.03
4	30.49	33.54	35.89	38.04
5	41.68	45.85	49.06	52.00
6	53.86	59.25	63.39	67.20

Manhours include checking out of job storage, handling, job hauling, layout, and installing conduit as outlined.

Manhours exclude fabrication of offset bends, installation of fittings and wire, and placement of scaffolding. See respective tables for these time frames.

For heights above 25 feet increase manhours 10% for each additional 5 feet or fraction thereof.

If parallel runs are to be installed at the same time apply the following percentages of the manhours for each additional run:

For Second Parallel Run—97%
For Third Parallel Run—94%
For Fourth Parallel Run—90%
For Fifth Parallel Run—85%
For Each Additional Run—82%

RIGID ALUMINUM CONDUIT
Underground Construction
(Concealed Installation)

MANHOURS PER HUNDRED LINEAR FEET

Size (Inches)	Manhours Underground
1/2	5.17
3/4	6.33
1	6.33
1-1/4	6.33
1-1/2	6.46
2	8.04
2-1/2	10.48
3	13.77
3-1/2	14.61
4	19.82
5	27.09
6	35.00

Manhours include checking out of job storage, handling, job hauling, laying out, and installing conduit as outlined.

Manhours exclude excavation, placing of concrete, backfill, fabrication of offset bends, installation of fittings and wire. See respective tables for these time frames.

If parallel runs are to be installed at the same time apply the following percentages of the manhours for each additional run:

For Second Parallel Run—97%
For Third Parallel Run—94%
For Fourth Parallel Run—90%
For Fifth Parallel Run—85%
For Each Additional Run—82%

RIGID ALUMINUM CONDUIT

Flat Concrete Slab or Concrete Pan Form Construction

(Concealed Installation)

MANHOURS PER HUNDRED LINEAR FEET

Size	Manhours for Height to			
(Inches)	10'	15'	20'	25'
1/2	5.74	6.31	6.76	7.16
3/4	7.03	7.73	8.27	8.77
1	7.20	7.92	8.47	8.98
1-1/4	7.23	7.95	8.51	9.02
1-1/2	7.66	8.43	9.02	9.56
2	8.86	9.75	10.43	11.05
2-1/2	11.58	12.74	13.63	14.45
3	15.30	16.83	18.01	19.09
3-1/2	16.23	17.85	19.10	20.25
4	22.02	24.22	25.92	27.47
5	30.10	33.11	35.43	37.53
6	38.90	42.79	45.79	48.53

Concrete Walls (Concealed Installation)

MANHOURS PER HUNDRED LINEAR FEET

Size	Manhours for Height to			
(Inches)	10'	15'	20'	25'
1/2	5.97	6.57	7.03	7.45
3/4	7.31	8.04	8.61	9.12
1	7.49	8.24	8.81	9.34
1-1/4	7.52	8.27	8.85	9.38
1-1/2	7.97	8.76	9.38	9.94
2	9.21	10.14	10.85	11.50
2-1/2	12.04	13.25	14.17	15.03
3	15.91	17.50	18.73	19.85
3-1/2	16.88	18.57	19.87	21.06
4	22.90	25.19	26.95	28.57
5	31.30	34.43	36.84	39.06
6	40.46	44.50	47.62	50.47

Manhours include checking out of job storage, handling, job hauling, layout, and installing conduit as outlined.

Manhours exclude fabrication of offset bends, installation of fittings and wire, and placement of scaffolding. See respective tables for these time frames.

For heights above 25 feet increase manhours 10% for each additional 5 feet or fraction thereof.

If parallel runs are to be installed at the same time apply the following percentages of the manhours for each additional run:

For Second Parallel Run—97%
For Third Parallel Run—94%
For Fourth Parallel Run—90%
For Fifth Parallel Run—85%
For Each Additional Run—82%

RIGID ALUMINUM CONDUIT

Concrete Block, Brick, or Hollow Tile

(Concealed Installation)

MANHOURS PER HUNDRED LINEAR FEET

Size (Inches)	Manhours for Height to			
	10'	15'	20'	25'
1/2	6.31	6.95	7.43	7.88
3/4	7.73	8.51	9.10	9.65
1	7.92	8.71	9.32	9.88
1-1/4	7.95	8.75	9.36	9.92
1-1/2	8.43	9.27	9.92	10.51
2	9.75	10.72	11.47	12.16
2-1/2	12.74	14.01	14.99	15.89
3	16.83	18.51	19.81	21.00
3-1/2	17.85	19.64	21.01	22.27
4	24.22	26.64	28.51	30.22
5	33.11	36.42	39.97	41.31
6	42.79	47.07	50.36	53.39

Walls and Ceilings (Exposed Installation)

MANHOURS PER HUNDRED LINEAR FEET

Size (Inches)	Manhours for Height to			
	10'	15'	20'	25'
1/2	5.16	5.68	6.08	6.45
3/4	6.33	6.96	7.45	7.89
1	6.48	7.13	7.63	8.08
1-1/4	6.51	7.16	7.66	8.12
1-1/2	6.89	7.58	8.11	8.60
2	7.97	8.77	9.39	9.95
2-1/2	10.42	11.46	12.27	13.00
3	13.77	15.15	16.21	17.18
3-1/2	14.61	16.07	17.19	18.22
4	19.82	21.80	23.33	24.73
5	27.09	29.80	31.88	33.80
6	35.01	38.51	41.21	43.68

Manhours include checking out of job storage, handling, job hauling, layout, and installing conduit as outlined.

Manhours exclude fabrication of offset bends, installation of fittings and wire, and placement of scaffolding. See respective tables for these time frames.

For heights above 25 feet increase manhours 10% for each additional 5 feet or fraction thereof.

If parallel runs are to be installed at the same time apply the following percentages of the manhours for each additional run:

For Second Parallel Run—97%
For Third Parallel Run—94%
For Fourth Parallel Run—90%
For Fifth Parallel Run—85%
For Each Additional Run—82%

RIGID ALUMINUM CONDUIT

Wood Frame Construction (Concealed Installation)

MANHOURS PER HUNDRED LINEAR FEET

Size	Manhours for Height to			
(Inches)	10'	15'	20'	25'
1/2	4.88	5.37	5.74	6.09
3/4	5.98	6.57	7.03	7.46
1	6.12	6.73	7.20	7.64
1-1/4	6.15	6.76	7.23	7.67
1-1/2	6.51	7.16	7.66	8.12
2	7.53	8.28	8.86	9.40
2-1/2	9.84	10.83	11.59	12.28
3	13.01	14.31	15.31	16.23
3-1/2	13.80	15.18	16.24	17.21
4	18.72	20.59	22.03	23.35
5	25.59	28.14	30.11	31.92
6	33.07	36.37	38.92	41.25

Suspended Ceiling Construction (Concealed Installation)

MANHOURS PER HUNDRED LINEAR FEET

Size	Manhours for Height to			
(Inches)	10'	15'	20'	25'
1/2	5.97	6.57	7.03	7.45
3/4	7.31	8.04	8.61	9.12
1	7.49	8.24	8.81	9.34
1-1/4	7.52	8.27	8.85	9.38
1-1/2	7.97	8.76	9.38	9.94
2	9.21	10.14	10.85	11.50
2-1/2	12.04	13.25	14.17	15.03
3	15.91	17.50	18.73	19.85
3-1/2	16.88	18.57	19.87	21.06
4	22.90	25.19	26.95	28.57
5	31.30	34.43	36.84	39.06
6	40.46	44.50	47.62	50.47

Manhours include checking out of job storage, handling, job hauling, layout, and installing conduit as outlined.

Manhours exclude fabrication of offset bends, installation of fittings and wire, and placement of scaffolding. See respective tables for these time frames.

For heights above 25 feet increase manhours 10% for each additional 5 feet or fraction thereof.

If parallel runs are to be installed at the same time apply the following percentages of the manhours for each additional run:

For Second Parallel Run—97%
For Third Parallel Run—94%
For Fourth Parallel Run—90%
For Fifth Parallel Run—85%
For Each Additional Run—82%

RIGID ALUMINUM CONDUIT

Equipment Connections

(Exposed Installation)

MANHOURS PER HUNDRED LINEAR FEET

Size	Manhours for Height to			
(Inches)	10'	15'	20'	25'
1/2	5.16	5.68	6.08	6.45
3/4	6.33	6.96	7.45	7.89
1	6.48	7.13	7.63	8.08
1-1/4	6.51	7.16	7.66	8.12
1-1/2	6.89	7.58	8.11	8.60
2	7.97	8.77	9.39	9.95
2-1/2	10.42	11.46	12.27	13.00
3	13.77	15.15	16.21	17.18
3-1/2	14.61	16.07	17.19	18.22
4	19.82	21.80	23.33	24.73
5	27.09	29.80	31.88	33.80
6	35.01	38.51	41.21	43.68

Manhours include checking out of job storage, handling, job hauling, layout, and installing conduit as outlined.

Manhours exclude fabrication of offset bends, installation of fittings and wire, and placement of scaffolding. See respective tables for these time frames.

For heights above 25 feet increase manhours 10% for each additional 5 feet or fraction thereof.

If parallel runs are to be installed at the same time apply the following percentages of the manhours for each additional run:

For Second Parallel Run—97%
For Third Parallel Run—94%
For Fourth Parallel Run—90%
For Fifth Parallel Run—85%
For Each Additional Run—82%

PLASTIC CONDUIT
Underground Construction
(Concealed Installation)

MANHOURS PER HUNDRED LINEAR FEET

Size	Manhours
(Inches)	Underground
1/2	5.17
3/4	6.33
1	6.33
1-1/4	6.33
1-1/2	6.46
2	8.61
2-1/2	12.10
3	15.89
3-1/2	16.85
4	22.87
5	31.26
6	40.40

Manhours include checking out of job storage, handling, job hauling, laying out, and installing conduit as outlined.

Manhours exclude excavation, placing of concrete, backfill, fabrication of offset bends, installation of fittings and wire. See respective tables for these time frames.

If parallel runs are to be installed at the same time apply the following percentages of the manhours for each additional run:

For Second Parallel Run—97%
For Third Parallel Run—94%
For Fourth Parallel Run—90%
For Fifth Parallel Run—85%
For Each Additional Run—82%

PLASTIC CONDUIT
Flat Concrete Slab or Concrete Pan Form
(Concealed Installation)

MANHOURS PER HUNDRED LINEAR FEET

Size (Inches)	Manhours for Height to			
	10'	15'	20'	25'
1/2	5.74	6.31	6.76	7.16
3/4	7.03	7.23	8.27	8.77
1	7.20	7.92	8.47	8.98
1-1/4	7.23	7.95	8.51	9.02
1-1/2	7.66	8.43	9.02	9.56
2	9.57	10.53	11.26	11.94
2-1/2	16.04	17.64	18.88	20.01
3	17.66	19.43	20.79	22.03
3-1/2	18.73	20.60	22.05	23.37
4	25.41	27.95	29.91	31.70
5	34.73	38.20	40.88	43.33
6	44.88	49.37	52.82	55.99

Concrete Walls (Concealed Installation)

MANHOURS PER HUNDRED LINEAR FEET

Size (Inches)	Manhours for Height to			
	10'	15'	20'	25'
1/2	5.97	6.57	7.03	7.45
3/4	7.31	8.04	8.61	9.12
1	7.49	8.24	8.81	9.34
1-1/4	7.52	8.27	8.85	9.38
1-1/2	7.97	8.76	9.38	9.94
2	9.95	10.95	11.71	12.41
2-1/2	16.68	18.35	19.63	20.81
3	18.37	20.20	21.62	22.91
3-1/2	19.48	21.43	22.93	24.30
4	26.43	29.07	31.10	32.97
5	36.12	39.73	42.51	45.06
6	46.68	51.34	54.94	58.23

Manhours include checking out of job storage, handling, job hauling, layout, and installing conduit as outlined.

Manhours exclude fabrication of offset bends, installation of fittings and wire, and placement of scaffolding. See respective tables for these time frames.

For heights above 25 feet increase manhours 10% for each additional 5 feet or fraction thereof.

If parallel runs are to be installed at the same time apply the following percentages of the manhours for each additional run:

  For Second Parallel Run—97%
  For Third Parallel Run—94%
  For Fourth Parallel Run—90%
  For Fifth Parallel Run—85%
  For Each Additional Run—82%

PLASTIC CONDUIT

Concrete Block, Brick, or Hollow Tile

(Concealed Installation)

MANHOURS PER HUNDRED LINEAR FEET

Size (Inches)	Manhours for Height to			
	10'	15'	20'	25'
1/2	6.31	6.95	7.43	7.88
3/4	7.73	8.51	9.10	9.65
1	7.92	8.71	9.32	9.88
1-1/4	7.95	8.75	9.36	9.92
1-1/2	8.43	9.27	9.92	10.51
2	10.52	11.58	12.39	13.13
2-1/2	17.64	19.41	20.76	22.00
3	19.43	21.37	22.86	24.24
3-1/2	20.60	22.66	24.25	25.70
4	27.95	30.75	32.90	34.87
5	38.20	42.03	44.96	47.66
6	49.37	54.30	58.11	61.59

Walls and Ceilings (Exposed Installation)

MANHOURS PER HUNDRED LINEAR FEET

Size (Inches)	Manhours for Height to			
	10'	15'	20'	25'
1/2	5.17	5.68	6.08	6.45
3/4	6.33	6.96	7.45	7.89
1	6.48	7.13	7.63	8.08
1-1/4	6.51	7.16	7.67	8.12
1-1/2	6.89	7.58	8.11	8.60
2	8.61	9.47	10.14	10.75
2-1/2	14.44	15.88	16.99	18.00
3	15.89	17.48	18.71	19.83
3-1/2	16.86	18.54	19.84	21.93
4	22.87	25.16	26.92	28.53
5	31.26	34.38	36.79	39.00
6	40.39	44.93	47.54	50.39

Manhours include checking out of job storage, handling, job hauling, layout, and installing conduit as outlined.

Manhours exclude fabrication of offset bends, installation of fittings and wire, and placement of scaffolding. See respective tables for these time frames.

For heights above 25 feet increase manhours 10% for each additional 5 feet or fraction thereof.

If parallel runs are to be installed at the same time apply the following percentages of the manhours for each additional run:

For Second Parallel Run—97%
For Third Parallel Run—94%
For Fourth Parallel Run—90%
For Fifth Parallel Run—85%
For Each Additional Run—82%

PLASTIC CONDUIT

Wood Frame Construction (Concealed Installation)

MANHOURS PER HUNDRED LINEAR FEET

Size (Inches)	Manhours for Height to			
	10'	15'	20'	25'
1/2	4.88	5.37	5.74	6.09
3/4	5.98	6.57	7.03	7.46
1	6.12	6.73	7.20	7.64
1-1/4	6.15	6.76	7.23	7.64
1-1/2	6.51	7.16	7.66	8.12
2	8.13	8.95	9.57	10.15
2-1/2	13.63	15.00	16.05	17.01
3	15.01	16.51	17.67	18.73
3-1/2	15.92	17.51	18.74	19.86
4	21.60	23.76	25.42	26.95
5	29.52	32.47	34.75	36.83
6	38.15	41.96	44.90	47.59

Suspended Ceiling Construction (Concealed Installation)

MANHOURS PER HUNDRED LINEAR FEET

Size (Inches)	Manhours for Height to			
	10'	15'	20'	25'
1/2	5.97	6.57	7.03	7.45
3/4	7.31	8.04	8.61	9.12
1	7.49	8.24	8.81	9.34
1-1/4	7.52	8.27	8.85	9.38
1-1/2	7.97	8.76	9.38	9.94
2	9.95	10.95	11.71	12.41
2-1/2	16.68	18.35	19.63	20.81
3	18.37	20.20	21.62	22.91
3-1/2	19.48	21.43	22.93	24.30
4	26.43	29.07	31.10	32.97
5	36.12	39.73	42.51	45.06
6	46.68	51.34	54.94	58.23

Manhours include checking out of job storage, handling, job hauling, layout, and installing conduit as outlined.

Manhours exclude fabrication of offset bends, installation of fittings and wire, and placement of scaffolding. See respective tables for these time frames.

For heights above 25 feet increase manhours 10% for each additional 5 feet or fraction thereof.

If parallel runs are to be installed at the same time apply the following percentages of the manhours for each additional run:

For Second Parallel Run—97%
For Third Parallel Run—94%
For Fourth Parallel Run—90%
For Fifth Parallel Run—85%
For Each Additional Run—82%

PLASTIC CONDUIT

Equipment Connections

(Exposed Installation)

MANHOURS PER HUNDRED LINEAR FEET

Size	Manhours for Height to			
(Inches)	10'	15'	20'	25'
1/2	5.17	5.68	6.08	6.45
3/4	6.33	6.96	7.45	7.89
1	6.48	7.13	7.63	8.08
1-1/4	6.51	7.16	7.67	8.12
1-1/2	6.89	7.58	8.11	8.60
2	8.61	9.47	10.14	10.75
2-1/2	14.44	15.88	16.99	18.00
3	15.89	17.48	18.71	19.83
3-1/2	16.86	18.54	19.84	21.03
4	22.87	25.16	26.92	28.53
5	31.26	34.38	36.79	39.00
6	40.39	44.43	47.54	50.39

Manhours include checking out of job storage, handling, job hauling, layout, and installing conduit as outlined.

Manhours exclude fabrication of offset bends, installation of fittings and wire, and placement of scaffolding. See respective tables for these time frames.

For heights above 25 feet increase manhours 10% for each additional 5 feet or fraction thereof.

If parallel runs are to be installed at the same time apply the following percentages of the manhours for each additional run:

For Second Parallel Run—97%
For Third Parallel Run—94%
For Fourth Parallel Run—90%
For Fifth Parallel Run—85%
For Each Additional Run—82%

ELECTRIC METALLIC TUBING (THINWALL)
Underground Construction
(Concealed Installation)

MANHOURS PER HUNDRED LINEAR FEET

Size (Inches)	Manhours Underground
1/2	4.82
3/4	5.76
1	5.98
1-1/4	5.98
1-1/2	6.02
2	6.77
2-1/2	9.52
3	12.50
3-1/2	13.26
4	17.07
5	23.34
6	30.16

Manhours include checking out of job storage, handling, job hauling, laying out, and installing conduit as outlined.

Manhours exclude excavation, placing of concrete, backfill, fabrication of offset bends, installation of fittings and wire. See respective tables for these time frames.

If parallel runs are to be installed at the same time apply the following percentages of the manhours for each additional run:

For Second Parallel Run—97%
For Third Parallel Run—94%
For Fourth Parallel Run—90%
For Fifth Parallel Run—85%
For Each Additional Run—82%

ELECTRIC METALLIC TUBING (THINWALL)

Flat Concrete Slab or Concrete Pan Form

MANHOURS PER HUNDRED LINEAR FEET

Size	Manhours for Height to			
(Inches)	10'	15'	20'	25'
1/2	5.36	5.90	6.31	6.69
3/4	6.40	7.04	7.53	7.98
1	6.80	7.48	8.00	8.48
1-1/4	6.83	7.51	8.04	8.52
1-1/2	6.99	7.69	8.23	8.72
2	8.68	9.55	10.22	10.83
2-1/2	10.51	11.56	12.37	13.11
3	13.89	15.28	16.35	17.33
3-1/2	13.98	15.38	16.45	17.44
4	18.97	20.87	22.33	23.67

Concrete Walls (Concealed Installation)

MANHOURS PER HUNDRED LINEAR FEET

Size	Manhours for Height to			
(Inches)	10'	15'	20'	25'
1/2	5.57	6.13	6.57	6.96
3/4	6.66	7.32	7.83	8.30
1	7.07	7.78	8.32	8.82
1-1/4	7.10	7.81	8.36	8.86
1-1/2	7.27	8.00	8.56	9.07
2	9.03	9.93	10.62	11.26
2-1/2	10.93	12.02	12.87	13.64
3	14.46	15.89	17.00	18.02
3-1/2	14.54	15.99	17.11	18.14
4	19.73	21.70	23.22	24.61

Manhours include checking out of job storage, handling, job hauling, layout, and installing conduit as outlined.

Manhours exclude fabrication of offset bends, installation of fittings and wire, and placement of scaffolding. See respective tables for these time frames.

For heights above 25 feet increase manhours 10% for each additional 5 feet or fraction thereof.

If parallel runs are to be installed at the same time apply the following percentages of the manhours for each additional run:

For Second Parallel Run—97%
For Third Parallel Run—94%
For Fourth Parallel Run—90%
For Fifth Parallel Run—85%
For Each Additional Run—82%

ELECTRIC METALLIC TUBING (THINWALL)

Concrete Block, Brick, or Hollow Tile

(Concealed Installation)

MANHOURS PER HUNDRED LINEAR FEET

Size (Inches)	Manhours for Height to			
	10'	15'	20'	25'
1/2	5.90	6.49	6.94	7.36
3/4	7.04	7.74	8.29	8.78
1	7.48	8.23	8.80	9.33
1-1/4	7.51	8.26	8.84	9.37
1-1/2	7.69	8.46	9.05	9.59
2	9.55	10.50	11.24	11.91
2-1/2	11.56	12.72	13.61	14.42
3	15.28	16.81	17.98	19.06
3-1/2	15.38	16.92	18.10	19.19
4	20.87	22.95	24.56	26.03

Walls and Ceilings (Exposed Installation)

MANHOURS PER HUNDRED LINEAR FEET

Size (Inches)	Manhours for Height to			
	10'	15'	20'	25'
1/2	4.82	5.31	5.68	6.02
3/4	5.76	6.34	6.78	7.19
1	6.12	6.73	7.20	7.64
1-1/4	6.15	6.76	7.24	7.67
1-1/2	6.29	6.92	7.40	7.85
2	7.81	8.59	9.19	9.74
2-1/2	9.46	10.40	11.13	11.80
3	12.50	13.75	14.71	15.60
3-1/2	12.58	13.84	14.81	15.70
4	17.07	18.78	20.09	21.30

Manhours include checking out of job storage, handling, job hauling, layout, and installing conduit as outlined.

Manhours exclude fabrication of offset bends, installation of fittings and wire, and placement of scaffolding. See respective tables for these time frames.

For heights above 25 feet increase manhours 10% for each additional 5 feet or fraction thereof.

If parallel runs are to be installed at the same time apply the following percentages of the manhours for each additional run:

For Second Parallel Run—97%
For Third Parallel Run—94%
For Fourth Parallel Run—90%
For Fifth Parallel Run—85%
For Each Additional Run—82%

ELECTRIC METALLIC TUBING (THINWALL)
Wood Frame Construction
(Concealed Installation)

MANHOURS PER HUNDRED LINEAR FEET

Size	Manhours for Height to			
(Inches)	10'	15'	20'	25'
1/2	4.56	5.01	5.36	5.68
3/4	5.44	5.98	6.40	6.79
1	5.78	6.36	6.80	7.21
1-1/4	5.81	6.39	6.83	7.24
1-1/2	5.94	6.54	6.99	7.41
2	7.38	8.12	8.68	9.20
2-1/2	8.93	8.93	10.51	11.15
3	11.81	12.99	13.90	14.73
3-1/2	11.88	13.07	13.99	14.83
4	16.12	17.74	18.98	20.12

Suspended Ceiling Construction
(Concealed Installation)

MANHOURS PER HUNDRED LINEAR FEET

Size	Manhours for Height to			
(Inches)	10'	15'	20'	25'
1/2	5.57	6.13	6.52	6.96
3/4	6.66	7.32	7.83	8.30
1	7.07	7.78	8.32	8.82
1-1/4	7.10	7.81	8.36	8.86
1-1/2	7.27	8.00	8.56	9.07
2	9.03	9.93	10.62	11.26
2-1/2	10.93	12.02	12.87	13.64
3	14.46	15.89	17.00	18.02
3-1/2	14.54	15.99	17.11	18.14
4	19.73	21.70	23.22	24.61

Manhours include checking out of job storage, handling, job hauling, layout, and installing conduit as outlined.

Manhours exclude fabrication of offset bends, installation of fittings and wire, and placement of scaffolding. See respective tables for these time frames.

For heights above 25 feet increase manhours 10% for each additional 5 feet or fraction thereof.

If parallel runs are to be installed at the same time apply the following percentages of the manhours for each additional run:

For Second Parallel Run—97%
For Third Parallel Run—94%
For Fourth Parallel Run—90%
For Fifth Parallel Run—85%
For Each Additional Run—82%

ELECTRIC METALLIC TUBING (THINWALL)
Equipment Connections
(Exposed Installation)

MANHOURS PER HUNDRED LINEAR FEET

Size	Manhours for Height to			
(Inches)	10'	15'	20'	25'
1/2	4.82	5.31	5.68	6.02
3/4	5.76	6.34	6.78	7.19
1	6.12	6.73	7.20	7.64
1-1/4	6.15	6.76	7.24	7.67
1-1/2	6.29	6.92	7.40	7.85
2	7.81	8.59	9.19	9.74
2-1/2	9.46	10.40	11.13	11.80
3	12.50	13.75	14.71	15.60
3-1/2	12.58	13.84	14.81	15.70
4	17.07	18.78	20.09	21.30

Manhours include checking out of job storage, handling, job hauling, layout, and installing conduit as outlined.

Manhours exclude fabrication of offset bends, installation of fittings and wire, and placement of scaffolding. See respective tables for these time frames.

For heights above 25 feet increase manhours 10% for each additional 5 feet or fraction thereof.

If parallel runs are to be installed at the same time apply the following percentages of the manhours for each additional run:

For Second Parallel Run—97%
For Third Parallel Run—94%
For Fourth Parallel Run—90%
For Fifth Parallel Run—85%
For Each Additional Run—82%

FLEXIBLE STEEL CONDUIT

Concrete Block, Brick, or Hollow Tile

(Concealed Installation)

MANHOURS PER HUNDRED LINEAR FEET

Size	Manhours for Height to			
(Inches)	10'	15'	20'	25'
3/8	4.75	5.23	5.59	5.93
1/2	5.42	5.96	6.38	6.76
3/4	7.03	7.73	8.27	8.77
1	7.20	7.92	8.47	8.98
1-1/4	8.83	9.71	10.39	11.02
1-1/2	11.00	12.10	12.95	13.72
2	16.59	18.25	19.53	20.70
2-1/2	23.17	25.49	27.27	28.91
3	34.13	37.54	40.17	42.58

Walls and Ceilings

(Exposed Installation)

MANHOURS PER HUNDRED LINEAR FEET

Size	Manhours for Height to			
(Inches)	10'	15'	20'	25'
3/8	3.80	4.18	4.47	4.74
1/2	4.34	4.77	5.10	5.41
3/4	5.62	6.19	6.62	7.02
1	5.76	6.34	6.78	7.19
1-1/4	7.06	7.77	8.31	8.81
1-1/2	8.80	9.68	10.36	10.98
2	13.27	14.60	15.62	16.56
2-1/2	18.54	20.39	21.82	23.13
3	27.30	30.03	32.14	34.07

Manhours include checking out of job storage, handling, job hauling, layout, and installing flexible conduit as outlined.

Manhours exclude installation of fittings and wire, and placement of scaffolding. See respective tables for these time frames

For heights above 25 feet increase manhours 10% for each additional 5 feet or fraction thereof.

If parallel runs are to be installed at the same time apply the following percentages of the manhours for each additional run:

For Second Parallel Run—97%
For Third Parallel Run—94%
For Fourth Parallel Run—90%
For Fifth Parallel Run—85%
For Each Run Above Five—82%

FLEXIBLE STEEL CONDUIT

Wood Frame Construction
(Concealed Installation)

MANHOURS PER HUNDRED LINEAR FEET

Size (Inches)	Manhours for Height to			
	10'	15'	20'	25'
3/8	3.56	3.92	4.19	4.44
1/2	4.07	4.47	4.78	5.07
3/4	5.27	5.80	6.21	6.58
1	5.40	5.94	6.36	6.74
1-1/4	6.62	7.28	7.79	8.26
1-1/2	8.25	9.08	9.71	10.29
2	12.44	13.69	14.64	15.52
2-1/2	17.38	19.12	20.46	21.69
3	25.60	28.16	30.13	31.94

Suspended Ceiling Construction
(Concealed Installation)

MANHOURS PER HUNDRED LINEAR FEET

Size (Inches)	Manhours for Height to			
	10'	15'	20'	25'
3/8	4.56	5.02	5.37	5.69
1/2	5.20	5.72	6.12	6.49
3/4	6.75	7.42	7.94	8.42
1	6.91	7.60	8.14	8.62
1-1/4	8.48	9.32	9.98	10.58
1-1/2	10.56	11.62	12.43	13.17
2	15.93	17.52	18.75	19.87
2-1/2	22.24	24.47	26.18	27.75
3	32.76	36.04	38.56	40.88

Manhours include checking out of job storage, handling, job hauling, layout, and installing flexible conduit as outlined.

Manhours exclude installation of fittings and wire, and placement of scaffolding. See respective tables for these time frames

For heights above 25 feet increase manhours 10% for each additional 5 feet or fraction thereof.

If parallel runs are to be installed at the same time apply the following percentages of the manhours for each additional run:

For Second Parallel Run—97%
For Third Parallel Run—94%
For Fourth Parallel Run—90%
For Fifth Parallel Run—85%
For Each Run Above Five—82%

FLEXIBLE STEEL CONDUIT

Equipment Connections

(Exposed Installation)

MANHOURS PER HUNDRED LINEAR FEET

Size	Manhours for Height to			
(Inches)	10'	15'	20'	25'
3/8	3.80	4.18	4.47	4.74
1/2	4.34	4.77	5.10	5.41
3/4	5.62	6.19	6.62	7.02
1	5.76	6.34	6.78	7.19
1-1/4	7.06	7.77	8.31	8.81
1-1/2	8.80	9.68	10.36	10.98
2	13.27	14.60	15.62	16.56
2-1/2	18.54	20.39	21.82	23.13
3	27.30	30.03	32.14	34.07

Manhours include checking out of job storage, handling, job hauling, layout, and installing flexible conduit as outlined.

Manhours exclude installation of fittings and wire, and placement of scaffolding. See respective tables for these time frames.

For heights above 25 feet, increase manhours 10% for each additional 5 feet or fraction thereof.

If parallel runs are to be installed at the same time apply the following percentages of the manhours for each additional run:

For Second Parallel Run—97%
For Third Parallel Run—94%
For Fourth Parallel Run—90%
For Fifth Parallel Run—85%
For Each Run Above Five—82%

FABRICATED CONDUIT BENDS IN SHOP AND FIELD GALVANIZED RIGID STEEL OR ALUMINUM CONDUIT

MANHOURS EACH

Size (Inches)	Type of Bend						
	In Shop					In Field	
	Nos. 1, 2, 3 and 4	No. 5	No. 6	No. 7	No. 8	Nos. 1, 2, 3 and 4	No. 7
1/2	.18	–	.37	.28	–	.27	.37
3/4	.23	–	.42	.33	–	.33	.42
1	.27	–	.60	.47	–	.51	.56
1-1/4	.33	1.21	.70	.56	1.21	.58	.65
1-1/2	.42	1.35	.84	.70	1.35	.74	.84
2	.65	1.58	1.02	.88	1.58	.93	1.12
2-1/2	.79	1.81	1.26	1.12	1.81	1.16	1.49
3	.90	1.98	1.49	1.26	1.98	1.35	1.67
3-1/2	1.22	2.34	1.62	1.44	2.34	1.62	1.80
4	1.62	2.57	2.03	1.71	2.57	2.25	1.98
5	2.03	3.21	2.54	2.13	3.21	2.81	2.47
6	2.43	3.86	3.05	2.56	3.85	3.37	2.97

Manhours include checking out of storage, handling, job hauling to fabrication shop or field as required, checking sketches, bending with shoe, and checking.

Shop manhours include piece marking or coding. Field manhours for bends 1-inch and smaller are for bends made with a general use hickey.

Manhours exclude cutting, reaming, threading, joint make-up, and installation. See respective tables for these time frames.

STANDARD TYPES OF BENDS

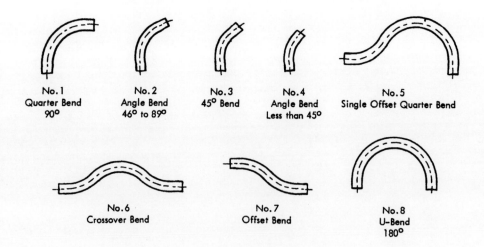

No. 1
Quarter Bend
90°

No. 2
Angle Bend
46° to 89°

No. 3
45° Bend

No. 4
Angle Bend
Less than 45°

No. 5
Single Offset Quarter Bend

No. 6
Crossover Bend

No. 7
Offset Bend

No. 8
U–Bend
180°

CUTTING, REAMING, AND THREADING GALVANIZED RIGID STEEL OR ALUMINUM CONDUIT

In Fabricating Shop or Field

MANHOURS EACH

| Size (Inches) | Operation and Location | | | | | |
| | Shop | | | Field | | |
	Cutting Only	Ream and Thread	Combined Operation	Cutting Only	Ream and Thread	Combined Operation
1/2	.11	.17	.28	.18	.27	.45
3/4	.11	.17	.28	.18	.27	.45
1	.12	.18	.30	.19	.28	.47
1-1/4	.13	.20	.33	.21	.33	.54
1-1/2	.14	.24	.38	.23	.40	.63
2	.17	.25	.42	.27	.41	.68
2-1/2	.18	.26	.44	.28	.42	.70
3	.22	.27	.49	.36	.45	.81
3-1/2	.24	.32	.56	.38	.54	.92
4	.27	.43	.70	.45	.72	1.17
5	.33	.53	.86	.56	.90	1.46
6	.41	.65	1.01	.68	1.08	1.76

Manhours include checking out of job storage, handling, job hauling to and from fabrication shop if required, measuring, checking, cutting, reaming, and threading or conduit as outlined.

Shop manhours are for use of power tools for all operations. Field manhours are for use of hand tools for all operations.

Manhours exclude joint make-up, field installation, and scaffolding. See respective tables for these time frames.

CUTTING AND END PREPARATION OF PLASTIC CONDUIT, ELECTRIC METALLIC TUBING, AND FLEXIBLE CONDUIT

MANHOURS EACH

Size (Inches)	Plastic Conduit	EMT (Thinwall)	Flexible Conduit
1/2	.34	.38	.50
3/4	.34	.38	.50
1	.35	.40	.52
1-1/4	.41	.46	.59
1-1/2	.47	.54	.69
2	.51	.58	.75
2-1/2	53	.60	.77
3	.61	.69	.89
3-1/2	.69	.78	1.00
4	.88	.99	1.29
5	1.10	1.24	1.61
6	1.32	1.50	1.94

Manhours include cutting and preparation of ends as required for items as outlined

Manhours exclude make-up of joint and installation. See respective tables for these time frames.

THREAD-ONS FOR
STEEL AND ALUMINUM FITTINGS

MANHOURS EACH

Size (Inches)	In Fab. Shop	Under-ground	Concrete Ground Slab	Manhours for Following Locations							
				Horizontal Runs in Height to				Vertical Runs in Height to			
				10′	15′	20′	25′	10′	15′	20′	25′
1/2	.35	.42	.28	.33	.35	.37	.40	.33	.37	.41	.44
3/4	.35	.42	.28	.33	.35	.37	.40	.33	.37	.41	.46
1	.56	.70	.46	.55	.59	.60	.62	.55	.60	.64	.68
1-1/4	.57	.74	.48	.57	.64	.66	.67	.57	.66	.69	.74
1-1/2	.68	.81	.54	.67	.68	.69	.74	.67	.69	.76	.80
2	.79	.95	.63	.76	.79	.81	.84	.76	.81	.86	.88
2-1/2	.82	.96	.65	.79	.82	.84	.87	.79	.84	.90	.96
3	.90	1.01	.72	.86	.90	.92	.96	.86	.92	1.00	1.02
3-1/2	.95	1.06	.76	.91	.95	.98	1.01	.91	.98	1.05	1.12
4	1.00	1.12	.80	.96	1.00	1.02	1.20	.96	1.02	1.25	1.30
5	1.25	1.40	1.00	1.20	1.25	1.28	1.50	1.20	1.28	1.60	1.72
6	1.50	1.68	1.20	1.44	1.50	1.53	1.80	1.44	1.53	1.90	2.00

Manhours include checking rigid steel or aluminum conduit fittings out of job storage, handling, job hauling, applying joint sealer, and making up of single thread-on. It should be remembered that a coupling or elbow has two thread-ons, a cap one thread-on, etc.

Manhours exclude cutting, reaming and threading of conduit and the placement of scaffolding. See respective tables for these time frames.

An allowance is included under shop fabrication thread-ons for hauling from fabrication site to erection site.

Manhours do not include installation of conduit bodies. See respective tables for these time frames.

See other tables for time requirements for electric metallic tube, plastic PVC, and flexible conduit make-ups.

MAKE-UPS FOR ELECTRIC METALLIC TUBING AND PLASTIC PVC CONDUIT

MANHOURS EACH

Size (Inches)	Manhours for Following Types and Locations															
	Electric Metallic Tubing								Plastic PVC							
	Horizontal Runs in Height to				Vertical Runs in Height to				Horizontal Runs in Height to				Vertical Runs in Height to			
	10'	15'	20'	25'	10'	15'	20'	25'	10'	15'	20'	25'	10'	15'	20'	25'
1/2	.28	.30	.31	.34	.28	.31	.35	.37	.25	.26	.28	.30	.25	.28	.31	.33
3/4	.28	.30	.31	.34	.28	.31	.35	.39	.25	.26	.28	.30	.25	.28	.31	.35
1	47	.50	.51	.53	.47	.51	.54	.58	41	.44	.45	.47	.42	.45	.48	.51
1-1/4	48	.54	.56	.57	48	.56	.59	.63	.42	.48	.49	.51	.43	.50	.52	.56
1-1/2	.57	.58	.59	.63	.57	.59	.65	.68	.50	.51	.52	.56	.50	.52	.57	.60
2	.65	.67	.69	.71	.65	.69	73	.75	.57	.59	.61	.63	.57	.61	.65	.66
2-1/2	.67	.70	71	.74	.67	.71	77	.82	.59	.62	.63	.65	.59	.63	.68	.72
3	73	.77	.78	.82	.73	.78	.85	.87	.65	.68	.69	.72	.65	.69	.75	.77
3-1/2	77	.81	.83	.86	.77	.83	.89	.95	.68	.71	.74	.76	.68	.74	.79	.84
4	.82	.85	.87	1.02	.82	.87	1.06	1.11	.72	.75	.77	.90	.72	.77	.94	.98
5	–	–	–	–	–	–	–	–	.90	.94	.96	1.13	.90	.96	1.20	1.29
6	–	–	–	–	–	–	–	–	1.08	1.13	1.15	1.37	1.08	1.15	1.43	1.50

Manhours include checking fittings out of job storage, handling, job hauling, applying sealer where necessary and one make-up.

Where fitting is slipped on and secured with set screws, allow one half make-up per screw.

Manhours do not include installation of conduit bodies, cutting and cleaning of tube or conduit or scaffolding. See respective tables for these time frames.

See other tables for rigid galvanized steel and aluminum conduit thread-ons, cuts, threads, and bevels.

MAKE-UPS OF
FLEXIBLE CONDUIT FITTINGS

MANHOURS EACH

Size (Inches)	Equipment Connections	Manhours for Following Locations							
		Horizontal Runs in Height to				Vertical Runs in Height to			
		10′	15′	20′	25′	10′	15′	20′	25′
3/8	.31	.36	.39	.41	.44	.36	.41	.45	.48
1/2	.31	.36	.39	.41	.44	.36	.41	.45	.48
3/4	.31	.36	.39	.41	.44	.36	.41	.45	.51
1	.51	.61	.65	.66	.68	.61	.66	.70	.75
1-1/4	.53	.63	.70	.73	.74	.63	.73	.76	.81
1-1/2	.59	.74	.75	.76	.81	.74	.76	.84	.88
2	.69	.84	.87	.89	.92	.84	.89	.95	.97
2-1/2	.72	.87	.90	.92	.96	.87	.92	.99	1.06
3	.79	.95	.99	1.01	1.06	.95	1.01	1.10	1.12
3-1/2	–	–	–	–	–	–	–	–	–
4	–	–	–	–	–	–	–	–	–

Manhours include checking fittings out of job storage, handling, job hauling, and complete assembling of make-up of fitting.

Manhours exclude installation of flexible conduit, and the placing and removal of scaffolding. See respective tables for these time frames.

BRANCH, SERVICE, AND FEEDER WIRING

This section includes manhour tables for the installation of branch, service, and feeder wire and cable and their connections as may be related to a process or industrial complex.

It is the intent of the manhour units to cover all normal directly related labor operations as may be involved for the complete installation of the items as outlined.

The manhours are averages of many projects installed under varied conditions and therefore should be adjusted to fit the needs of a specific project in accordance with the Introduction of this manual.

BRANCH CIRCUIT WIRING
SINGLE CONDUCTOR—600 VOLTS
Size #8 and Smaller—Solid Copper Wire
For Heights to 10 Feet

MANHOURS PER LINEAR FOOT AND NUMBER OF WIRES LISTED

Type of Insulation	Wire Size	Runs in Length											
		To 15'				16' to 30'				31' to 50'			
		Number of Wires				Number of Wires				Number of Wires			
		2	3	4	6	2	3	4	6	2	3	4	6
Rubber	#16	.015	.020	.028	.039	.014	.019	.025	.036	.013	.017	.024	.033
	#14	.016	.021	.029	.041	.014	.019	.025	.036	.014	.019	.025	.036
	#12	.019	.025	.035	.049	.017	.023	.031	.044	.016	.021	.029	.041
	#10	.028	.037	.051	.072	.025	.033	.045	.064	.024	.032	.044	.062
	# 8	.033	.044	.060	.085	.030	.040	.055	.077	.028	.037	.051	.072
Thermo-Plastic	#16	.014	.019	.025	.036	.013	.017	.024	.033	.012	.016	.022	.031
	#14	.015	.020	.028	.039	.014	.019	.025	.036	.013	.017	.024	.033
	#12	.018	.024	.033	.046	.016	.021	.029	.041	.015	.020	.027	.038
	#10	.026	.035	.047	.067	.023	.031	.042	.059	.022	.029	.040	.056
	# 8	.031	.041	.056	.080	.028	.037	.051	.072	.026	.035	.047	.067

For Heights to 15 Feet

MANHOURS PER LINEAR FOOT AND HUMBER OF WIRES LISTED

Type of Insulation	Wire Size	Runs in Length											
		To 15'				16' to 30'				31' to 50'			
		Number of Wires				Number of Wires				Number of Wires			
		2	3	4	6	2	3	4	6	2	3	4	6
Rubber	#16	.016	.021	.029	.041	.014	.019	.025	.036	.014	.019	.025	.036
	#14	.017	.023	.031	.044	.015	.020	.028	.039	.014	.019	.025	.036
	#12	.021	.028	.038	.054	.019	.025	.035	.049	.018	.024	.033	.046
	#10	.030	.040	.040	.077	.027	.036	.049	.069	.026	.035	.047	.067
	# 8	.036	.048	.048	.092	.032	.043	.058	.082	.031	.041	.056	.080
Thermo-Plastic	#16	.015	.020	.028	.039	.014	.019	.025	.036	.013	.017	.024	.033
	#14	.016	.021	.029	.041	.014	.019	.025	.036	.014	.019	.025	.036
	#12	.019	.025	.035	.049	.017	.023	.031	.044	.016	.021	.029	.041
	#10	.028	.037	.051	.072	.025	.033	.045	.064	.024	.032	.044	.062
	# 8	.033	.044	.060	.085	.030	.040	.055	.077	.028	.037	.051	.072

Manhours include checking out of job storage, handling, job hauling, placing wire coils or spools on reel stands or racks or opening wire cartons, fishing raceway runs, attaching wire to fish tape or other pull-in means, pulling wire into conduit type raceway and duct or pull-out and laying wire into wireways or channels, identifying or "polling out" the circuit conductors, and splicing and tagging for connections to panelboards, equipment, or outlet device.

If duplicate pulls above 6 are made reduce manhours 4% for each 2-, 3-, 4-, or 6-wire installation to compensate for equipment set-up time.

Manhours apply for all types of rubber or thermoplastic insulated wire.

Manhours exclude extensive identification, "ringing-out," connection to panelboard, equipment or outlet device and scaffolding. See respective tables for these time frames.

BRANCH CIRCUIT WIRING
SINGLE CONDUCTOR—600 VOLTS
Size #8 and Smaller—Solid Copper Wire
For Heights to 20 Feet

MANHOURS PER LINEAR FOOT AND NUMBER OF WIRES LISTED

Type of Insulation	Wire Size	Runs in Length											
		To 15'				16' to 30'				31' to 50'			
		Number of Wires				Number of Wires				Number of Wires			
		2	3	4	6	2	3	4	6	2	3	4	6
Rubber	#16	.017	.023	.031	.044	.015	.020	.028	.039	.014	.019	.025	.036
	#14	.018	.024	.033	.046	.016	.021	.029	.041	.015	.020	.028	.039
	#12	.021	.028	.038	.054	.019	.025	.035	.049	.018	.024	.033	.046
	#10	.031	.041	.056	.080	.028	.037	.051	.072	.026	.035	.047	.067
	# 8	.037	.049	.067	.095	.033	.044	.060	.085	.031	.041	.056	.080
Thermo-Plastic	#16	.016	.021	.029	.041	.014	.019	.025	.036	.014	.019	.025	.036
	#14	.017	.023	.031	.044	.015	.020	.028	.039	.014	.019	.025	.036
	#12	.020	.027	.036	.051	.018	.024	.033	.046	.017	.023	.031	.044
	#10	.029	.039	.053	.074	.026	.035	.047	.067	.025	.033	.045	.065
	# 8	.035	.047	.064	.090	.032	.043	.058	.082	.030	.040	.055	.077

For Heights to 25 Feet

MANHOURS PER LINEAR FOOT AND NUMBER OF WIRES LISTED

Type of Insulation	Wire Size	Runs in Length											
		To 15'				16' to 30'				31' to 50'			
		Number of Wires				Number of Wires				Number of Wires			
		2	3	4	6	2	3	4	6	2	3	4	6
Rubber	#16	.017	.023	.031	.044	.015	.020	.028	.039	.014	.019	.025	.036
	#14	.018	.024	.033	.046	.016	.021	.029	.041	.015	.020	.028	.039
	#12	.022	.029	.040	.057	.020	.027	.036	.051	.019	.025	.035	.049
	#10	.032	.043	.058	.082	.029	.039	.053	.075	.027	.036	.049	.069
	# 8	.038	.051	.069	.099	.034	.045	.062	.087	.032	.043	.058	.082
Thermo-Plastic	#16	.016	.021	.029	.041	.014	.019	.025	.036	.014	.019	.025	.036
	#14	.017	.023	.031	.044	.015	.020	.028	.039	.014	.019	.025	.036
	#12	.021	.028	.038	.054	.019	.025	.035	.049	.018	.024	.033	.046
	#10	.030	.040	.055	.077	.027	.036	.049	.069	.026	.025	.035	.049
	# 8	.036	.048	.065	.092	.032	.043	.058	.082	.031	.041	.056	.080

Manhours include checking out of job storage, handling, job hauling, placing wire coils or spools on reel stands or racks or opening wire cartons, fishing raceway runs, attaching wire to fish tape or other pull-in means, pulling wire into conduit type raceway and duct or pull-out and laying wire into wireways or channels, identifying or "polling out" the circuit conductors, and splicing and tagging for connections to panelboards, equipment, or outlet device.

If duplicate pulls above 6 are made reduce manhours 4% for each 2-, 3-, 4-, or 6-wire installation to compensate for equipment set-up time.

Manhours apply for all types of rubber or thermoplastic insulated wire.

Manhours exclude extensive identification, "ringing-out," connection to panelboard, equipment or outlet device and scaffolding. See respective tables for these time frames.

BRANCH CIRCUIT WIRING
SINGLE CONDUCTOR—600 VOLTS
Size #8 and Smaller—Stranded Copper Wire
For Heights to 10 Feet

MANHOURS PER LINEAR FOOT AND NUMBER OF WIRES LISTED

Type of Insulation	Wire Size	Runs in Length											
		To 15'				16' to 30'				31' to 50'			
		Number of Wires				Number of Wires				Number of Wires			
		2	3	4	6	2	3	4	6	2	3	4	6
Rubber	#16	.017	.023	.031	.044	.015	.020	.028	.039	.014	.019	.025	.036
	#14	.019	.025	.035	.049	.017	.023	.031	.044	.016	.021	.029	.041
	#12	.021	.028	.038	.054	.019	.025	.035	.049	.018	.024	.033	.046
	#10	.030	.040	.055	.077	.027	.036	.049	.069	.026	.035	.047	.067
	# 8	.035	.047	.064	.090	.032	.043	.058	.082	.030	.040	.055	.077
Thermo-Plastic	#16	.016	.021	.029	.041	.014	.019	.025	.036	.014	.019	.025	.036
	#14	.017	.023	.031	.044	.015	.020	.028	.039	.014	.019	.025	.036
	#12	.020	.027	.036	.051	.018	.024	.033	.046	.017	.023	.031	.044
	#10	.028	.037	.051	.072	.025	.033	.045	.064	.024	.032	.044	.062
	# 8	.033	.044	.060	.085	.030	.040	.055	.077	.028	.037	.051	.072

For Heights to 15 Feet

MANHOURS PER LINEAR FOOT AND NUMBER OF WIRES LISTED

Type of Insulation	Wire Size	Runs in Length											
		To 15'				16' to 30'				31' to 50'			
		Number of Wires				Number of Wires				Number of Wires			
		2	3	4	6	2	3	4	6	2	3	4	6
Rubber	#16	.018	.024	.033	.046	.016	.021	.029	.041	.015	.020	.028	.039
	#14	.021	.028	.038	.054	.019	.025	.035	.049	.018	.024	.033	.046
	#12	.023	.031	.042	.059	.021	.028	.038	.054	.020	.027	.036	.051
	#10	.032	.043	.058	.082	.029	.039	.053	.074	.027	.036	.049	.069
	# 8	.038	.051	.069	.099	.034	.045	.062	.087	.032	.043	.058	.082
Thermo-Plastic	#16	.017	.023	.031	.044	.015	.020	.028	.039	.014	.019	.025	.036
	#14	.018	.024	.033	.046	.016	.021	.029	.041	.015	.020	.028	.039
	#12	.022	.029	.040	.056	.020	.027	.036	.051	.019	.025	.035	.049
	#10	.030	.040	.055	.077	.027	.036	.049	.069	.026	.035	.047	.067
	# 8	.036	.048	.065	.092	.032	.043	.058	.082	.031	.041	.056	.080

Manhours include checking out of job storage, handling, job hauling, placing wire coils or spools on reel stands or racks or opening wire cartons, fishing raceway runs, attaching wire to fish tape or other pull-in means, pulling wire into conduit type raceway and duct or pull-out and laying wire into wireways or channels, identifying or "polling out" the circuit conductors, and splicing and tagging for connections to panelboards, equipment, or outlet device.

If duplicate pulls above 6 are made reduce manhours 4% for each 2-, 3-, 4-, or 6-wire installation to compensate for equipment set-up time.

Manhours apply for all types of rubber or thermoplastic insulated wire.

Manhours exclude extensive identification, "ringing-out," connection to panelboard, equipment or outlet device and scaffolding. See respective tables for these time frames.

BRANCH CIRCUIT WIRING
SINGLE CONDUCTOR—600 VOLTS
Size #8 and Smaller—Stranded Copper Wire
For Heights to 20 Feet

MANHOURS PER LINEAR FOOT AND NUMBER OF WIRES LISTED

Type of Insulation	Wire Size	Runs in Length											
		To 15'				16' to 30'				31' to 50'			
		Number of Wires				Number of Wires				Number of Wires			
		2	3	4	6	2	3	4	6	2	3	4	6
Rubber	#16	.019	.025	.035	.049	.017	.023	.031	.044	.016	.021	.029	.041
	#14	.021	.028	.038	.054	.019	.025	.035	.049	.018	.024	.033	.046
	#12	.024	.032	.044	.062	.022	.029	.040	.056	.020	.027	.036	.051
	#10	.034	.045	.062	.087	.031	.041	.056	.080	.029	.039	.053	.074
	# 8	.039	.052	.071	.100	.035	.047	.064	.090	.033	.044	.060	.085
Thermo-Plastic	#16	.018	.024	.033	.046	.016	.021	.029	.041	.015	.020	.028	.039
	#14	.019	.025	.035	.049	.017	.023	.031	.044	.016	.021	.029	.041
	#12	.022	.029	.040	.056	.020	.027	.036	.051	.019	.025	.035	.049
	#10	.031	.041	.056	.080	.028	.037	.051	.072	.026	.035	.047	.067
	# 8	.037	.049	.067	.095	.033	.044	.060	.085	.031	.041	.056	.080

For Heights to 25 Feet

MANHOURS PER LINEAR FOOT AND NUMBER OF WIRES LISTED

Type of Insulation	Wire Size	Runs in Length											
		To 15'				16' to 30'				31' to 50'			
		Number of Wires				Number of Wires				Number of Wires			
		2	3	4	6	2	3	4	6	2	3	4	6
Rubber	#16	.020	.027	.036	.051	.018	.024	.033	.046	.017	.023	.031	.044
	#14	.022	.029	.040	.056	.020	.027	.036	.051	.019	.025	.035	.049
	#12	.024	.032	.044	.062	.022	.029	.040	.056	.020	.027	.036	.051
	#10	.035	.047	.064	.090	.032	.043	.058	.082	.030	.040	.055	.077
	# 8	.040	.053	.073	.103	.036	.048	.065	.092	.034	.044	.062	.087
Thermo-Plastic	#16	.018	.024	.033	.046	.016	.021	.029	.041	.015	.020	.029	.039
	#14	.020	.027	.036	.051	.018	.024	.033	.046	.017	.023	.031	.044
	#12	.023	.031	.042	.059	.021	.028	.038	.054	.020	.027	.036	.051
	#10	.032	.043	.058	.082	.029	.039	.053	.074	.027	.036	.049	.069
	# 8	.038	.051	.069	.099	.034	.045	.062	.087	.032	.043	.058	.082

Manhours include checking out of job storage, handling, job hauling, placing wire coils or spools on reel stands or racks or opening wire cartons, fishing raceway runs, attaching wire to fish tape or other pull-in means, pulling wire into conduit type raceway and duct or pull-out and laying wire into wireways or channels, identifying or "polling out" the circuit conductors, and splicing and tagging for connections to panelboards, equipment, or outlet device.

If duplicate pulls above 6 are made reduce manhours 4% for each 2-, 3-, 4-, or 6-wire installation to compensate for equipment set-up time.

Manhours apply for all types of rubber or thermoplastic insulated wire.

Manhours exclude extensive identification, "ringing-out," connection to panelboard, equipment or outlet device and scaffolding. See respective tables for these time frames.

BRANCH CIRCUIT WIRING
SINGLE CONDUCTOR—600 VOLTS
Size #8 and Smaller—Solid Aluminum Wire
For Heights to 10 Feet

MANHOURS PER LINEAR FOOT AND NUMBER OF WIRES LISTED

Type of Insulation	Wire Size	Runs in Length											
		To 15'				16' to 30'				31' to 50'			
		Number of Wires				Number of Wires				Number of Wires			
		2	3	4	6	2	3	4	6	2	3	4	6
Rubber	#16	–	–	–	–	–	–	–	–	–	–	–	–
	#14	–	–	–	–	–	–	–	–	–	–	–	–
	#12	.019	.025	.035	.049	.017	.023	.031	.044	.016	.021	.029	.041
	#10	.028	.037	.051	.072	.025	.033	.045	.064	.024	.032	.044	.062
	# 8	.033	.044	.060	.085	.030	.040	.055	.077	.028	.037	.051	.072
Thermo-Plastic	#16	–	–	–	–	–	–	–	–	–	–	–	–
	#14	–	–	–	–	–	–	–	–	–	–	–	–
	#12	.018	.024	.033	.046	.016	.021	.029	.041	.015	.020	.029	.039
	#10	.026	.035	.047	.067	.023	.031	.042	.059	.022	.029	.040	.056
	# 8	.031	.041	.056	.080	.028	.037	.051	.072	.026	.035	.047	.067

For Heights to 15 Feet

MANHOURS PER LINEAR FOOT AND NUMBER OF WIRES LISTED

Type of Insulation	Wire Size	Runs in Length											
		To 15'				16' to 30'				31' to 50'			
		Number of Wires				Number of Wires				Number of Wires			
		2	3	4	6	2	3	4	6	2	3	4	6
Rubber	#16	–	–	–	–	–	–	–	–	–	–	–	–
	#14	–	–	–	–	–	–	–	–	–	–	–	–
	#12	.021	.028	.038	.054	.019	.025	.035	.049	.018	.024	.033	.046
	#10	.039	.040	.055	.077	.027	.036	.049	.069	.026	.035	.047	.067
	# 8	.036	.048	.065	.092	.032	.043	.058	.082	.031	.041	.056	.080
Thermo-Plastic	#16	–	–	–	–	–	–	–	–	–	–	–	–
	#14	–	–	–	–	–	–	–	–	–	–	–	–
	#12	.019	.025	.035	.049	.017	.023	.031	.044	.016	.021	.029	.041
	#10	.028	.037	.051	.072	.025	.033	.045	.064	.024	.032	.044	.062
	# 8	.033	.044	.060	.085	.030	.040	.055	.077	.028	.037	.051	.072

Manhours include checking out of job storage, handling, job hauling, placing wire coils or spools on reel stands or racks or opening wire cartons, fishing raceway runs, attaching wire to fish tape or other pull-in means, pulling wire into conduit type raceway and duct or pull-out and laying wire into wireways or channels, identifying or "polling out" the circuit conductors, and splicing and tagging for connections to panelboards, equipment, or outlet device.

If duplicate pulls above 6 are made reduce manhours 4% for each 2-, 3-, 4-, or 6-wire installation to compensate for equipment set-up time.

Manhours apply for all types of rubber or thermoplastic insulated wire.

Manhours exclude extensive identification, "ringing-out," connection to panelboard, equipment or outlet device and scaffolding. See respective tables for these time frames.

BRANCH CIRCUIT WIRING
SINGLE CONDUCTOR—600 VOLTS
Size #8 and Smaller—Solid Aluminum Wire
For Heights to 20 Feet

MANHOURS PER LINEAR FOOT AND NUMBER OF WIRES LISTED

Type of Insulation	Wire Size	Runs in Length											
		To 15'				16' to 30'				31' to 50'			
		Number of Wires				Number of Wires				Number of Wires			
		2	3	4	6	2	3	4	6	2	3	4	6
Rubber	#16	–	–	–	–	–	–	–	–	–	–	–	–
	#14	–	–	–	–	–	–	–	–	–	–	–	–
	#12	.021	.028	.038	.054	.019	.025	.035	.049	.018	.024	.033	.046
	#10	.031	.041	.056	.080	.028	.037	.051	.072	.026	.035	.047	.067
	# 8	.037	.049	.067	.095	.033	.044	.060	.085	.031	.041	.056	.080
Thermo-Plastic	#16	–	–	–	–	–	–	–	–	–	–	–	–
	#14	–	–	–	–	–	–	–	–	–	–	–	–
	#12	.020	.027	.036	.051	.018	.024	.033	.046	.017	.023	.031	.044
	#10	.029	.039	.053	.074	.026	.035	.047	.067	.025	.033	.045	.064
	# 8	.035	.047	.064	.090	.032	.043	.058	.082	.030	.040	.055	.077

For Heights to 25 Feet

MANHOURS PER LINEAR FOOT AND NUMBER OF WIRES LISTED

Type of Insulation	Wire Size	Runs in Length											
		To 15'				16' to 30'				31' to 50'			
		Number of Wires				Number of Wires				Number of Wires			
		2	3	4	6	2	3	4	6	2	3	4	6
Rubber	#16	–	–	–	–	–	–	–	–	–	–	–	–
	#14	–	–	–	–	–	–	–	–	–	–	–	–
	#12	.022	.029	.040	.056	.020	.027	.036	.051	.019	.025	.035	.049
	#10	.032	.043	.058	.082	.029	.039	.053	.074	.027	.036	.049	.069
	# 8	.038	.051	.069	.099	.034	.045	.062	.087	.032	.043	.058	.082
Thermo-Plastic	#16	–	–	–	–	–	–	–	–	–	–	–	–
	#14	–	–	–	–	–	–	–	–	–	–	–	–
	#12	.021	.028	.038	.054	.019	.025	.035	.049	.0218	.024	.033	.046
	#10	.030	.040	.055	.077	.027	.036	.049	.069	.026	.035	.047	.067
	# 8	.036	.048	.065	.092	.032	.043	.058	.082	.031	.041	.056	.080

Manhours include checking out of job storage, handling, job hauling, placing wire coils or spools on reel stands or racks or opening wire cartons, fishing raceway runs, attaching wire to fish tape or other pull-in means, pulling wire into conduit type raceway and duct or pull-out and laying wire into wireways or channels, identifying or "polling out" the circuit conductors, and splicing and tagging for connections to panelboards, equipment, or outlet device.

If duplicate pulls above 6 are made reduce manhours 4% for each 2-, 3-, 4-, or 6-wire installation to compensate for equipment set-up time.

Manhours apply for all types of rubber or thermoplastic insulated wire.

Manhours exclude extensive identification, "ringing-out," connection to panelboard, equipment or outlet device and scaffolding. See respective tables for these time frames.

BRANCH CIRCUIT WIRING
SINGLE CONDUCTOR—600 VOLTS
Size #8 and Smaller—Stranded Aluminum Wire
For Heights to 10 Feet

MANHOURS PER LINEAR FOOT AND NUMBER OF WIRES LISTED

Type of Insulation	Wire Size	Runs in Length											
		To 15'				16' to 30'				31' to 50'			
		Number of Wires				Number of Wires				Number of Wires			
		2	3	4	6	2	3	4	6	2	3	4	6
Rubber	#16	–	–	–	–	–	–	–	–	–	–	–	–
	#14	–	–	–	–	–	–	–	–	–	–	–	–
	#12	.020	.027	.036	.051	.019	.025	.035	.049	.018	.024	.033	.046
	#10	.030	.040	.055	.077	.027	.036	.049	.069	.026	.035	.047	.067
	# 8	.035	.047	.064	.090	.032	.043	.058	.082	.030	.040	.055	.077
Thermo-Plastic	#16	–	–	–	–	–	–	–	–	–	–	–	–
	#14	–	–	–	–	–	–	–	–	–	–	–	–
	#12	.020	.027	.036	.051	.018	.024	.033	.046	.017	.023	.031	.044
	#10	.028	.037	.051	.072	.025	.033	.045	.064	.024	.032	.044	.062
	# 8	.033	.044	.060	.085	.030	.040	.055	.077	.028	.037	.051	.072

For Heights to 15 Feet

MANHOURS PER LINEAR FOOT AND NUMBER OF WIRES LISTED

Type of Insulation	Wire Size	Runs in Length											
		To 15'				16' to 30'				31' to 50'			
		Number of Wires				Number of Wires				Number of Wires			
		2	3	4	6	2	3	4	6	2	3	4	6
Rubber	#16	–	–	–	–	–	–	–	–	–	–	–	–
	#14	m	–	–	–	–	–	–	–	–	–	–	–
	#12	.022	.029	.040	.056	.021	.028	.038	.054	.020	.027	.036	.051
	#10	.032	.043	.058	.082	.029	.039	.053	.074	.027	.036	.049	.069
	# 8	.038	.051	.069	.099	.034	.045	.062	.087	.032	.043	.058	.082
Thermo-Plastic	#16	–	–	–	–	–	–	–	–	–	–	–	–
	#14	–	–	–	–	–	–	–	–	–	–	–	–
	#12	.022	.029	.040	.056	.020	.027	.036	.051	.019	.025	.035	.049
	#10	.030	.040	.055	.077	.027	.036	.049	.069	.026	.035	.047	.067
	# 8	.036	.048	.065	.092	.032	.043	.058	.082	.031	.041	.056	.080

Manhours include checking out of job storage, handling, job hauling, placing wire coils or spools on reel stands or racks or opening wire cartons, fishing raceway runs, attaching wire to fish tape or other pull-in means, pulling wire into conduit type raceway and duct or pull-out and laying wire into wireways or channels, identifying or "polling out" the circuit conductors, and splicing and tagging for connections to panelboards, equipment, or outlet device.

If duplicate pulls above 6 are made reduce manhours 4% for each 2-, 3-, 4-, or 6-wire installation to compensate for equipment set-up time.

Manhours apply for all types of rubber or thermoplastic insulated wire.

Manhours exclude extensive identification, "ringing-out," connection to panelboard, equipment or outlet device and scaffolding. See respective tables for these time frames.

BRANCH CIRCUIT WIRING
SINGLE CONDUCTOR—600 VOLTS
Size #8 and Smaller—Stranded Aluminum Wire
For Heights to 20 Feet

MANHOURS PER LINEAR FOOT AND NUMBER OF WIRES LISTED

Type of Insulation	Wire Size	To 15' Number of Wires				16' to 30' Number of Wires				31' to 50' Number of Wires			
		2	3	4	6	2	3	4	6	2	3	4	6
Rubber	#16	–	–	–	–	–	–	–	–	–	–	–	–
	#14	–	–	–	–	–	–	–	–	–	–	–	–
	#12	.022	.029	.029	.056	.020	.027	.036	.051	.019	.025	.035	.049
	#10	.034	.045	.062	.087	.031	.041	.056	.080	.029	.039	.053	.074
	# 8	.039	.052	.071	.100	.035	.047	.064	.090	.033	.044	.060	.085
Thermo-Plastic	#16	–	–	–	–	–	–	–	–	–	–	–	–
	#14	–	–	–	–	–	–	–	–	–	–	–	–
	#12	.022	.029	.040	.056	.020	.027	.036	.051	.019	.025	.035	.049
	#10	.031	.041	.056	.080	.028	.037	.051	.072	.026	.035	.047	.067
	# 8	.037	.049	.067	.095	.033	.044	.060	.085	.031	.041	.056	.080

For Heights to 25 Feet

MANHOURS PER LINEAR FOOT AND NUMBER OF WIRES LISTED

Type of Insulation	Wire Size	To 15' Number of Wires				16' to 30' Number of Wires				31' to 50' Number of Wires			
		2	3	4	6	2	3	4	6	2	3	4	6
Rubber	#16	–	–	–	–	–	–	–	–	–	–	–	–
	#14	–	–	–	–	–	–	–	–	–	–	–	–
	#12	.023	.031	.042	.059	.021	.028	.038	.054	.020	.027	.036	.051
	#10	.035	.047	.064	.090	.032	.043	.058	.082	.030	.040	.055	.077
	# 8	.040	.053	.073	.103	.036	.048	.065	.092	.034	.045	.062	.087
Thermo-Plastic	#16	–	–	–	–	–	–	–	–	–	–	–	–
	#14	–	–	–	–	–	–	–	–	–	–	–	–
	#12	.023	.031	.042	.059	.021	.028	.038	.054	.020	.027	.036	.051
	#10	.032	.043	.058	.082	.029	.039	.053	.074	.027	.036	.049	.069
	# 8	.038	.051	.069	.099	.034	.045	.062	.087	.032	.043	.058	.082

Manhours include checking out of job storage, handling, job hauling, placing wire coils or spools on reel stands or racks or opening wire cartons, fishing raceway runs, attaching wire to fish tape or other pull-in means, pulling wire into conduit type raceway and duct or pull-out and laying wire into wireways or channels, identifying or "polling out" the circuit conductors, and splicing and tagging for connections to panelboards, equipment, or outlet device.

If duplicate pulls above 6 are made reduce manhours 4% for each 2-, 3-, 4-, or 6-wire installation to compensate for equipment set-up time.

Manhours apply for all types of rubber or thermoplastic insulated wire.

Manhours exclude extensive identification, "ringing-out," connection to panelboard, equipment or outlet device and scaffolding. See respective tables for these time frames.

FIXTURE WIRING
SINGLE CONDUCTOR—600 VOLTS
Size #12 and Smaller—Stranded and Solid Copper

MANHOURS PER HUNDRED FEET OF SINGLE CONDUCTOR

Wire Type	Wire Size	Manhours for Height to			
		10'	15'	20'	25'
AF Plain Stranded	#18	2.00	2.16	2.24	2.30
	#16	2.20	2.38	2.47	2.55
	#14	2.40	2.59	2.70	2.78
	#12	2.62	2.81	2.92	3.00
AF Braid Stranded	#18	2.00	2.16	2.24	2.30
	#16	2.20	2.38	2.47	2.55
	#14	2.40	2.59	2.70	2.78
	#12	2.62	2.81	2.92	3.00
RF Stranded	#18	2.00	2.16	2.24	2.30
	#16	2.20	2.38	2.47	2.55
TF Stranded	#18	1.87	2.00	2.10	2.16
	#16	2.13	2.30	2.39	2.46
TF Solid	#18	1.87	2.00	2.10	2.16
	#16	2.06	2.22	2.31	2.38

Manhours include checking out of job storage, handling, job hauling, placing and opening wire cartons, fishing raceway runs, attaching wire to fish tape or other pull-in means, pulling wire into conduit type raceway or pull-out and laying wire into fixture or support channels, identifying or "polling out" the circuit conductors, and splicing and tagging for connections to lighting fixtures.

Manhours are for installation of single conductor only. If second wire or ground wire is installed at same time deduct 20% from manhours for second wire.

Manhours exclude connecting of wire to lighting device and scaffolding. See respective tables for these time frames.

For heights above 25 feet increase manhours 3% for each additional 5 feet or fraction thereof.

BOILER ROOM WIRING
SINGLE CONDUCTOR—300 VOLTS
Size #8 and Smaller—Solid and Stranded Copper

MANHOURS PER HUNDRED FEET OF SINGLE CONDUCTOR

Wire Type	Wire Size	Manhours for Height to			
		10'	15'	20'	25'
AVA Solid	#16	2.00	2.16	2.24	2.30
	#14	2.15	2.32	2.41	2.49
	#12	2.42	2.61	2.72	2.80
	#10	3.15	3.40	3.54	3.64
	# 8	4.00	4.32	4.49	4.63
AVA Stranded	#16	2.24	2.42	2.52	2.59
	#14	2.37	2.56	2.66	2.74
	#12	2.64	2.85	2.97	3.05
	#10	3.74	4.04	4.20	4.33
	# 8	4.29	4.63	4.82	4.96

Manhours include checking out of job storage, handling, job hauling, placing and opening wire cartons, fishing raceway runs, attaching wire to fish tape or other pull-in means, pulling wire into raceway, identifying or "polling out" the circuit conductors, splicing as required, and tagging for connections.

Manhours are for installation of single conductor only. If second or ground wire is installed at the same time deduct 20% from manhours for second wire.

Manhours exclude installation of raceway, connecting of wiring, and scaffolding. See respective tables for these time frames.

For heights above 25 feet increase manhours 5% for each additional 5 feet or fraction thereof.

FLEXIBLE METALLIC ARMORED CABLE FOR BRANCH CIRCUITS

BX and BXL Types—Wire Size #8 and Smaller

Wood or Metal Frame Construction (Concealed Work)

MANHOURS PER 100 LINEAR FEET

Type of Cable	Wire Size	Number of Conductors	Manhours for Height to			
			10'	15'	20'	25'
AC Solid BX	#14	2	3.71	4.00	4.17	4.29
	#14	3	4.33	4.68	4.86	5.00
	#14	4	4.95	5.35	5.56	5.73
	#12	2	4.33	4.68	4.86	5.00
	#12	3	4.95	5.35	5.56	5.73
	#12	4	5.50	5.94	6.18	6.36
	#10	2	6.05	6.53	6.80	7.00
	#10	3	6.67	7.20	7.49	7.72
	#10	4	7.56	8.16	8.49	8.75
	# 8	2	7.29	7.87	8.19	8.43
	# 8	3	8.53	9.21	9.58	9.87
	# 8	4	9.90	10.69	11.12	11.45
ACL Solid BXL	#14	2	3.85	4.16	4.32	4.45
	#14	3	4.47	4.83	5.02	5.17
	#12	2	4.47	4.83	5.02	5.17
	#12	3	4.95	5.35	5.56	5.73
	#10	2	5.98	6.46	6.72	6.92
	#10	3	6.74	7.28	7.57	7.80
	# 8	2	7.84	8.47	8.81	9.07
	# 8	3	8.94	9.66	10.04	10.34
Solid Armored	# 8	1	4.40	4.75	4.94	5.09
	# 6	1	5.50	5.94	6.18	6.36
	# 4	1	6.60	7.13	7.41	7.64

Manhours include checking out of job storage, handling, job hauling, placing cable reels on stands or racks, layout, cutting, installing, placing normal cable clamps at termination, and normal securing of cable in position.

Manhours exclude installation of special connectors, fasteners and anchors, and scaffolding. See respective tables for these time frames.

For heights above 25 feet increase manhours 5% for each additional 5 feet or fraction thereof.

FLEXIBLE METALLIC ARMORED CABLE FOR BRANCH CIRCUITS

BX and BXL Types—Wire Size #8 and Smaller

Masonry Walls (Concealed Work)

MANHOURS PER 100 LINEAR FEET

Type of Cable	Wire Size	Number of Conductors	Manhours for Height to			
			10'	15'	20'	25'
AC Solid BX	#14	2	4.33	4.68	4.86	5.00
	#14	3	5.05	5.45	5.67	5.84
	#14	4	5.77	6.23	6.48	6.68
	#12	2	5.05	5.45	5.67	5.84
	#12	3	5.77	6.23	6.48	6.68
	#12	4	6.41	6.92	7.20	7.42
	#10	2	7.05	7.61	7.92	8.16
	#10	3	7.78	8.40	8.74	9.00
	#10	4	8.81	9.51	9.90	10.19
	# 8	2	8.50	9.18	9.55	9.83
	# 8	3	9.95	10.75	11.18	11.51
	# 8	4	11.54	12.46	12.96	13.35
ACL Solid BXL	#14	2	4.50	4.86	5.05	5.21
	#14	3	5.21	5.63	5.85	6.03
	#12	2	5.21	5.63	5.85	6.03
	#12	3	5.77	6.23	6.48	6.68
	#10	2	6.97	7.53	7.83	8.06
	#10	3	7.86	8.49	8.83	9.09
	# 8	2	9.14	9.87	10.27	10.57
	# 8	3	10.42	11.25	11.70	12.05
Solid Armored	# 8	1	5.13	5.54	5.76	5.93
	# 6	1	6.41	6.92	7.20	7.42
	# 4	1	7.70	8.32	8.65	8.91

Manhours include checking out of job storage, handling, job hauling, placing cable reels on stands or racks, layout, cutting, installing, placing normal cable clamps at termination, and normal securing of cable in position.

Manhours exclude installation of special connectors, fasteners and anchors, and scaffolding. See respective tables for these time frames.

For heights above 25 feet increase manhours 5% for each additional 5 feet or fraction thereof.

FLEXIBLE METALLIC ARMORED CABLE
FOR BRANCH CIRCUITS

BX and BXL Types—Wire Size #8 and Smaller

Suspended Ceilings (Concealed Work)

MANHOURS PER 100 LINEAR FEET

Type of Cable	Wire Size	Number of Conductors	Manhours for Height to			
			10′	15′	20′	25′
AC Solid BX	#14	2	4.12	4.45	4.63	4.77
	#14	3	4.81	5.19	5.40	5.56
	#14	4	5.49	5.93	6.17	6.35
	#12	2	4.81	5.19	5.40	5.56
	#12	3	5.49	5.93	6.17	6.35
	#12	4	6.11	6.60	6.86	7.07
	#10	2	6.72	7.26	7.55	7.77
	#10	3	7.40	7.99	8.31	8.56
	#10	4	8.39	9.06	9.42	9.71
	# 8	2	8.09	8.74	9.09	9.36
	# 8	3	9.47	10.23	10.64	10.96
	# 8	4	11.00	11.88	12.36	12.73
ACL Solid BXL	#14	2	4.27	4.61	4.80	4.94
	#14	3	4.96	5.36	5.57	5.74
	#12	2	4.96	5.36	5.57	5.74
	#12	3	5.49	5.93	6.17	6.35
	#10	2	6.64	7.17	7.45	7.68
	#10	3	7.48	8.07	8.40	8.65
	# 8	2	8.70	9.40	9.77	10.06
	# 8	3	9.92	10.71	11.14	11.48
Solid Armored	# 8	1	4.88	5.27	5.48	5.65
	# 6	1	6.11	6.60	6.86	7.07
	# 4	1	7.33	7.92	8.23	8.48

Manhours include checking out of job storage, handling, job hauling, placing cable reels on stands or racks, layout, cutting, installing, placing normal cable clamps at termination, and normal securing of cable in position.

Manhours exclude installation of special connectors, fasteners and anchors, and scaffolding. See respective tables for these time frames.

For heights above 25 feet increase manhours 5% for each additional 5 feet or fraction thereof.

FLEXIBLE METALLIC ARMORED CABLE FOR BRANCH CIRCUITS

BX and BXL Types—Wire Size #8 and Smaller

Flat Ceilings or Walls (Exposed Work)

MANHOURS PER 100 LINEAR FEET

Type of Cable	Wire Size	Number of Conductors	Manhours for Height to			
			10'	15'	20'	25'
AC Solid BX	#14	2	3.91	4.22	4.39	4.52
	#14	3	4.57	4.94	5.13	5.29
	#14	4	5.22	5.64	5.86	6.04
	#12	2	4.57	4.94	5.13	5.29
	#12	3	5.22	5.64	5.86	6.04
	#12	4	5.80	6.26	6.51	6.71
	#10	2	6.38	6.89	7.17	7.38
	#10	3	7.04	7.60	7.91	8.14
	#10	4	7.98	8.62	8.96	9.23
	# 8	2	7.69	8.31	8.64	8.90
	# 8	3	9.00	8.72	10.11	10.41
	# 8	4	10.44	11.28	11.73	12.08
ACL Solid BXL	#14	2	4.06	4.38	4.56	4.70
	#14	3	4.72	5.10	5.30	5.46
	#12	2	4.72	5.10	5.30	5.46
	#12	3	5.22	5.64	5.86	6.04
	#10	2	6.31	6.81	7.09	7.30
	#10	3	7.11	7.68	7.99	8.23
	# 8	2	8.27	8.93	9.29	9.57
	# 8	3	8.43	10.18	10.59	10.91
Solid Armored	# 8	1	4.64	5.00	5.21	5.37
	# 6	1	5.80	6.26	6.51	6.71
	# 4	1	6.96	7.52	7.82	8.05

Manhours include checking out of job storage, handling, job hauling, placing cable reels on stands or racks, layout, cutting, installing, placing normal cable clamps at termination, and normal securing of cable in position.

Manhours exclude installation of special connectors, fasteners and anchors, and scaffolding. See respective tables for these time frames.

For heights above 25 feet increase manhours 5% for each additional 5 feet or fraction thereof.

NONMETALLIC SHEATHED CABLE
FOR BRANCH CIRCUITS

Type NM Braid Sheathed Copper Conductor (Romex)
Type NM Plastic Sheathed Copper Conductor (Romex)
Type UF Thermoplastic Sheathed Copper Conductor
Type SE Unarmored 2 Insulated, 1 Uninsulated Copper Conductor
Wood or Metal Frame Construction (Concealed Work)

MANHOURS PER 100 LINEAR FEET

Type of Cable	Wire Size	Number of Conductors	Manhours for Height to			
			10'	15'	20'	25'
	#14	1	1.10	1.19	1.24	1.27
	#14	2	2.97	3.21	3.34	3.44
	#14	3	3.41	3.68	3.83	3.95
	#12	1	1.38	1.49	1.55	1.60
Types NM & UF	#12	2	3.41	3.68	3.83	3.95
Braid Cable–No. Grd.	#12	3	3.91	4.22	4.39	4.52
Plastic Cable–No. Grd.	#10	1	1.65	1.78	1.85	1.91
Thermoplastic–No. Grd.	#10	2	4.79	5.17	5.38	5.54
	#10	3	5.45	5.89	6.12	6.31
	# 8	1	2.20	2.38	2.47	2.55
	# 8	2	5.34	5.77	6.00	6.18
	# 8	3	6.00	6.48	6.74	6.94
	#14	2	3.47	3.75	3.90	4.00
	#14	3	3.91	4.22	4.39	4.52
Types NM & UF	#12	2	4.46	4.22	4.39	4.52
Braid Cable W/Grd.	#12	3	5.46	4.82	5.00	5.16
Plastic Cable W/Grd.	#10	2	6.00	5.90	6.13	6.32
Thermoplastic W/Grd.	#10	3	5.83	6.48	6.74	6.94
	# 8	2	6.38	6.30	6.55	6.74
	# 8	3		6.89	7.17	7.38
	#12	3	4.84	5.23	5.44	5.60
Type SE Unarmored	#10	3	6.27	6.77	7.04	7.25
	# 8	3	7.37	7.96	8.28	8.52

Manhours include checking out of job storage, handling, job hauling, placing cable reels on stands or racks, layout, cutting, installing, placing normal cable clamps at termination, and normal securing of cable in position.

Manhours exclude installation of special connectors, fasteners and anchors, and scaffolding. See respective tables for these time frames.

For heights above 25 feet increase manhours 5% for each additional 5 feet or fraction thereof.

NONMETALLIC SHEATHED CABLE FOR BRANCH CIRCUITS

Type NM Braid Sheathed Copper Conductor (Romex)
Type NM Plastic Sheathed Copper Conductor (Romex)
Type UF Thermoplastic Sheathed Copper Conductor
Type SE Unarmored 2 Insulated, 1 Uninsulated Copper Conductor
Masonry Walls (Concealed Work)

MANHOURS PER 100 LINEAR FEET

Type of Cable	Wire Size	Number of Conductors	Manhours for Height to			
			10'	15'	20'	25'
Types NM & UF Braid Cable—No. Grd. Plastic Cable—No. Grd. Thermoplastic—No. Grd.	#14	1	1.28	1.38	1.44	1.48
	#14	2	3.46	3.74	3.89	4.00
	#14	3	3.98	4.30	4.47	4.60
	#12	1	1.61	1.74	1.81	1.86
	#12	2	3.98	4.30	4.47	4.60
	#12	3	4.56	4.92	5.12	5.28
	#10	1	1.92	2.07	2.16	2.22
	#10	2	5.59	6.04	6.28	6.47
	#10	3	6.35	6.86	7.13	7.35
	# 8	1	2.57	2.78	2.89	2.97
	# 8	2	6.23	6.73	7.00	7.21
	# 8	3	7.00	7.56	7.86	8.10
Types NM & UF Braid Cable W/Grd. Plastic Cable W/Grd. Thermoplastic W/Grd.	#14	2	4.05	4.37	4.55	4.69
	#14	3	4.56	4.92	5.12	5.28
	#12	2	4.56	4.92	5.12	5.28
	#12	3	5.20	5.62	5.84	6.02
	#10	2	6.37	6.88	7.15	7.40
	#10	3	7.00	7.56	7.86	8.10
	# 8	2	6.80	7.34	7.64	7.87
	# 8	3	7.44	8.04	8.36	8.61
Type SE Unarmored	#12	3	5.64	6.09	6.33	6.52
	#10	3	7.31	7.89	8.21	8.46
	# 8	3	8.59	9.28	9.65	9.94

Manhours include checking out of job storage, handling, job hauling, placing cable reels on stands or racks, layout, cutting, installing, placing normal cable clamps at termination, and normal securing of cable in position.

Manhours exclude installation of special connectors, fasteners and anchors, and scaffolding. See respective tables for these time frames.

For heights above 25 feet increase manhours 5% for each additional 5 feet or fraction thereof.

NONMETALLIC SHEATHED CABLE
FOR BRANCH CIRCUITS

Type NM Braid Sheathed Copper Conductor (Romex)
Type NM Plastic Sheathed Copper Conductor (Romex)
Type UF Thermoplastic Sheathed Copper Conductor
Type SE Unarmored 2 Insulated, 1 Uninsulated Copper Conductor
Suspended Ceilings (Concealed Work)

MANHOURS PER 100 LINEAR FEET

Type of Cable	Wire Size	Number of Conductors	Manhours for Height to			
			10'	15'	20'	25'
Types NM & UF Braid Cable–No. Grd. Plastic Cable–No. Grd. Thermoplastic–No. Grd.	#14	1	1.22	1.32	1.37	1.41
	#14	2	3.30	3.56	3.71	3.82
	#14	3	3.79	4.09	4.26	4.38
	#12	1	1.53	1.65	1.72	1.77
	#12	2	3.79	4.09	4.26	4.38
	#12	3	4.34	4.69	4.87	5.02
	#10	1	1.83	1.98	2.06	2.12
	#10	2	5.32	5.75	5.98	6.15
	#10	3	6.05	6.53	6.80	7.00
	# 8	1	2.44	2.64	2.74	2.82
	# 8	2	5.93	6.40	6.66	6.86
	# 8	3	6.66	7.19	7.48	7.70
Types NM & UF Braid Cable W/Grd. Plastic Cable W/Grd. Thermoplastic W/Grd.	#14	2	3.85	4.16	4.32	4.45
	#14	3	4.34	4.69	4.87	5.02
	#12	2	4.34	4.69	4.87	5.02
	#12	3	4.95	5.35	5.56	5.73
	#10	2	6.06	6.54	6.81	7.01
	#10	3	6.66	7.19	7.48	7.70
	# 8	2	6.47	6.99	7.27	7.49
	# 8	3	7.08	7.65	7.95	8.19
Type SE Unarmored	#12	3	5.37	5.80	6.03	6.21
	#10	3	6.96	7.52	7.82	8.05
	# 8	3	8.18	8.83	9.19	9 46

Manhours include checking out of job storage, handling, job hauling, placing cable reels on stands or racks, layout, cutting, installing, placing normal cable clamps at termination, and normal securing of cable in position.

Manhours exclude installation of special connectors, fasteners and anchors, and scaffolding. See respective tables for these time frames.

For heights above 25 feet increase manhours 5% for each additional 5 feet or fraction thereof.

NONMETALLIC SHEATHED CABLE FOR BRANCH CIRCUITS

Type NM Braid Sheathed Copper Conductor (Romex)
Type NM Plastic Sheathed Copper Conductor (Romex)
Type UF Thermoplastic Sheathed Copper Conductor
Type SE Unarmored 2 Insulated, 1 Uninsulated Copper Conductor
Flat Ceilings or Walls (Exposed Work)

MANHOURS PER 100 LINEAR FEET

Type of Cable	Wire Size	Number of Conductors	Manhours for Height to			
			10'	15'	20'	25'
	#14	1	1.17	1.26	1.31	1.35
	#14	2	3.15	3.40	3.54	3.64
	#14	3	3.61	3.90	4.05	4.18
	#12	1	1.46	1.58	1.64	1.69
Types NM & UF	#12	2	3.61	3.90	4.05	4.18
Braid Cable—No. Grd.	#12	3	4.14	4.47	4.65	4.79
Plastic Cable—No. Grd.	#10	1	1.75	1.89	1.97	2.02
Thermoplastic—No. Grd.	#10	2	5.08	5.49	5.71	5.88
	#10	3	5.78	6.24	6.49	6.69
	# 8	1	2.33	2.52	2.62	2.70
	# 8	2	5.66	6.11	6.36	6.55
	# 8	3	6.36	6.87	7.14	7.36
	#14	2	3.68	3.97	4.13	4.26
	#14	3	4.14	4.47	4.65	4.79
Types NM & UF	#12	2	4.14	4.47	4.65	4.79
Braid Cable W/Grd.	#12	3	4.73	5.11	5.31	5.47
Plastic Cable W/Grd.	#10	2	5.79	6.25	6.50	6.70
Thermoplastic W/Grd.	#10	3	6.36	6.87	7.14	7.36
	# 8	2	6.18	6.67	6.94	7.15
	# 8	3	6.76	7.30	7.59	7.82
	#12	3	5.13	5.54	5.76	5.93
Type SE Unarmored	#10	3	6.65	7.18	7.47	7.69
	# 8	3	7.81	8.43	8.77	9.04

Manhours include checking out of job storage, handling, job hauling, placing cable reels on stands or racks, layout, cutting, installing, placing normal cable clamps at termination, and normal securing of cable in position.

Manhours exclude installation of special connectors, fasteners and anchors, and scaffolding. See respective tables for these time frames.

For heights above 25 feet increase manhours 5% for each additional 5 feet or fraction thereof.

CONNECTORS FOR
FLEXIBLE METALLIC ARMORED CABLE
AND
.NONMETALLIC SHEATHED CABLE

MANHOURS EACH

Item and Description	Size (Inches)	Manhours for Height to			
		10'	15'	20'	25'
Flexible Metallic Armored Cable					
Straight Connector	3/8	.044	.048	.049	.051
Straight Connector	1/2	.143	.154	.161	.165
Straight Connector	3/4	.180	.194	.202	.208
Duplex Connector	3/8	.056	.060	.063	.065
Angle Connector	3/8	.050	.054	.056	.058
Angle Connector	1/2	.147	.184	.191	.197
Angle Connector	3/4	.190	.205	.213	.220
Nonmetallic Sheathed Cable					
Snap Connector	3/8	.041	.044	.046	.047
Screw Connector	3/8	.050	.054	.056	.058
Screw Connector	1/2	.143	.154	.161	.165
Screw Connector	3/4	.180	.194	.202	.208

Manhours include checking out of job storage, handling, job hauling, placing and connecting, connectors as outlined.

Manhours exclude installation of cable, fasteners, and scaffolding. See respective tables for these time frames.

For heights above 25 feet increase manhours 5% for each additional 5 feet or fraction thereof.

LOW VOLTAGE THERMOSTAT AND LIGHTING CONTROL

Branch Cable—Size #18

Wood or Metal Frame Construction (Concealed)

MANHOURS PER 100 LINEAR FEET

Type of Cable	Wire Size	Number of Conductors	Manhours for Height to			
			10'	15'	20'	25'
Plastic Insulated Twisted, No Braid Overall Jacket Waxed Braid Jacket	#18	2	0.89	0.96	1.00	1.03
		3	1.05	1.13	1.18	1.21
		4	1.28	1.38	1.44	1.48
		5	1.31	1.41	1.47	1.52
		6	1.46	1.58	1.64	1.69
		7	1.57	1.70	1.76	1.82
Plastic Insulated Steel Armored	#18	2	1.78	1.92	2.00	2.06
		3	2.10	2.27	2.36	2.43
		4	2.33	2.52	2.62	2.70
		5	2.63	2.84	2.95	3.04
		6	2.96	3.20	3.32	3.42
Plastic Insulated Parallel	#18	2	0.89	0.96	1.00	1.03
		3	1.05	1.13	1.18	1.21
Plastic Insulated Parallel, Indoor	#18	1	0.84	0.91	0.94	0.97
		2	0.89	0.96	1.00	1.03
		3	1.05	1.13	1.18	1.21
Fabric Insulated Plastic Jacket, Indoor	#18	14	2.33	2.52	2.62	2.70
		19	2.90	3.13	3.26	3.35

Manhours include checking out of job storage, handling, job hauling, opening wire cartons, fishing raceway runs and attaching cable to fish tape or other pull-in means where required, necessary measuring and cutting, pulling in or placing wire, identifying and tagging for connections.

Manhours exclude extensive identification, making connections, and scaffolding. See respective tables for these time frames.

For heights above 25 feet increase manhours 3% for each additional 5 feet or fraction thereof.

LOW VOLTAGE THERMOSTAT AND LIGHTING CONTROL

Branch Cable—Size # 18

Masonry Wall Construction (Concealed)

MANHOURS PER 100 LINEAR FEET

Type of Cable	Wire Size	Number of Conductors	Manhours for Height to			
			10'	15'	20'	25'
Plastic Insulated Twisted, No Braid Overall Jacket Waxed Braid Jacket	#18	2	1.04	1.12	1.17	1.20
		3	1.22	1.32	1.37	1.41
		4	1.49	1.61	1.67	1.72
		5	1.53	1.65	1.72	1.77
		6	1.70	1.84	1.91	1.97
		7	1.83	1.98	2.06	2.12
Plastic Insulated Steel Armored	#18	2	2.08	2.25	2.34	2.41
		3	2.45	2.65	2.75	2.83
		4	2.72	2.94	3.06	3.15
		5	3.07	3.32	3.45	3.55
		6	3.45	3.73	3.88	3.99
Plastic Insulated Parallel	#18	2	1.04	1.12	1.17	1.20
		3	1.22	1.32	1.37	1.41
Plastic Insulated Parallel, Indoor	#18	1	0.98	1.06	1.10	1.13
		2	1.04	1.12	1.17	1.20
		3	1.22	1.32	1.37	1.41
Fabric Insulated Plastic Jacket, Indoor	#18	14	2.72	2.94	3.06	3.15
		19	3.38	3.65	3.80	3.91

Manhours include checking out of job storage, handling, job hauling, opening wire cartons, fishing raceway runs and attaching cable to fish tape or other pull-in means where required, necessary measuring and cutting, pulling in or placing wire, identifying and tagging for connections.

Manhours exclude extensive identification, making connections, and scaffolding. See respective tables for these time frames.

For heights above 25 feet increase manhours 3% for each additional 5 feet or fraction thereof.

LOW VOLTAGE THERMOSTAT AND LIGHTING CONTROL

Branch Cable—Size #18

Suspended Ceilings (Concealed)

MANHOURS PER 100 LINEAR FEET

Type of Cable	Wire Size	Number of Conductors	Manhours for Height to			
			10′	15′	20′	25′
Plastic Insulated Twisted, No Braid Overall Jacket Waxed Braid Jacket	#18	2	0.99	1.07	1.11	1.15
		3	1.17	1.26	1.31	1.35
		4	1.42	1.53	1.59	1.64
		5	1.45	1.57	1.63	1.68
		6	1.62	1.75	1.82	1.87
		7	1.74	1.88	1.95	2.00
Plastic Insulated Steel Armored	#18	2	1.98	2.14	2.22	2.29
		3	2.33	2.52	2.62	2.70
		4	2.59	2.80	2.91	3.00
		5	2.92	3.15	3.28	3.38
		6	3.29	3.55	3.70	3.81
Plastic Insulated Parallel	#18	2	0.99	1.07	1.11	1.15
		3	1.17	1.26	1.31	1.35
Plastic Insulated Parallel, Indoor	#18	1	0.93	1.00	1.04	1.08
		2	0.99	1.07	1.11	1.15
		3	1.17	1.26	1.31	1.35
Fabric Insulated Plastic Jacket, Indoor	#18	14	2.59	2.80	2.91	3.00
		19	3.22	3.48	3.62	3.73

Manhours include checking out of job storage, handling, job hauling, opening wire cartons, fishing raceway runs and attaching cable to fish tape or other pull-in means where required, necessary measuring and cutting, pulling in or placing wire, identifying and tagging for connections.

Manhours exclude extensive identification, making connections, and scaffolding. See respective tables for these time frames.

For heights above 25 feet increase manhours 3% for each additional 5 feet or fraction thereof.

LOW VOLTAGE THERMOSTAT AND LIGHTING CONTROL

Branch Cable—Size #18

Flat Ceilings and Walls (Exposed)

MANHOURS PER 100 LINEAR FEET

Type of Cable	Wire Size	Number of Conductors	Manhours for Height to			
			10′	15′	20′	25′
Plastic Insulated Twisted, No Braid Overall Jacket Waxed Braid Jacket	#18	2	0.94	1.02	1.06	1.09
		3	1.11	1.20	1.25	1.28
		4	1.35	1.46	1.52	1.56
		5	1.38	1.49	1.55	1.60
		6	1.54	1.66	1.73	1.78
		7	1.66	1.79	1.86	1.92
Plastic Insulated Steel Armored	#18	2	1.88	2.03	2.11	2.17
		3	2.22	2.40	2.49	2.57
		4	2.46	2.66	2.76	2.85
		5	2.77	2.99	3.11	3.20
		6	3.12	3.37	3.50	3.61
Plastic Insulated Parallel	#18	2	0.94	1.02	1.06	1.09
		3	1.11	1.20	1.25	1.28
Plastic Insulated Parallel, Indoor	#18	1	0.89	0.96	1.00	1.03
		2	0.94	1.02	1.06	1.09
		3	1.11	1.20	1.25	1.28
Fabric Insulated Plastic Jacket, Indoor	#18	14	2.46	2.66	2.76	2.85
		19	3.06	3.30	3.44	3.54

Manhours include checking out of job storage, handling, job hauling, opening wire cartons, fishing raceway runs and attaching cable to fish tape or other pull-in means where required, necessary measuring and cutting, pulling in or placing wire, identifying and tagging for connections.

Manhours exclude extensive identification, making connections, and scaffolding. See respective tables for these time frames.

For heights above 25 feet increase manhours 3% for each additional 5 feet or fraction thereof.

LOW VOLTAGE THERMOSTAT AND LIGHTING CONTROL

Branch Cable—Size #18

Installed in Conduit

MANHOURS PER 100 LINEAR FEET

Type of Cable	Wire Size	Number of Conductors	Manhours for Height to			
			10′	15′	20′	25′
Plastic Insulated Twisted, No Braid Overall Jacket Waxed Braid Jacket	#18	2	1.13	1.22	1.27	1.31
		3	1.33	1.44	1.49	1.54
		4	1.63	1.76	1.83	1.89
		5	1.66	1.79	1.86	1.92
		6	1.85	2.00	2.08	2.14
		7	1.99	2.15	2.24	2.30
Plastic Insulated Steel Armored	#18	2	2.26	2.44	2.54	2.61
		3	2.67	2.88	3.00	3.09
		4	2.96	3.20	3.32	3.42
		5	3.34	3.61	3.75	3.86
		6	3.76	4.06	4.22	4.35
Plastic Insulated Parallel	#18	2	1.13	1.22	1.27	1.31
		3	1.33	1 44	1.49	1.54
Plastic Insulated Parallel, Indoor	#18	1	1.07	1.16	1.20	1.24
		2	1.13	1.22	1.27	1.31
		3	1.33	1.44	1.49	1.54
Fabric Insulated Plastic Jacket, Indoor	#18	14	2.96	3.20	3.32	3.42
		19	3.68	3.97	4.13	4.26

Manhours include checking out of job storage, handling, job hauling, opening wire cartons, fishing raceway runs and attaching cable to fish tape or other pull-in means where required, necessary measuring and cutting, pulling in or placing wire, identifying and tagging for connections.

Manhours exclude extensive identification, making connections, and scaffolding. See respective tables for these time frames.

For heights above 25 feet increase manhours 3% for each additional 5 feet or fraction thereof.

MINERAL INSULATED CABLE AND TERMINATORS
Wire Size #8 and Smaller
Concrete Slab Construction (Concealed)

MANHOURS PER FOOT OF CABLE AND FOR EACH OF TERMINATORS

Cable Size	Number of Conductors	Cable Installation				Terminator Installation			
		Manhours for Height to				Manhours for Height to			
		10′	15′	20′	25′	10′	15′	20′	25′
16	1	.037	.040	.042	.043	.33	.36	.37	.38
16	2	.043	.046	.048	.050	.39	.42	.44	.45
16	3	.049	.053	.055	.057	.44	.48	.49	.51
16	4	.055	.059	.062	.064	.50	.54	.56	.58
16	7	.091	.098	.102	.105	.76	.82	.85	.88
14	1	.037	.040	.042	.043	.33	.36	.37	.38
14	2	.043	.046	.048	.050	.39	.42	.44	.45
14	3	.049	.053	.055	.057	.44	.48	.49	.51
14	4	.055	.059	.062	.064	.50	.54	.56	.58
14	7	.091	.098	.102	.105	.76	.82	.85	.88
12	1	.043	.046	.048	.050	.33	.36	.37	.38
12	2	.048	.052	.054	.056	.39	.42	.44	.45
12	3	.054	.058	.061	.062	.44	.48	.49	.51
12	4	.061	.066	.069	.071	.50	.54	.56	.58
12	7	.108	.117	.121	.125	.76	.82	.85	.88
10	1	.049	.053	.055	.057	.39	.42	.44	.45
10	2	.055	.059	.062	.064	.44	.48	.49	.51
10	3	.062	.067	.069	.072	.50	.54	.56	.58
10	4	.067	.072	.075	.078	.55	.59	.62	.64
10	7	.105	.113	.118	.121	.94	1.02	1.06	1.09
8	1	.062	.067	.069	.072	.44	.48	.49	.51
8	2	.067	.072	.075	.078	.50	.54	.56	.58
8	3	.074	.080	.083	.086	.55	.59	.62	.64
8	4	.080	.086	.090	.093	.73	.79	.82	.84

All manhours include checking out of job storage, handling, and job hauling.

Cable installation manhours include placing cable coils or spools on reel stands or racks or opening cable cartons, measuring and cutting cable, normal securing of cable in position, identifying or "polling out" the conductors, and tagging for connections to branch circuits, panel boards, or equipment.

Terminator manhours include stripping of cable sheath up to 36 inches, placing terminator, and making connection.

Manhours exclude extensive identification and scaffolding. See respective tables for these time frames.

For heights above 25 feet increase manhours 5% for each additional 5 feet or fraction thereof.

MINERAL INSULATED CABLE AND TERMINATORS

Wire Size #8 and Smaller

Masonry Walls (Concealed)

MANHOURS PER FOOT OF CABLE AND FOR EACH OF TERMINATORS

Cable Size	Number of Conductors	Cable Installation Manhours for Height to				Terminator Installation Manhours for Height to			
		10'	15'	20'	25'	10'	15'	20'	25'
16	1	.039	.042	.044	.045	.35	.38	.39	.40
16	2	.045	.049	.051	.052	.41	.44	.46	.47
16	3	.051	.055	.057	.059	.46	.50	.52	.53
16	4	.058	.063	.065	.067	.53	.57	.60	.61
16	7	.096	.104	.108	.111	.80	.86	.90	.93
14	1	.039	.042	.044	.045	.35	.38	.39	.40
14	2	.045	.049	.051	.052	.41	.44	.46	.47
14	3	.051	.055	.057	.059	.46	.50	.52	.53
14	4	.058	.063	.065	.067	.53	.57	.60	.61
14	7	.096	.104	.108	.111	.80	.86	.90	.93
12	1	.045	.049	.051	.052	.35	.38	.39	.40
12	2	.050	.054	.056	.058	.41	.44	.46	.47
12	3	.057	.062	.064	.066	.46	.50	.52	.53
12	4	.064	.069	.072	.074	.53	.57	.60	.61
12	7	.113	.122	.127	.131	.80	.86	.90	.93
10	1	.051	.055	.057	.059	.41	.44	.46	.47
10	2	.058	.063	.065	.067	.46	.50	.52	.53
10	3	.065	.070	.073	.075	.53	.57	.60	.61
10	4	.070	.076	.079	.081	.58	.63	.65	.67
10	7	.110	.119	.124	.127	.99	1.07	1.11	1.15
8	1	.065	.070	.073	.075	.46	.50	.52	.53
8	2	.070	.076	.079	.081	.53	.57	.60	.61
8	3	.078	.084	.088	.090	.58	.63	.65	.67
8	4	.084	.091	.094	.097	.77	.83	.86	.89

All manhours include checking out of job storage, handling, and job hauling.

Cable installation manhours include placing cable coils or spools on reel stands or racks or opening cable cartons, measuring and cutting cable, normal securing of cable in position, identifying or "polling out" the conductors, and tagging for connections to branch circuits, panel boards, or equipment.

Terminator manhours include stripping of cable sheath up to 36 inches, placing terminator, and making connection.

Manhours exclude extensive identification and scaffolding. See respective tables for these time frames.

For heights above 25 feet increase manhours 5% for each additional 5 feet or fraction thereof.

MINERAL INSULATED CABLE AND TERMINATORS

Wire Size #8 and Smaller
Flat Ceilings and Walls (Exposed)

MANHOURS PER FOOT OF CABLE AND FOR EACH OF TERMINATORS

Cable Size	Number of Conductors	Cable Installation Manhours for Height to				Terminator Installation Manhours for Height to			
		10'	15'	20'	25'	10'	15'	20'	25'
16	1	.035	.038	.039	.040	.31	.33	.35	.36
16	2	.040	.432	.449	.463	.37	.40	.42	.43
16	3	.047	.051	.053	.054	.42	.45	.47	.49
16	4	.052	.056	.058	.060	.48	.52	.54	.56
16	7	.086	.093	.097	.099	.72	.78	.81	.83
14	1	.035	.038	.039	.040	.31	.33	.35	.36
14	2	.040	.432	.449	.463	.37	.40	.42	.43
14	3	.047	.051	.053	.054	.42	.45	.47	.49
14	4	.052	.056	.058	.060	.48	.52	.54	.56
14	7	.086	.093	.097	.099	.72	.78	.81	.83
12	1	.040	.432	.449	.463	.31	.33	.35	.36
12	2	.046	.050	.052	.053	.37	.40	.42	.43
12	3	.051	.055	.057	.059	.42	.45	.47	.49
12	4	.058	.063	.065	.067	.48	.52	.54	.56
12	7	.103	.111	.116	.119	.72	.78	.81	.83
10	1	.047	.051	.053	.054	.37	.40	.42	.43
10	2	.052	.056	.058	.060	.42	.45	.47	.49
10	3	.059	.064	.066	.068	.48	.52	.54	.56
10	4	.064	.069	.072	.074	.52	.56	.58	,60
10	7	.100	.108	.112	.116	.89	.96	1.00	1.03
8	1	.059	.064	.066	.068	.42	.45	.47	.49
8	2	.064	.069	.072	.074	.48	.52	.54	.56
8	3	.070	.076	.079	.081	.52	.56	.58	.60
8	4	.076	.082	.085	.088	.69	.75	.78	.80

All manhours include checking out of job storage, handling, and job hauling.

Cable installation manhours include placing cable coils or spools on reel stands or racks or opening cable cartons, measuring and cutting cable, normal securing of cable in position, identifying or "polling out" the conductors, and tagging for connections to branch circuits, panel boards, or equipment.

Terminator manhours include stripping of cable sheath up to 36 inches, placing terminator, and making connection.

Manhours exclude extensive identification and scaffolding. See respective tables for these time frames.

For heights above 25 feet increase manhours 5% for each additional 5 feet or fraction thereof.

MINERAL INSULATED CABLE AND TERMINATORS

Wire Size #8 and Smaller

Open Truss or Bar Joist (Exposed)

MANHOURS PER FOOT OF CABLE AND FOR EACH OF TERMINATORS

Cable Size	Number of Conductors	Cable Installation Manhours for Height to				Terminator Installation Manhours for Height to			
		10′	15′	20′	25′	10′	15′	20′	25′
16	1	.033	.036	.037	.038	.30	.32	.34	.35
16	2	.039	.042	.044	.045	.35	.38	.39	.40
16	3	.044	.048	.049	.051	.40	.43	.45	.46
16	4	.050	.054	.056	.058	.45	.49	.51	.52
16	7	.082	.089	.092	.095	.68	.73	.76	.79
14	1	.033	.036	.037	.038	.30	.32	.34	.35
14	2	.039	.042	.044	.045	.35	.38	.39	.40
14	3	.044	.048	.049	.051	.40	.43	.45	.46
14	4	.050	.054	.056	.058	.45	.49	.51	.52
14	7	.082	.089	.092	.095	.68	.73	.76	.79
12	1	.039	.042	.044	.045	.30	.32	.34	.35
12	2	.043	.046	.048	.050	.35	.38	.39	.40
12	3	.049	.053	.055	.057	.40	.43	.45	.46
12	4	.055	.059	.062	.064	.45	.49	.51	.52
12	7	.097	.105	.109	.112	.68	.73	.76	.79
10	1	.044	.048	.049	.051	.35	.38	.39	.40
10	2	.050	.054	.056	.058	.40	.43	.45	.46
10	3	.056	.060	.063	.065	.45	.49	.51	.52
10	4	.060	.065	.067	.069	.50	.54	.56	.58
10	7	.095	.103	.107	.110	.85	.92	.95	.98
8	1	.056	.060	.063	.065	.40	.43	.45	.46
8	2	.060	.065	.067	.069	.45	.49	.51	.52
8	3	.067	.072	.075	.078	.50	.54	.56	.58
8	4	.072	.078	.081	.083	.66	.71	.74	.76

All manhours include checking out of job storage, handling, and job hauling.

Cable installation manhours include placing cable coils or spools on reel stands or racks or opening cable cartons, measuring and cutting cable, normal securing of cable in position, identifying or "polling out" the conductors, and tagging for connections to branch circuits, panel boards, or equipment.

Terminator manhours include stripping of cable sheath up to 36 inches, placing terminator, and making connection.

Manhours exclude extensive identification and scaffolding. See respective tables for these time frames.

For heights above 25 feet increase manhours 5% for each additional 5 feet or fraction thereof.

MINERAL INSULATED CABLE AND TERMINATORS
Wire Size #8 and Smaller
Connect Equipment (Exposed)

MANHOURS PER FOOT OF CABLE AND FOR EACH OF TERMINATORS

Cable Size	Number of Conductors	Cable Installation				Terminator Installation			
		Manhours for Height to				Manhours for Height to			
		10'	15'	20'	25'	10'	15'	20'	25'
16	1	.040	.432	.449	.463	.36	.39	.40	.42
16	2	.046	.050	.052	.053	.42	.45	.47	.49
16	3	.053	.057	.060	.061	.48	.52	.54	.56
16	4	.059	.064	.066	.068	.54	.58	.61	.62
16	7	.098	.106	.110	.113	.82	.89	.92	.95
14	1	.040	.432	.449	.463	.36	.39	.40	.42
14	2	.046	.050	.052	.053	.42	.45	.47	.49
14	3	.053	.057	.060	.061	.48	.52	.54	.56
14	4	.059	.064	.066	.068	.54	.58	.61	.62
14	7	.098	.106	.110	.113	.82	.89	.92	.95
12	1	.046	.050	.052	.053	.36	.39	.40	.42
12	2	.052	.056	.058	.060	.42	.45	.47	.49
12	3	.058	.063	.065	.067	.48	.52	.54	.56
12	4	.066	.071	.074	.076	.54	.58	.61	.62
12	7	.117	.126	.131	.135	.82	.89	.92	.95
10	1	.053	.057	.060	.061	.42	.45	.47	.49
10	2	.059	.064	.066	.068	.48	.52	.54	.56
10	3	.067	.072	.075	.078	.54	.58	.61	.62
10	4	.072	.078	.081	.083	.59	.64	.66	.68
10	7	.113	.122	.127	.131	1.02	1.10	1.15	1.18
8	1	.067	.072	.075	.078	.48	.52	.54	.56
8	2	.072	.078	.081	.083	.54	.58	.61	.62
8	3	.080	.086	.090	.093	.59	.64	.66	.68
8	4	.086	.093	.097	.099	.79	.85	.89	.91

All manhours include checking out of job storage, handling, and job hauling.

Cable installation manhours include placing cable coils or spools on reel stands or racks or opening cable cartons, measuring and cutting cable, normal securing of cable in position, identifying or "polling out" the conductors, and tagging for connections to branch circuits, panel boards, or equipment.

Terminator manhours include stripping of cable sheath up to 36 inches, placing terminator, and making connection.

Manhours exclude extensive identification and scaffolding. See respective tables for these time frames.

For heights above 25 feet increase manhours 5% for each additional 5 feet or fraction thereof.

ALUMINUM SHEATHED CABLE AND CONNECTORS

Wire Size #8 and Smaller

Wood or Metal Frame Construction (Concealed)

MANHOURS PER FOOT OF CABLE AND EACH OF CONNECTORS

Cable Size	Number of Conductors	Cable Installation				Connector Installation			
		Manhours for Height to				Manhours for Height to			
		10'	15'	20'	25'	10'	15'	20'	25'
14	2	.043	.046	.048	.050	.33	.36	.37	.38
14	3	.049	.053	.055	.057	.41	.44	.46	.47
14	4	.055	.059	.062	.064	.44	.48	.49	.51
14	5	.067	.072	.075	.078	.51	.55	.57	.59
14	6	.079	.085	.089	.091	.60	.65	.67	.69
14	7	.091	.098	.102	.105	.69	.75	.78	.80
14	9	.108	.117	.121	.125	.78	.84	.88	.90
14	12	.124	.134	.139	.143	.85	.92	.95	.98
12	2	.048	.052	.054	.056	.33	.36	.37	.38
12	3	.054	.058	.061	.062	.41	.44	.46	.47
12	4	.061	.066	.069	.071	.44	.48	.49	.51
12	5	.073	.079	.082	.084	.51	.55	.57	.59
12	6	.090	.097	.101	.104	.60	.65	.67	.69
12	7	.108	.117	.121	.125	.69	.75	.78	.80
12	9	.130	.140	.146	.150	.78	.84	.88	.90
12	12	.152	.164	.171	.176	.85	.92	.95	.98
10	2	.055	.059	.062	.064	.39	.42	44	.45
10	3	.061	.066	.069	.071	.44	.48	.49	.51
10	4	.067	.072	.075	.078	.50	.54	.56	.59
10	5	.080	.086	.090	.093	.58	.63	.65	.67
10	7	.116	.125	.130	.134	.87	.94	.98	1.01
8	3	.073	.079	.082	.084	.56	.60	.63	.65

All manhours include checking out of job storage, handling, and job hauling.

Cable installation manhours include placing cable coils or spools on reel stands or racks or opening cable cartons, measuring and cutting cable, normal securing of cable in position, identifying or "polling out" the conductors and tagging for connection to branch circuits, panel boards, or equipment.

Connector manhours include stripping of cable sheath up to 36 inches, placing connector, and making connection.

Manhours exclude extensive identification, and scaffolding. See respective tables for these time frames.

For heights above 25 feet increase manhours 5% for each additional 5 feet or fraction thereof.

ALUMINUM SHEATHED CABLE AND CONNECTORS

Wire Size #8 and Smaller
Masonry Walls (Concealed)

MANHOURS PER FOOT OF CABLE AND EACH OF CONNECTORS

Cable Size	Number of Conductors	Cable Installation				Connector Installation			
		Manhours for Height to				Manhours for Height to			
		10'	15'	20'	25'	10'	15'	20'	25'
14	2	.050	.054	.056	.058	.38	.41	.43	.44
14	3	.057	.062	.064	.066	.48	.52	.54	.56
14	4	.064	.069	.072	.074	.51	.55	.57	.59
14	5	.078	.084	.088	.090	.59	.64	.66	.68
14	6	.092	.099	.103	.106	.70	.76	.79	.81
14	7	.106	.114	.119	.123	.80	.86	.90	.93
14	9	.126	.136	.142	.146	.91	.98	1.02	1.05
14	12	.145	.157	.163	.168	.99	1.07	1.11	1.15
12	2	.056	.060	.063	.065	.38	.41	.43	.44
12	3	.063	.068	.071	.073	.48	.52	.54	.56
12	4	.071	.077	.080	.082	.51	.55	.57	.59
12	5	.085	.092	.095	.098	.59	.64	.66	.68
12	6	.105	.113	.118	.121	.70	.76	.79	.81
12	7	.126	.136	.142	.146	.80	.86	.90	.93
12	9	.152	.164	.171	.176	.91	.98	1.02	1.05
12	12	.177	.191	.199	.205	.99	1.07	1.11	1.15
10	2	.064	.069	.072	.074	.45	.49	.51	.52
10	3	.071	.077	.080	.082	.51	.55	.57	.59
10	4	.078	.084	.088	.090	.58	.63	.65	.67
10	5	.093	.100	.104	.108	.68	.73	.76	.79
10	7	.135	.146	.152	.156	1.01	1.09	1.13	1.17
8	3	.085	.092	.095	.098	.65	.70	.73	.75

All manhours include checking out of job storage, handling, and job hauling.

Cable installation manhours include placing cable coils or spools on reel stands or racks or opening cable cartons, measuring and cutting cable, normal securing of cable in position, identifying or "polling out" the conductors and tagging for connection to branch circuits, panel boards, or equipment.

Connector manhours include stripping of cable sheath up to 36 inches, placing connector, and making connection.

Manhours exclude extensive identification, and scaffolding. See respective tables for these time frames.

For heights above 25 feet increase manhours 5% for each additional 5 feet or fraction thereof.

ALUMINUM SHEATHED CABLE AND CONNECTORS

Wire Size #8 and Smaller

Suspended Ceilings (Concealed)

MANHOURS PER FOOT OF CABLE AND EACH OF CONNECTORS

Cable Size	Number of Conductors	Cable Installation				Connector Installation			
		Manhours for Height to				Manhours for Height to			
		10'	15'	20'	25'	10'	15'	20'	25'
14	2	.048	.052	.054	.056	.37	.40	.42	.43
14	3	.054	.058	.061	.062	.46	.50	.52	.53
14	4	.061	.066	.069	.071	.49	.53	.55	.57
14	5	.074	.080	.083	.086	.57	.62	.64	.66
14	6	.088	.095	.099	.102	.67	.72	.75	.78
14	7	.101	.109	.113	.117	.77	.83	.86	.89
14	9	.120	.130	.135	.139	.87	.94	.98	1.01
14	12	.138	.149	.155	.160	.94	1.02	1.06	1.09
12	2	.053	.057	.060	.061	.37	.40	.42	.43
12	3	.060	.065	.067	.069	46	.50	.52	.53
12	4	.068	.073	.076	.079	.49	.53	.55	.57
12	5	.081	.087	.091	.094	.57	.62	.64	.66
12	6	.100	.108	.112	.116	.67	.72	.75	.78
12	7	.120	.130	.135	.139	.77	.83	.86	.89
12	9	.144	.156	.162	.167	.87	.94	.98	1.01
12	12	.169	.183	.190	.196	.94	1.02	1.06	1.09
10	2	.061	.066	.069	.071	43	.46	.48	.50
10	3	.068	.073	.076	.079	49	.53	.55	.57
10	4	.074	.080	.083	.086	.56	.60	.63	.65
10	5	.089	.096	.100	.103	.64	.69	.72	.74
10	7	.129	.139	.145	.149	.97	1.05	1.09	1.12
8	3	.081	.087	.091	.094	.62	.67	.70	.72

All manhours include checking out of job storage, handling, and job hauling.

Cable installation manhours include placing cable coils or spools on reel stands or racks or opening cable cartons, measuring and cutting cable, normal securing of cable in position, identifying or "polling out" the conductors and tagging for connection to branch circuits, panel boards, or equipment.

Connector manhours include stripping of cable sheath up to 36 inches, placing connector, and making connection.

Manhours exclude extensive identification, and scaffolding. See respective tables for these time frames.

For heights above 25 feet increase manhours 5% for each additional 5 feet or fraction thereof.

ALUMINUM SHEATHED CABLE AND CONNECTORS

Wire Size #8 and Smaller

Flat Ceilings and Walls (Exposed)

MANHOURS PER FOOT OF CABLE AND EACH OF CONNECTORS

Cable Size	Number of Conductors	Cable Installation				Connector Installation			
		Manhours for Height to				Manhours for Height to			
		10′	15′	20′	25′	10′	15′	20′	25′
14	2	.046	.050	.052	.053	.35	.38	.39	.40
14	3	.052	.056	.058	.060	.43	.46	.48	.50
14	4	.058	.063	.065	.067	.47	.51	.53	.54
14	5	.071	.077	.080	.082	.54	.58	.61	.62
14	6	.084	.091	.094	.097	.64	.69	.72	.74
14	7	.096	.104	.108	.111	.73	.79	.82	.84
14	9	.114	.123	.128	.132	.83	.90	.93	.96
14	12	.131	.141	.147	.152	.90	.97	1.02	1.04
12	2	.051	.055	.057	.059	.35	.38	.39	.40
12	3	.057	.062	.064	.066	.43	.46	.48	.50
12	4	.065	.070	.073	.075	.47	.51	.53	.54
12	5	.077	.083	.086	.089	.54	.58	.61	.62
12	6	.095	.103	.107	.110	.64	.69	.72	.74
12	7	.114	.123	.128	.132	.73	.79	.82	.84
12	9	.138	.149	.155	.160	.83	.90	.93	.96
12	12	.161	.174	.181	.186	.90	.97	1.01	1.04
10	2	.058	.063	.065	.067	.41	.44	.46	.47
10	3	.065	.070	.073	.075	.47	.51	.53	.54
10	4	.071	.077	.080	.082	.53	.57	.60	.61
10	5	.085	.092	.095	.098	.61	.66	.69	.71
10	7	.123	.133	.138	.142	.92	.99	1.03	1.06
8	3	.077	.083	.086	.089	.59	.64	.66	.68

All manhours include checking out of job storage, handling, and job hauling.

Cable installation manhours include placing cable coils or spools on reel stands or racks or opening cable cartons, measuring and cutting cable, normal securing of cable in position, identifying or "polling out" the conductors and tagging for connection to branch circuits, panel boards, or equipment.

Connector manhours include stripping of cable sheath up to 36 inches, placing connector, and making connection.

Manhours exclude extensive identification, and scaffolding. See respective tables for these time frames.

For heights above 25 feet increase manhours 5% for each additional 5 feet or fraction thereof.

ALUMINUM SHEATHED CABLE AND CONNECTORS

Wire Size #8 and Smaller

Open Truss or Bar Joist (Exposed)

MANHOURS PER FOOT OF CABLE AND EACH OF CONNECTORS

Cable Size	Number of Conductors	Cable Installation				Connector Installation			
		Manhours for Height to				Manhours for Height to			
		10'	15'	20'	25'	10'	15'	20'	25'
14	2	.043	.046	.048	.050	.33	.36	.37	.38
14	3	.049	.053	.055	.057	.41	.44	.46	.47
14	4	.055	.059	.062	.064	.44	.48	.49	.51
14	5	.067	.072	.075	.078	.51	.55	.57	.59
14	6	.079	.085	.089	.091	.60	.65	.67	.69
14	7	.091	.098	.102	.105	.69	.75	.78	.80
14	9	.108	.117	.121	.125	.78	.84	.88	.90
14	12	.124	.134	.139	.143	.85	.92	.95	.98
12	2	.048	.052	.054	.056	.33	.36	.37	.38
12	3	.054	.058	.061	.062	.41	.44	.46	.47
12	4	.061	.066	.069	.071	.44	.48	.49	.51
12	5	.073	.079	.082	.084	.51	.55	.57	.59
12	6	.090	.097	.101	.104	.60	.65	.67	.69
12	7	.108	.117	.121	.125	.69	.75	.78	.80
12	9	.130	.140	.146	.150	.78	.84	.88	.90
12	12	.152	.164	.171	.176	.85	.92	.95	.98
10	2	.055	.059	.062	.064	.39	.42	.44	.45
10	3	.061	.066	.069	.071	.44	.48	.49	.51
10	4	.067	.072	.075	.078	.50	.54	.56	.58
10	5	.080	.086	.090	.093	.58	.63	.65	.67
10	7	.116	.125	.130	.134	.87	.94	.98	1.01
8	3	.073	.079	.082	.084	.56	.60	.63	.65

All manhours include checking out of job storage, handling, and job hauling.

Cable installation manhours include placing cable coils or spools on reel stands or racks or opening cable cartons, measuring and cutting cable, normal securing of cable in position, identifying or "polling out" the conductors and tagging for connection to branch circuits, panel boards, or equipment.

Connector manhours include stripping of cable sheath up to 36 inches, placing connector, and making connection.

Manhours exclude extensive identification, and scaffolding. See respective tables for these time frames.

For heights above 25 feet increase manhours 5% for each additional 5 feet or fraction thereof.

ALUMINUM SHEATHED CABLE AND CONNECTORS

Wire Size #8 and Smaller

Connect Equipment (Exposed)

MANHOURS PER FOOT OF CABLE AND EACH OF CONNECTORS

Cable Size	Number of Conductors	Cable Installation				Connector Installation			
		Manhours for Height to				Manhours for Height to			
		10'	15'	20'	25'	10'	15'	20'	25'
14	2	.046	.050	.052	.053	.36	.39	.40	.42
14	3	.053	.057	.060	.061	.44	.48	.49	.51
14	4	.059	.064	.066	.068	.48	.52	.54	.56
14	5	.072	.078	.081	.083	.55	.59	.62	.64
14	6	.085	.092	.095	.098	.65	.70	.73	.75
14	7	.098	.106	.110	.113	.75	.81	.84	.87
14	9	.117	.126	.131	.135	.84	.91	.94	.97
14	12	.134	.145	.151	.155	.92	.99	1.03	1.06
12	2	.052	.056	.058	.060	.36	.39	.40	.42
12	3	.058	.063	.065	.067	.44	.48	.49	.51
12	4	.066	.071	.074	.076	.48	.52	.54	.56
12	5	.079	.085	.089	.091	.55	.59	.62	.64
12	6	.097	.105	.109	.112	.65	.70	.73	.75
12	7	.117	.126	.131	.135	.75	.81	.84	.87
12	9	.140	.151	.157	.162	.84	.91	.94	.97
12	12	.164	.177	.184	.190	.92	.99	1.03	1.06
10	2	.059	.064	.066	.068	.42	.45	.47	.49
10	3	.066	.071	.074	.076	.48	.52	.54	.56
10	4	.072	.078	.081	.083	.54	.59	.61	.62
10	5	.086	.093	.097	.099	.63	.68	.71	.73
10	7	.125	.135	.140	.145	.94	1.02	1.06	1.09
8	3	.079	.085	.089	.091	.60	.65	.67	.69

All manhours include checking out of job storage, handling, and job hauling.

Cable installation manhours include placing cable coils or spools on reel stands or racks or opening cable cartons, measuring and cutting cable, normal securing of cable in position, identifying or "polling out" the conductors and tagging for connection to branch circuits, panel boards, or equipment.

Connector manhours include stripping of cable sheath up to 36 inches, placing connector, and making connection.

Manhours exclude extensive identification, and scaffolding. See respective tables for these time frames.

For heights above 25 feet increase manhours 5% for each additional 5 feet or fraction thereof.

SERVICE AND FEEDER WIRING
SINGLE CONDUCTOR—600 VOLTS

Size #6 and Larger—Stranded Copper Wire

Rubber Insulated

For Heights to 10 Feet

MANHOURS PER LINEAR FOOT AND NUMBER OF WIRES LISTED

Wire Size	Runs in Length											
	To 50'				51' to 100'				101' to 200'			
	Number of Wires				Number of Wires				Number of Wires			
	2	3	4	6	2	3	4	6	2	3	4	6
6	.031	.036	.043	.056	.029	.032	.036	.046	.027	.029	.033	.040
4	.037	.043	.051	.066	.034	.037	.043	.054	.032	.035	.039	.047
2	.041	.047	.057	.074	.038	.042	.048	.060	.035	.038	.042	.051
1	.047	.054	.065	.084	.043	.047	.054	.068	.040	.043	.048	.059
1/0	.057	.066	.079	.102	.052	.057	.066	.082	.049	.053	.059	.072
2/0	.066	.076	.091	.118	.061	.067	.077	.096	.057	.062	.069	.083
3/0	.077	.089	.106	.138	.071	.078	.090	.112	.066	.071	.080	.097
4/0	.089	.102	.123	.160	.082	.090	.104	.130	.077	.083	.093	.107
250 MCM	.094	.108	.130	.169	.086	.095	.109	.136	.081	.087	.098	.119
300 MCM	.100	.115	.138	.179	.092	.101	.116	.145	.086	.093	.104	.126
350 MCM	.106	.122	.146	.190	.098	.108	.124	.155	.091	.098	.110	.133
400 MCM	.115	.132	.159	.206	.106	.117	.134	.168	.099	.107	.120	.145
500 MCM	.124	.143	.171	.222	.114	.125	.144	.180	.107	.116	.129	.157
750 MCM	.165	.190	.228	.296	.152	.167	.192	.240	.142	.153	.172	.208
1000 MCM	.201	.231	.277	.361	.185	.204	.234	.293	.173	.187	.209	.253

Manhours include checking out of job storage, handling, job hauling, placing wire coils or spools on reel stands or racks or opening wire cartons, fishing raceway runs, attaching wire to fish tape or other pull-in means, measuring and cutting wire, pulling wire into conduit type raceway or duct, or pull-out and lay wire into wireways or channels, identifying or "polling out" the circuit conductors, and tagging for connections to branch circuits, panel boards, or equipment.

If duplicate pulls above 6 are made reduce manhours 4% for each 2-, 3-, 4-, or 6-wire installation to compensate for equipment set-up time.

Manhours exclude extensive identification, "ringing-out," connection to branch circuit, panel board or equipment, and scaffolding. See respective tables for these time requirements.

SERVICE AND FEEDER WIRING
SINGLE CONDUCTOR—600 VOLTS
Size #6 and Larger—Stranded Copper Wire
Rubber Insulated
For Heights to 15 Feet

MANHOURS PER LINEAR FOOT AND NUMBER OF WIRES LISTED

Wire Size	Runs in Length											
	To 50'				51' to 100'				101' to 200'			
	Number of Wires				Number of Wires				Number of Wires			
	2	3	4	6	2	3	4	6	2	3	4	6
6	.033	.038	.046	.059	.030	.033	.038	.041	.028	.030	.034	.041
4	.040	.046	.055	.072	.037	.041	.047	.059	.034	.037	.041	.050
2	.044	.051	.061	.079	.040	.044	.051	.063	.038	.041	.046	.056
1	.051	.059	.070	.091	.047	.052	.059	.074	.044	.048	.053	.064
1/0	.062	.071	.086	.111	.057	.063	.072	.090	.053	.057	.064	.078
2/0	.071	.082	.098	.127	.065	.072	.082	.103	.061	.066	.074	.089
3/0	.083	.095	.115	.149	.076	.084	.096	.120	.071	.077	.086	.104
4/0	.096	.110	.132	.172	.088	.097	.111	.139	.083	.090	.100	.121
250 MCM	.102	.117	.141	.183	.094	.103	.119	.149	.088	.095	.106	.129
300 MCM	.108	.124	.149	.194	.099	.109	.125	.157	.093	.100	.112	.136
350 MCM	.114	.131	.157	.205	.105	.116	.133	.166	.098	.106	.119	.143
400 MCM	.124	.143	.171	.222	.114	.125	.144	.180	.107	.116	.129	.157
500 MCM	.134	.154	.185	.240	.123	.135	.156	.194	.115	.124	.139	.168
750 MCM	.178	.205	.246	.319	.164	.180	.207	.259	.153	.165	.185	.224
1000 MCM	.217	.250	.299	.389	.200	.220	.253	.316	.187	.202	.226	.274

Manhours include checking out of job storage, handling, job hauling, placing wire coils or spools on reel stands or racks or opening wire cartons, fishing raceway runs, attaching wire to fish tape or other pull-in means, measuring and cutting wire, pulling wire into conduit type raceway or duct, or pull-out and lay wire into wireways or channels, identifying or "polling out" the circuit conductors, and tagging for connections to branch circuits, panel boards, or equipment.

If duplicate pulls above 6 are made reduce manhours 4% for each 2-, 3-, 4-, or 6-wire installation to compensate for equipment set-up time.

Manhours exclude extensive identification, "ringing-out," connection to branch circuit, panel board or equipment, and scaffolding. See respective tables for these time requirements.

SERVICE AND FEEDER WIRING
SINGLE CONDUCTOR—600 VOLTS
Size #6 and Larger—Stranded Copper Wire
Rubber Insulated
For Heights to 20 Feet

MANHOURS PER LINEAR FOOT AND NUMBER OF WIRES LISTED

Wire Size	Runs in Length											
	To 50'				51' to 100'				101' to 200'			
	Number of Wires				Number of Wires				Number of Wires			
	2	3	4	6	2	3	4	6	2	3	4	6
6	.034	.039	.047	.061	.031	.034	.039	.049	.029	.031	.035	.042
4	.042	.048	.058	.075	.039	.043	.049	.062	.036	.039	.044	.053
2	.046	.053	.063	.083	.042	.046	.053	.066	.040	.043	.048	.059
1	.053	.061	.073	.095	.049	.054	.062	.077	.046	.050	.056	.067
1/0	.064	.074	.088	.115	.059	.065	.075	.093	.055	.059	.067	.080
2/0	.074	.085	.102	.133	.068	.075	.086	.108	.064	.069	.077	.094
3/0	.086	.099	.119	.154	.079	.087	.100	.125	.074	.080	.090	.108
4/0	.100	.115	.138	.179	.092	.101	.116	.145	.086	.093	.104	.126
250 MCM	.106	.122	.146	.190	.098	.108	.124	.155	.091	.098	.110	.133
300 MCM	.112	.129	.155	.201	.103	.113	.130	.163	.096	.104	.116	.141
350 MCM	.119	.137	.164	.213	.109	.120	.138	.172	.102	.110	.123	.149
400 MCM	.129	.148	.178	.231	.119	.131	.151	.188	.111	.120	.134	.162
500 MCM	.139	.160	.192	.249	.128	.141	.162	.202	.120	.130	.145	.176
750 MCM	.177	.204	.244	.318	.163	.179	.206	.258	.152	.164	.184	.222
1000 MCM	.226	.260	.312	.405	.208	.229	.263	.329	.194	.210	.235	.284

Manhours include checking out of job storage, handling, job hauling, placing wire coils or spools on reel stands or racks or opening wire cartons, fishing raceway runs, attaching wire to fish tape or other pull-in means, measuring and cutting wire, pulling wire into conduit type raceway or duct, or pull-out and lay wire into wireways or channels, identifying or "polling out" the circuit conductors, and tagging for connections to branch circuits, panel boards, or equipment.

If duplicate pulls above 6 are made reduce manhours 4% for each 2-, 3-, 4-, or 6-wire installation to compensate for equipment set-up time.

Manhours exclude extensive identification, "ringing-out," connection to branch circuit, panel board or equipment, and scaffolding. See respective tables for these time requirements.

SERVICE AND FEEDER WIRING
SINGLE CONDUCTOR—600 VOLTS
Size #6 and Larger—Stranded Copper Wire
Rubber Insulated
For Heights to 25 Feet

MANHOURS PER LINEAR FOOT AND NUMBER OF WIRES LISTED

Wire Size	Runs in Length											
	To 50'				51' to 100'				101' to 200'			
	Number of Wires				Number of Wires				Number of Wires			
	2	3	4	6	2	3	4	6	2	3	4	6
6	.035	.040	.048	.063	.032	.035	.040	.051	.030	.032	.036	.044
4	.043	.049	.059	.077	.040	.044	.051	.063	.037	.040	.045	.054
2	.047	.054	.065	.084	.043	.047	.054	.068	.040	.043	.048	.059
1	.055	.063	.076	.099	.051	.056	.065	.081	.047	.051	.057	.069
1/0	.066	.076	.091	.118	.061	.067	.077	.096	.057	.062	.069	.083
2/0	.076	.087	.105	.136	.070	.077	.089	.111	.065	.070	.079	.095
3/0	.089	.102	.123	.160	.082	.090	.104	.130	.077	.083	.093	.107
4/0	.103	.118	.142	.185	.095	.105	.120	.150	.089	.096	.108	.130
250 MCM	.109	.125	.150	.196	.100	.111	.127	.158	.094	.102	.114	.138
300 MCM	.115	.132	.159	.206	.106	.117	.134	.168	.099	.107	.120	.145
350 MCM	.123	.141	.170	.221	.113	.124	.143	.179	.106	.114	.128	.155
400 MCM	.133	.153	.184	.239	.122	.134	.154	.193	.114	.123	.138	.167
500 MCM	.143	.164	.197	.257	.132	.145	.167	.209	.123	.133	.149	.180
750 MCM	.182	.209	.251	.327	.167	.184	.211	.264	.157	.170	.190	.230
1000 MCM	.233	.268	.322	.418	.214	.235	.271	.338	.200	.216	.242	.293

Manhours include checking out of job storage, handling, job hauling, placing wire coils or spools on reel stands or racks or opening wire cartons, fishing raceway runs, attaching wire to fish tape or other pull-in means, measuring and cutting wire, pulling wire into conduit type raceway or duct, or pull-out and lay wire into wireways or channels, identifying or "polling out" the circuit conductors, and tagging for connections to branch circuits, panel boards, or equipment.

If duplicate pulls above 6 are made reduce manhours 4% for each 2-, 3-, 4-, or 6-wire installation to compensate for equipment set-up time.

Manhours exclude extensive identification, "ringing-out," connection to branch circuit, panel board or equipment, and scaffolding. See respective tables for these time requirements.

For heights above 25 feet increase manhours 5% for each additional 5 feet or fraction thereof.

SERVICE AND FEEDER WIRING
SINGLE CONDUCTOR—600 VOLTS
Size #6 and Larger—Stranded Aluminum Wire
Rubber Insulated
For Heights to 10 Feet

MANHOURS PER LINEAR FOOT AND NUMBER OF WIRES LISTED

Wire Size	Runs in Length											
	To 50′				51′ to 100′				101′ to 200′			
	Number of Wires				Number of Wires				Number of Wires			
	2	3	4	6	2	3	4	6	2	3	4	6
6	.029	.033	.040	.052	.027	.030	.034	.043	.025	.027	.030	.037
4	.035	.040	.048	.063	.032	.035	.040	.051	.030	.032	.036	.044
2	.039	.045	.054	.070	.036	.040	.046	.060	.034	.037	.041	.050
1	.042	.048	.058	.075	.039	.043	.049	.062	.036	.039	.044	.053
1/0	.053	.061	.073	.095	.049	.054	.062	.077	.046	.050	.056	.067
2/0	.059	.068	.081	.106	.054	.059	.068	.085	.051	.055	.062	.075
3/0	.069	.079	.095	.124	.063	.069	.080	.100	.059	.064	.071	.086
4/0	.080	.092	.110	.144	.074	.081	.094	.117	.069	.075	.083	.101
250 MCM	.084	.097	.116	.151	.077	.085	.097	.122	.072	.078	.087	.105
300 MCM	.090	.104	.124	.161	.083	.091	.105	.131	.077	.083	.093	.113
350 MCM	.096	.110	.132	.172	.088	.097	.111	.139	.083	.090	.100	.121
400 MCM	.101	.116	.139	.181	.092	.101	.116	.145	.087	.094	.105	.127
500 MCM	.104	.120	.144	.187	.096	.106	.121	.152	.089	.096	.108	.130
750 MCM	.142	.163	.196	.255	.131	.144	.166	.207	.122	.132	.148	.179
1000 MCM	.175	.201	.242	.314	.161	.177	.204	.255	.151	.163	.183	.221

Manhours include checking out of job storage, handling, job hauling, placing wire coils or spools on reel stands or racks or opening wire cartons, fishing raceway runs, attaching wire to fish tape or other pull-in means, measuring and cutting wire, pulling wire into conduit type raceway or duct, or pull-out and lay wire into wireways or channels, identifying or "polling out" the circuit conductors, and tagging for connections to branch circuits, panel boards, or equipment.

If duplicate pulls above 6 are made reduce manhours 4% for each 2-, 3-, 4-, or 6-wire installation to compensate for equipment set-up time.

Manhours exclude extensive identification, "ringing-out," connection to branch circuit, panel board or equipment, and scaffolding. See respective tables for these time requirements.

SERVICE AND FEEDER WIRING
SINGLE CONDUCTOR—600 VOLTS
Size #6 and Larger—Stranded Aluminum Wire
Rubber Insulated
For Heights to 15 Feet

MANHOURS PER LINEAR FOOT AND NUMBER OF WIRES LISTED

Wire Size	Runs in Length											
	To 50'				51' to 100'				101' to 200'			
	Number of Wires				Number of Wires				Number of Wires			
	2	3	4	6	2	3	4	6	2	3	4	6
6	.031	.036	.043	.056	.029	.032	.036	.046	.027	.029	.033	.040
4	.038	.044	.052	.068	.035	.039	.044	.055	.033	.036	.040	.048
2	.042	.048	.058	.075	.039	.043	.049	.062	.036	.039	.044	.053
1	.045	.052	.062	.081	.041	.045	.052	.065	.039	.042	.047	.057
1/0	.057	.066	.079	.102	.052	.057	.066	.082	.049	.053	.059	.072
2/0	.064	.074	.088	.115	.059	.065	.075	.093	.055	.059	.067	.080
3/0	.075	.086	.104	.135	.069	.076	.087	.109	.065	.070	.079	.095
4/0	.086	.099	.119	.154	.079	.087	.100	.125	.074	.080	.090	.108
250 MCM	.091	.105	.126	.163	.084	.092	.106	.133	.078	.084	.094	.114
300 MCM	.097	.112	.134	.174	.089	.098	.113	.141	.083	.090	.100	.121
350 MCM	.104	.120	.144	.187	.096	.106	.121	.152	.089	.096	.108	.130
400 MCM	.109	.125	.150	.196	.100	.111	.127	.158	.094	.102	.114	.138
500 MCM	.112	.129	.155	.201	.103	.113	.130	.163	.096	.104	.116	.141
750 MCM	.153	.176	.211	.274	.141	.155	.178	.223	.132	.143	.160	.193
1000 MCM	.189	.217	.261	.339	.174	.191	.220	.275	.163	.176	.197	.239

Manhours include checking out of job storage, handling, job hauling, placing wire coils or spools on reel stands or racks or opening wire cartons, fishing raceway runs, attaching wire to fish tape or other pull-in means, measuring and cutting wire, pulling wire into conduit type raceway or duct, or pull-out and lay wire into wireways or channels, identifying or "polling out" the circuit conductors, and tagging for connections to branch circuits, panel boards, or equipment.

If duplicate pulls above 6 are made reduce manhours 4% for each 2-, 3-, 4-, or 6-wire installation to compensate for equipment set-up time.

Manhours exclude extensive identification, "ringing-out," connection to branch circuit, panel board or equipment, and scaffolding. See respective tables for these time requirements.

SERVICE AND FEEDER WIRING
SINGLE CONDUCTOR—600 VOLTS
Size #6 and Larger—Stranded Aluminum Wire
Rubber Insulated
For Heights to 20 Feet

MANHOURS PER LINEAR FOOT AND NUMBER OF WIRES LISTED

Wire Size	Runs in Length											
	To 50'				51' to 100'				101' to 200'			
	Number of Wires				Number of Wires				Number of Wires			
	2	3	4	6	2	3	4	6	2	3	4	6
6	.032	.037	.044	.057	.029	.032	.037	.046	.028	.030	.034	.041
4	.040	.046	.055	.072	.037	.041	.047	.059	.034	.037	.041	.050
2	.044	.051	.061	.079	.040	.044	.051	.063	.038	.041	.046	.056
1	.047	.054	.065	.084	.043	.047	.054	.068	.040	.043	.048	.059
1/0	.059	.068	.081	.106	.054	.059	.068	.085	.051	.055	.062	.075
2/0	.067	.077	.092	.120	.062	.068	.078	.098	.058	.063	.070	.085
3/0	.078	.090	.108	.140	.072	.079	.091	.114	.067	.072	.081	.098
4/0	.089	.102	.123	.160	.082	.090	.104	.130	.077	.083	.093	.107
250 MCM	.095	.109	.131	.170	.087	.096	.110	.138	.082	.089	.099	.120
300 MCM	.101	.116	.139	.181	.092	.101	.116	.145	.087	.094	.105	.127
350 MCM	.108	.124	.149	.194	.099	.109	.125	.157	.093	.100	.112	.136
400 MCM	.113	.130	.156	.203	.104	.114	.132	.164	.097	.105	.117	.142
500 MCM	.116	.133	.160	.208	.108	.119	.137	.171	.100	.108	.121	.146
750 MCM	.159	.183	.219	.285	.146	.161	.185	.231	.137	.148	.166	.201
1000 MCM	.197	.227	.272	.353	.181	.199	.229	.286	.169	.183	.204	.247

Manhours include checking out of job storage, handling, job hauling, placing wire coils or spools on reel stands or racks or opening wire cartons, fishing raceway runs, attaching wire to fish tape or other pull-in means, measuring and cutting wire, pulling wire into conduit type raceway or duct, or pull-out and lay wire into wireways or channels, identifying or "polling out" the circuit conductors, and tagging for connections to branch circuits, panel boards, or equipment.

If duplicate pulls above 6 are made reduce manhours 4% for each 2-, 3-, 4-, or 6-wire installation to compensate for equipment set-up time.

Manhours exclude extensive identification, "ringing-out," connection to branch circuit, panel board or equipment, and scaffolding. See respective tables for these time requirements.

SERVICE AND FEEDER WIRING
SINGLE CONDUCTOR—600 VOLTS
Size #6 and Larger—Stranded Aluminum Wire
Rubber Insulated
For Heights to 25 Feet

MANHOURS PER LINEAR FOOT AND NUMBER OF WIRES LISTED

Wire Size	Runs in Length											
	To 50'				51' to 100'				101' to 200'			
	Number of Wires				Number of Wires				Number of Wires			
	2	3	4	6	2	3	4	6	2	3	4	6
6	.033	.038	.046	.059	.030	.033	.038	.041	.028	.030	.034	.041
4	.041	.047	.057	.074	.038	.042	.048	.060	.035	.038	.042	.051
2	.045	.052	.062	.081	.041	.045	.052	.065	.039	.042	.047	.057
1	.048	.055	.066	.086	.044	.048	.056	.070	.041	.044	.050	.060
1/0	.061	.070	.084	.109	.056	.062	.071	.089	.052	.056	.063	.076
2/0	.069	.079	.095	.124	.063	.069	.080	.100	.059	.064	.071	.086
3/0	.080	.092	.110	.144	.074	.081	.094	.117	.069	.075	.083	.101
4/0	.092	.106	.127	.165	.085	.094	.108	.134	.079	.085	.096	.116
250 MCM	.098	.113	.135	.176	.090	.099	.114	.142	.084	.091	.102	.123
300 MCM	.104	.120	.144	.187	.096	.106	.121	.152	.089	.096	.108	.130
350 MCM	.111	.128	.153	.199	.102	.112	.129	.161	.095	.103	.115	.139
400 MCM	.116	.133	.160	.208	.108	.119	.137	.171	.100	.108	.121	.146
500 MCM	.119	.137	.164	.213	.109	.120	.138	.172	.102	.110	.123	.149
750 MCM	.164	.189	.226	.294	.151	.166	.191	.239	.141	.152	.171	.206
1000 MCM	.203	.233	.280	.364	.187	.206	.237	.296	.175	.189	.212	.256

Manhours include checking out of job storage, handling, job hauling, placing wire coils or spools on reel stands or racks or opening wire cartons, fishing raceway runs, attaching wire to fish tape or other pull-in means, measuring and cutting wire, pulling wire into conduit type raceway or duct, or pull-out and lay wire into wireways or channels, identifying or "polling out" the circuit conductors, and tagging for connections to branch circuits, panel boards, or equipment.

If duplicate pulls above 6 are made reduce manhours 4% for each 2-, 3-, 4-, or 6-wire installation to compensate for equipment set-up time.

Manhours exclude extensive identification, "ringing-out," connection to branch circuit, panel board or equipment, and scaffolding. See respective tables for these time requirements.

For heights above 25 feet increase manhours 5% for each additional 5 feet or fraction thereof.

SERVICE AND FEEDER WIRING
SINGLE CONDUCTOR—600 VOLTS
Size #6 and Larger—Stranded Copper Wire
Thermoplastic Insulated
For Heights to 10 Feet

MANHOURS PER LINEAR FOOT AND NUMBER OF WIRES LISTED

Wire Size	Runs in Length											
	To 50'				51' to 100'				101' to 200'			
	Number of Wires				Number of Wires				Number of Wires			
	2	3	4	6	2	3	4	6	2	3	4	6
6	.029	.033	.040	.052	.027	.030	.034	.043	.025	.027	.030	.037
4	.035	.040	.048	.063	.032	.035	.040	.051	.030	.032	.036	.044
2	.037	.043	.051	.066	.034	.037	.043	.054	.032	.035	.039	.047
1	.042	.048	.058	.075	.039	.043	.049	.062	.036	.039	.044	.053
1/0	.053	.061	.073	.095	.049	.054	.062	.077	.046	.050	.056	.067
2/0	.062	.071	.086	.111	.057	.063	.072	.090	.053	.057	.064	.078
3/0	.072	.083	.099	.129	.066	.073	.083	.104	.062	.067	.075	.091
4/0	.084	.097	.116	.151	.077	.085	.097	.122	.072	.078	.087	.105
250 MCM	.090	.104	.124	.161	.083	.091	.105	.131	.077	.083	.093	.113
300 MCM	.095	.109	.131	.170	.087	.096	.110	.138	.082	.089	.099	.120
350 MCM	.100	.115	.138	.179	.092	.101	.116	.145	.086	.093	.104	.126
400 MCM	.110	.127	.152	.197	.101	.111	.128	.160	.095	.103	.115	.139
500 MCM	.117	.135	.161	.210	.108	.119	.137	.171	.101	.109	.122	.148
750 MCM	.158	.182	.218	.283	.145	.160	.183	.229	.136	.147	.165	.199
1000 MCM	.190	.219	.262	.341	.175	.193	.221	.277	.163	.176	.197	.239

Manhours include checking out of job storage, handling, job hauling, placing wire coils or spools on reel stands or racks or opening wire cartons, fishing raceway runs, attaching wire to fish tape or other pull-in means, measuring and cutting wire, pulling wire into conduit type raceway or duct, or pull-out and lay wire into wireways or channels, identifying or "polling out" the circuit conductors, and tagging for connections to branch circuits, panel boards, or equipment.

If duplicate pulls above 6 are made reduce manhours 4% for each 2-, 3-, 4-, or 6-wire installation to compensate for equipment set-up time.

Manhours exclude extensive identification, "ringing-out," connection to branch circuit, panel board or equipment, and scaffolding. See respective tables for these time requirements.

SERVICE AND FEEDER WIRING
SINGLE CONDUCTOR—600 VOLTS
Size # 6 and Larger—Stranded Copper Wire
Thermoplastic Insulated
For Heights to 15 Feet

MANHOURS PER LINEAR FOOT AND NUMBER OF WIRES LISTED

Wire Size	Runs in Length											
	To 50'				51' to 100'				101' to 200'			
	Number of Wires				Number of Wires				Number of Wires			
	2	3	4	6	2	3	4	6	2	3	4	6
6	.031	.036	.043	.056	.029	.032	.036	.046	.027	.029	.033	.040
4	.038	.044	.052	.068	.035	.039	.044	.055	.033	.036	.040	.048
2	.040	.046	.055	.072	.037	.041	.047	.059	.034	.037	.041	.050
1	.045	.052	.062	.081	.041	.045	.052	.065	.039	.042	.047	.057
1/0	.057	.066	.079	.102	.052	.057	.066	.082	.049	.053	.059	.072
2/0	.067	.077	.092	.120	.062	.068	.078	.098	.058	.063	.070	.085
3/0	.078	.090	.108	.140	.072	.079	.091	.114	.067	.072	.081	.098
4/0	.091	.105	.126	.163	.084	.092	.106	.133	.078	.084	.094	.114
250 MCM	.097	.112	.134	.174	.089	.098	.113	.141	.083	.090	.100	.121
300 MCM	.103	.118	.142	.185	.095	.105	.120	.150	.089	.096	.108	.130
350 MCM	.108	.124	.149	.194	.099	.109	.125	.157	.093	.100	.112	.136
400 MCM	.119	.137	.164	.213	.109	.120	.138	.172	.102	.110	.123	.149
500 MCM	.126	.145	.174	.226	.116	.128	.147	.183	.108	.117	.131	.158
750 MCM	.171	.197	.236	.307	.157	.173	.199	.248	.147	.159	.178	.215
1000 MCM	.205	.236	.283	.368	.189	.208	.239	.299	.176	.190	.213	.258

Manhours include checking out of job storage, handling, job hauling, placing wire coils or spools on reel stands or racks or opening wire cartons, fishing raceway runs, attaching wire to fish tape or other pull-in means, measuring and cutting wire, pulling wire into conduit type raceway or duct, or pull-out and lay wire into wireways or channels, identifying or "polling out" the circuit conductors, and tagging for connections to branch circuits, panel boards, or equipment.

If duplicate pulls above 6 are made reduce manhours 4% for each 2-, 3-, 4-, or 6-wire installation to compensate for equipment set-up time.

Manhours exclude extensive identification, "ringing-out," connection to branch circuit, panel board or equipment, and scaffolding. See respective tables for these time requirements.

SERVICE AND FEEDER WIRING
SINGLE CONDUCTOR—600 VOLTS
Size #6 and Larger—Stranded Copper Wire
Thermoplastic Insulated
For Heights to 20 Feet

MANHOURS PER LINEAR FOOT AND NUMBER OF WIRES LISTED

Wire Size	Runs in Length											
	To 50'				51' to 100'				101' to 200'			
	Number of Wires				Number of Wires				Number of Wires			
	2	3	4	6	2	3	4	6	2	3	4	6
6	.032	.037	.044	.057	.029	.032	.037	.046	.028	.030	.034	.041
4	.040	.046	.055	.072	.037	.041	.047	.059	.034	.037	.041	.050
2	.042	.048	.058	.075	.039	.043	.049	.062	.036	.039	.044	.053
1	.047	.054	.065	.084	.043	.047	.054	.068	.040	.043	.048	.059
1/0	.059	.068	.081	.106	.054	.059	.068	.085	.051	.055	.062	.075
2/0	.070	.081	.097	.126	.064	.070	.081	.101	.060	.065	.073	.088
3/0	.081	.093	.112	.145	.075	.083	.095	.119	.070	.076	.085	.102
4/0	.095	.109	.131	.170	.087	.096	.110	.138	.082	.089	.099	.120
250 MCM	.101	.116	.139	.181	.092	.101	.116	.145	.087	.094	.105	.127
300 MCM	.107	.123	.148	.192	.098	.108	.124	.155	.092	.099	.111	.135
350 MCM	.112	.129	.155	.201	.103	.113	.130	.163	.096	.104	.116	.141
400 MCM	.124	.143	.171	.222	.114	.125	.144	.180	.107	.116	.129	.157
500 MCM	.131	.151	.181	.235	.121	.133	.153	.191	.113	.122	.137	.165
750 MCM	.178	.205	.246	.319	.164	.180	.207	.259	.153	.165	.185	.224
1000 MCM	.213	.245	.294	.382	.196	.216	.248	.310	.183	.198	.221	.268

Manhours include checking out of job storage, handling, job hauling, placing wire coils or spools on reel stands or racks or opening wire cartons, fishing raceway runs, attaching wire to fish tape or other pull-in means, measuring and cutting wire, pulling wire into conduit type raceway or duct, or pull-out and lay wire into wireways or channels, identifying or "polling out" the circuit conductors, and tagging for connections to branch circuits, panel boards, or equipment.

If duplicate pulls above 6 are made reduce manhours 4% for each 2-, 3-, 4-, or 6-wire installation to compensate for equipment set-up time.

Manhours exclude extensive identification, "ringing-out," connection to branch circuit, panel board or equipment, and scaffolding. See respective tables for these time requirements.

SERVICE AND FEEDER WIRING
SINGLE CONDUCTOR—600 VOLTS
Size #6 and Larger—Stranded Copper Wire
Thermoplastic Insulated
For Heights to 25 Feet

MANHOURS PER LINEAR FOOT AND NUMBER OF WIRES LISTED

Wire Size	Runs in Length											
	To 50'				51' to 100'				101' to 200'			
	Number of Wires				Number of Wires				Number of Wires			
	2	3	4	6	2	3	4	6	2	3	4	6
6	.033	.038	.046	.059	.030	.033	.038	.041	.028	.030	.034	.041
4	.041	.047	.057	.074	.038	.042	.048	.060	.035	.038	.042	.051
2	.043	.049	.059	.077	.040	.044	.051	.063	.037	.040	.045	.054
1	.048	.055	.066	.086	.044	.048	.056	.070	.041	.044	.050	.060
1/0	.061	.070	.084	.109	.056	.062	.071	.089	.052	.056	.063	.076
2/0	.072	.083	.099	.129	.066	.073	.083	.104	.062	.067	.075	.091
3/0	.083	.095	.115	.149	.076	.084	.096	.120	.071	.077	.086	.104
4/0	.098	.113	.135	.176	.090	.099	.114	.142	.084	.091	.102	.123
250 MCM	.104	.120	.144	.187	.096	.106	.121	.152	.089	.096	.108	.130
300 MCM	.110	.127	.152	.197	.101	.111	.128	.160	.095	.103	.115	.139
350 MCM	.115	.132	.159	.206	.106	.117	.134	.168	.099	.107	.120	.145
400 MCM	.128	.147	.177	.230	.118	.130	.149	.187	.110	.119	.133	.161
500 MCM	.135	.155	.186	.242	.124	.136	.157	.196	.116	.125	.140	.170
750 MCM	.183	.210	.253	.328	.168	.185	.213	.266	.157	.170	.190	.230
1000 MCM	.219	.252	.302	.393	.201	.221	.254	.318	.188	.203	.227	.275

Manhours include checking out of job storage, handling, job hauling, placing wire coils or spools on reel stands or racks or opening wire cartons, fishing raceway runs, attaching wire to fish tape or other pull-in means, measuring and cutting wire, pulling wire into conduit type raceway or duct, or pull-out and lay wire into wireways or channels, identifying or "polling out" the circuit conductors, and tagging for connections to branch circuits, panel boards, or equipment.

If duplicate pulls above 6 are made reduce manhours 4% for each 2-, 3-, 4-, or 6-wire installation to compensate for equipment set-up time.

Manhours exclude extensive identification, "ringing-out," connection to branch circuit, panel board or equipment, and scaffolding. See respective tables for these time requirements.

For heights avove 25 feet increase manhours 5% for each additional 5 feet or fraction thereof.

SERVICE AND FEEDER WIRING
SINGLE CONDUCTOR—600 VOLTS

Size #6 and Larger—Stranded Aluminum Wire
Thermoplastic Insulated
For Heights to 10 Feet

MANHOURS PER LINEAR FOOT AND NUMBER OF WIRES LISTED

Wire Size	Runs in Length											
	To 50'				51' to 100'				101' to 200'			
	Number of Wires				Number of Wires				Number of Wires			
	2	3	4	6	2	3	4	6	2	3	4	6
6	.027	.031	.037	.048	.025	.028	.032	.040	.023	.025	.029	.034
4	.033	.038	.046	.059	.030	.033	.038	.041	.028	.030	.034	.041
2	.035	.040	.048	.063	.032	.035	.040	.051	.030	.032	.036	.044
1	.038	.044	.052	.068	.035	.039	.044	.055	.033	.036	.040	.048
1/0	.049	.056	.068	.088	.045	.050	.057	.071	.042	.045	.051	.061
2/0	.055	.063	.076	.099	.051	.056	.065	.081	.047	.051	.057	.069
3/0	.065	.075	.090	.117	.060	.066	.076	.095	.056	.060	.068	.082
4/0	.076	.087	.105	.136	.070	.077	.089	.111	.065	.070	.079	.095
250 MCM	.081	.093	.112	.145	.075	.083	.095	.119	.070	.076	.085	.102
300 MCM	.085	.098	.117	.152	.078	.086	.099	.123	.073	.079	.088	.107
350 MCM	.092	.106	.127	.165	.085	.094	.108	.134	.079	.085	.096	.116
400 MCM	.096	.110	.132	.172	.088	.097	.111	.139	.083	.090	.100	.121
500 MCM	.098	.113	.135	.176	.090	.099	.114	.142	.084	.091	.102	.123
750 MCM	.133	.153	.184	.239	.122	.134	.154	.193	.114	.123	.138	.167
1000 MCM	.165	.190	.228	.296	.152	.167	.192	.240	.142	.153	.172	.208

Manhours include checking out of job storage, handling, job hauling, placing wire coils or spools on reel stands or racks or opening wire cartons, fishing raceway runs, attaching wire to fish tape or other pull-in means, measuring and cutting wire, pulling wire into conduit type raceway or duct, or pull-out and lay wire into wireways or channels, identifying or "polling out" the circuit conductors, and tagging for connections to branch circuits, panel boards, or equipment.

If duplicate pulls above 6 are made reduce manhours 4% for each 2-, 3-, 4-, or 6-wire installation to compensate for equipment set-up time.

Manhours exclude extensive identification, "ringing-out," connection to branch circuit, panel board or equipment, and scaffolding. See respective tables for these time requirements.

SERVICE AND FEEDER WIRING
SINGLE CONDUCTOR—600 VOLTS
Size #6 and Larger—Stranded Aluminum Wire
Thermoplastic Insulated
For Heights to 15 Feet

MANHOURS PER LINEAR FOOT AND NUMBER OF WIRES LISTED

Wire Size	Runs in Length											
	To 50'				51' to 100'				101' to 200'			
	Number of Wires				Number of Wires				Number of Wires			
	2	3	4	6	2	3	4	6	2	3	4	6
6	.029	.033	.040	.052	.027	.030	.034	.043	.025	.027	.030	.037
4	.036	.041	.050	.065	.033	.036	.042	.052	.031	.033	.037	.045
2	.038	.044	.052	.068	.035	.039	.044	.055	.033	.036	.040	.048
1	.041	.047	.057	.074	.038	.042	.048	.060	.035	.038	.042	.051
1/0	.053	.061	.073	.095	.049	.054	.062	.077	.046	.050	.056	.067
2/0	.059	.068	.081	.106	.054	.059	.068	.085	.051	.055	.062	.075
3/0	.070	.081	.097	.126	.064	.070	.081	.101	.060	.065	.073	.088
4/0	.082	.094	.113	.147	.075	.083	.095	.119	.071	.077	.086	.104
250 MCM	.087	.100	.120	.156	.080	.088	.101	.127	.075	.081	.091	.110
300 MCM	.092	.106	.127	.165	.085	.094	.108	.134	.079	.085	.096	.116
350 MCM	.099	.114	.137	.178	.091	.100	.115	.144	.085	.092	.103	.124
400 MCM	.104	.120	.144	.187	.096	.106	.121	.152	.089	.096	.108	.130
500 MCM	.106	.122	.146	.190	.098	.108	.124	.155	.091	.098	.110	.133
750 MCM	.144	.166	.199	.258	.132	.145	.167	.209	.124	.134	.150	.181
1000 MCM	.178	.205	.246	.319	.164	.180	.207	.259	.153	.165	.185	.224

Manhours include checking out of job storage, handling, job hauling, placing wire coils or spools on reel stands or racks or opening wire cartons, fishing raceway runs, attaching wire to fish tape or other pull-in means, measuring and cutting wire, pulling wire into conduit type raceway or duct, or pull-out and lay wire into wireways or channels, identifying or "polling out" the circuit conductors, and tagging for connections to branch circuits, panel boards, or equipment.

If duplicate pulls above 6 are made reduce manhours 4% for each 2-, 3-, 4-, or 6-wire installation to compensate for equipment set-up time.

Manhours exclude extensive identification, "ringing-out," connection to branch circuit, panel board or equipment, and scaffolding. See respective tables for these time requirements.

SERVICE AND FEEDER WIRING
SINGLE CONDUCTOR—600 VOLTS
Size #6 and Larger—Stranded Aluminum Wire
Thermoplastic Insulated
For Heights to 20 Feet

MANHOURS PER LINEAR FOOT AND NUMBER OF WIRES LISTED

Wire Size	Runs in Length											
	To 50'				51' to 100'				101' to 200'			
	Number of Wires				Number of Wires				Number of Wires			
	2	3	4	6	2	3	4	6	2	3	4	6
6	.030	.034	.041	.054	.028	.031	.035	.044	.026	.028	.031	.038
4	.037	.043	.051	.066	.034	.037	.043	.054	.032	.035	.039	.047
2	.039	.045	.054	.070	.036	.040	.046	.060	.034	.037	.041	.050
1	.043	.049	.059	.077	.040	.044	.051	.063	.037	.040	.045	.054
1/0	.055	.063	.076	.099	.051	.056	.065	.081	.047	.051	.057	.069
2/0	.061	.070	.084	.109	.056	.062	.071	.089	.052	.056	.063	.076
3/0	.073	.084	.101	.131	.067	.074	.085	.106	.063	.068	.076	.092
4/0	.085	.098	.117	.152	.078	.086	.099	.123	.073	.079	.088	.107
250 MCM	.090	.104	.124	.161	.083	.091	.105	.131	.077	.083	.093	.113
300 MCM	.096	.110	.132	.172	.088	.097	.111	.139	.083	.090	.100	.121
350 MCM	.103	.118	.142	.185	.095	.105	.120	.150	.089	.096	.108	.130
400 MCM	.108	.124	.149	.194	.099	.109	.125	.157	.093	.100	.112	.136
500 MCM	.110	.127	.152	.197	.101	.111	.128	.160	.095	.103	.115	.139
750 MCM	.150	.173	.207	.269	.138	.152	.175	.218	.129	.139	.156	.189
1000 MCM	.185	.213	.255	.332	.170	.187	.215	.269	.159	.172	.192	.233

Manhours include checking out of job storage, handling, job hauling, placing wire coils or spools on reel stands or racks or opening wire cartons, fishing raceway runs, attaching wire to fish tape or other pull-in means, measuring and cutting wire, pulling wire into conduit type raceway or duct, or pull-out and lay wire into wireways or channels, identifying or "polling out" the circuit conductors, and tagging for connections to branch circuits, panel boards, or equipment.

If duplicate pulls above 6 are made reduce manhours 4% for each 2-, 3-, 4-, or 6-wire installation to compensate for equipment set-up time.

Manhours exclude extensive identification, "ringing-out," connection to branch circuit, panel board or equipment, and scaffolding. See respective tables for these time requirements.

SERVICE AND FEEDER WIRING
SINGLE CONDUCTOR—600 VOLTS
Size #6 and Larger—Stranded Aluminum Wire
Thermoplastic Insulated
For Heights to 25 Feet

MANHOURS PER LINEAR FOOT AND NUMBER OF WIRES LISTED

Wire Size	Runs in Length											
	To 50'				51' to 100'				101' to 200'			
	Number of Wires				Number of Wires				Number of Wires			
	2	3	4	6	2	3	4	6	2	3	4	6
6	.031	.036	.043	.056	.029	.032	.036	.046	.027	.029	.033	.040
4	.038	.044	.052	.068	.035	.039	.044	.055	.033	.036	.040	.048
2	.040	.046	.055	.072	.037	.041	.047	.059	.034	.037	.041	.050
1	.044	.051	.061	.079	.040	.044	.051	.063	.038	.041	.046	.056
1/0	.057	.066	.079	.102	.052	.057	.066	.082	.049	.053	.059	.072
2/0	.063	.072	.087	.113	.058	.064	.073	.092	.054	.058	.065	.079
3/0	.075	.086	.104	.135	.069	.076	.087	.109	.065	.070	.092	.095
4/0	.088	.101	.121	.158	.081	.089	.102	.128	.076	.082	.092	.111
250 MCM	.093	.107	.128	.167	.086	.095	.109	.136	.080	.086	.097	.117
300 MCM	.099	.114	.137	.178	.091	.100	.115	.144	.085	.092	.103	.124
350 MCM	.106	.122	.146	.190	.098	.108	.124	.155	.091	.098	.110	.133
400 MCM	.111	.128	.153	.199	.102	.112	.129	.161	.095	.103	.115	.139
500 MCM	.113	.130	.156	.203	.104	.114	.132	.164	.097	.105	.117	.142
750 MCM	.155	.178	.214	.278	.143	.157	.181	.226	.133	.144	.161	.195
1000 MCM	.191	.220	.264	.343	.176	.194	.223	.278	.164	.177	.198	.240

Manhours include checking out of job storage, handling, job hauling, placing wire coils or spools on reel stands or racks or opening wire cartons, fishing raceway runs, attaching wire to fish tape or other pull-in means, measuring and cutting wire, pulling wire into conduit type raceway or duct, or pull-out and lay wire into wireways or channels, identifying or "polling out" the circuit conductors, and tagging for connections to branch circuits, panel boards, or equipment.

If duplicate pulls above 6 are made reduce manhours 4% for each 2-, 3-, 4-, or 6-wire installation to compensate for equipment set-up time.

Manhours exclude extensive identification, "ringing-out," connection to branch circuit, panel board or equipment, and scaffolding. See respective tables for these time requirements.

For heights above 25 feet increase manhours 5% for each additional 5 feet or fraction thereof.

WIRE CONNECTORS AND LUGS

MANHOURS EACH

Manhours Required for Items Listed

Wire Size	U-Bolt Connector Tape Wrapped	2-Way Compression Connector Tape Wrapped	Bolt Lug	Compression Lug
6	0.66	0.55	0.33	0.28
4	0.77	0.66	0.33	0.28
2	0.88	0.77	0.33	0.28
1	1.27	1.10	0.66	0.55
1/0	1.38	1.27	0.83	0.72
2/0	1.65	1.38	0.99	0.83
3/0	1.93	1.65	1.21	0.99
4/0	2.20	1.93	1.38	1.10
250 MCM	2.75	2.20	1.71	1.38
300 MCM	3.00	2.48	1.93	1.65
350 MCM	3.30	3.00	2.10	1.76
400 MCM	3.85	3.30	2.20	1.93
500 MCM	4.40	3.85	2.53	2.20
750 MCM	5.50	4.40	3.85	3.03
1000 MCM	6.60	5.50	4.40	3.58

Manhours include checking out of job storage, handling, job hauling, and installing items as listed.

Manhours exclude installation of wire, tagging, and scaffolding. See respective tables for these time frames.

WIRE VERTICAL RISER SUPPORTS

MANHOURS EACH

Size (Inches)	2–4-Wire Support Conduit Fitting	5- and More Wire Support Conduit Fitting	Wood Wedge Cable Support
1-1/4	0.94	1.10	0.55
1-1/2	1.21	1.38	0.83
2	1.82	1.98	1.10
2-1/2	2.42	2.75	1.65
3	3.03	3.30	2.20
3-1/2	3.63	4.13	2.75
4	4.68	5.23	3.85
5	5.78	6.60	4.95
6	7.15	8.25	6.60

Manhours include checking out of job storage, handling, job hauling, and installing items as listed on conduit termination, in pull boxes or panel cabinets.

Manhours exclude installation of wiring, tagging, and scaffolding. See respective tables for these time frames.

SERVICE AND FEEDER WIRE GROUNDING MATERIALS

MANHOURS FOR UNITS LISTED

Item and Description	Size	Unit	Manhours
Ground Clamps	1/2"-1"	Each	1.10
Ground Clamps	1-1/4"-2"	Each	1.93
Ground Clamps	2-1/2"-3-1/2"	Each	2.75
Ground Clamps	4"-5"	Each	3.85
Ground Clamps	6"	Each	4.40
Ground Straps	3/8"-1"	Each	0.55
Ground Straps	3/8"-2"	Each	0.83
Ground Straps—Perforated	3/4"	Each	0.83
Ground Clamp with 1/2" Conduit Hub	1/2"-1"	Each	1.65
Ground Clamp with 1/2" Conduit Hub	1-1/4"-2"	Each	2.48
Ground Clamp with 1/2" Conduit Hub	2-1/4"-4"	Each	3.30
Ground Clamp with 3/4" Conduit Hub	1/2"-1"	Each	1.93
Ground Clamp with 3/4" Conduit Hub	1-1/4"-2"	Each	2.75
Ground Clamp with 3/4" Conduit Hub	2-1/2"-4"	Each	3.58
Ground Clamp with 1" Conduit Hub	1/2"-1"	Each	2.20
Ground Clamp with 1" Conduit Hub	1-1/4"-2"	Each	3.03
Ground Clamp with 1" Conduit Hub	2-1/4"-4"	Each	3.85
Ground Rods	3/8"x5'0"	Each	1.10
Ground Rods	1/2"x5'0"	Each	1.10
Ground Rods	1/2"x8'0"	Each	1.65
Ground Rods	5/8"x8'0"	Each	1.65
Ground Rods	5/8"x10'0"	Each	2.20
Ground Rods	3/4"x10'0"	Each	2.30
Ground Rods	1"x10'0"	Each	4.40
Ground Rod Clamps	3/8"	Each	0.83
Ground Rod Clamps	1/2"	Each	1.10
Ground Rod Clamps	5/8"	Each	1.38
Ground Rod Clamps	3/4"	Each	1.65
Ground Rod Clamps	1"	Each	1.93
Solid Bare Copper Wire	#8	Lin. Ft.	0.02
Solid Bare Copper Wire	#6	Lin. Ft.	0.03
Stranded Bare Copper Wire	#4	Lin. Ft.	0.04
Stranded Bare Copper Wire	#2	Lin. Ft.	0.05
Stranded Bare Copper Wire	#1/0	Lin. Ft.	0.06
Stranded Bare Copper Wire	#2/0	Lin. Ft.	0.07
Stranded Bare Copper Wire	#3/0	Lin. Ft.	0.08

Manhours include checking out of job storage, handling, job hauling, and installing items as outlined for grounding conductors.

Manhours exclude installation of service and feeder wiring, tagging, and scaffolding. See respective tables for these time frames.

MINERAL INSULATED SHEATHED CABLE
Wire Size #6 and Larger
Wood or Metal Frame Construction (Concealed)

MANHOURS PER FOOT OF CABLE AND FOR EACH OF FITTINGS

Cable Size	Number of Conductors	Cable Installation				Service Entrance or Terminator Installation			
		Manhours for Height to				Manhours for Height to			
		10'	15'	20'	25'	10'	15'	20'	25'
6	1	.083	.090	.093	.096	1.49	1.61	1.67	1.72
6	2	.088	.095	.099	.102	1.71	1.85	1.92	1.98
6	3	.099	.107	.111	.115	1.98	2.14	2.22	2.29
6	4	.110	.119	.124	.127	2.15	2.32	2.41	2.49
4	1	.088	.095	.099	.102	1.60	1.73	1.80	1.85
4	2	.094	.102	.106	.109	2.04	2.20	2.29	2.36
4	3	.110	.119	.124	.127	2.20	2.38	2.47	2.55
2	1	.094	.102	.106	.109	1.76	1.90	1.98	2.04
1	1	.099	.107	.111	.115	1.87	2.02	2.10	2.16
0	1	.105	.113	.118	.121	2.04	2.20	2.29	2.36
00	1	.110	.119	.124	.127	2.20	2.38	2.47	2.55
000	1	.121	.131	.136	.140	2.42	2.61	2.72	2.80
0000	1	.138	.149	.155	.160	2.64	2.85	2.97	3.05
250 MCM	1	.149	.160	.166	.171	3.03	3.27	3.40	3.51

All manhours include checking out of job storage, handling, and job hauling.

Cable installation manhours include placing cable coils or spools on reel stands or racks or opening cable cartons, measuring and cutting cable, normal securing of cable in position, identifying or "polling-out" the conductors, and tagging for connections to branch circuits, panel boards, or equipment.

Service entrance or terminator manhours include stripping of cable sheath up to 36 inches, placing service entrance or terminator, and making connection.

Manhours exclude extensive identification, and scaffolding. See respective tables for these time frames.

For heights above 25 feet increase manhours 5% for each additional 5 feet or fraction thereof.

MINERAL INSULATED SHEATHED CABLE
Wire Size #6 and Larger
Suspended Ceilings (Concealed)

MANHOURS PER FOOT OF CABLE AND FOR EACH OF FITTINGS

Cable Size	Number of Conductors	Cable Installation				Service Entrance or Terminator Installation			
		Manhours for Height to				Manhours for Height to			
		10'	15'	20'	25'	10'	15'	20'	25'
6	1	.072	.078	.081	.083	1.34	1.45	1.51	1.55
6	2	.077	.083	.086	.089	1.54	1.66	1.73	1.78
6	3	.088	.095	.099	.102	1.78	1.92	2.00	2.06
6	4	.099	.107	.111	.115	1.94	2.10	2.18	2.24
4	1	.077	.083	.086	.089	1.44	1.56	1.62	1.67
4	2	.083	.090	.093	.096	1.84	1.99	2.07	2.13
4	3	.094	.102	.106	.109	1.98	2.14	2.22	2.29
2	1	.083	.090	.093	.096	1.58	1.71	1.77	1.83
1	1	.088	.095	.099	.102	1.68	1.81	1.87	1.94
0	1	.094	.102	.106	.109	1.84	1.99	2.07	2.13
00	1	.099	.107	.111	.115	1.98	2.14	2.22	2.29
000	1	.110	.119	.124	.127	2.18	2.35	2.45	2.52
0000	1	.127	.137	.143	.150	2.38	2.57	2.67	2.75
250 MCM	1	.132	.143	.148	.153	2.73	2.95	3.07	3.16

All manhours include checking out of job storage, handling, and job hauling.

Cable installation manhours include placing cable coils or spools on reel stands or racks or opening cable cartons, measuring and cutting cable, normal securing of cable in position, identifying or "polling-out" the conductors, and tagging for connections to branch circuits, panel boards, or equipment.

Service entrance or terminator manhours include stripping of cable sheath up to 36 inches, placing service entrance or terminator, and making connection.

Manhours exclude extensive identification, and scaffolding. See respective tables for these time frames.

For heights above 25 feet increase manhours 5% for each additional 5 feet or fraction thereof.

MINERAL INSULATED SHEATHED CABLE

Wire Size #6 and Larger

Concrete Floors and Walls (Concealed)

MANHOURS PER FOOT OF CABLE AND FOR EACH OF FITTINGS

Cable Size	Number of Conductors	Cable Installation Manhours for Height to				Service Entrance or Terminator Installation Manhours for Height to			
		10'	15'	20'	25'	10'	15'	20'	25'
6	1	.047	.051	.053	.054	0.89	0.96	1.00	1.03
6	2	.050	.054	.056	.058	1.03	1.11	1.16	1.19
6	3	.056	.060	.063	.065	1.19	1.29	1.34	1.38
6	4	.063	.068	.071	.073	1.29	1.39	1.45	1.49
4	1	.050	.054	.056	.058	0.96	1.03	1.08	1.11
4	2	.053	.057	.060	.061	1.22	1.32	1.37	1.41
4	3	.063	.068	.071	.073	1.32	1.43	1.48	1.53
2	1	.053	.057	.060	.061	1.06	1.14	1.19	1.23
1	1	.056	.060	.063	.065	1.12	1.21	1.26	1.30
0	1	.059	.064	.066	.068	1.22	1.32	1.37	1.41
00	1	.063	.068	.071	.073	1.32	1.43	1.48	1.53
000	1	.069	.075	.078	.080	1.45	1.57	1.63	1.68
0000	1	.078	.084	.088	.090	1.58	1.71	1.77	1.83
250 MCM	1	.085	.092	.095	.098	1.82	1.97	2.04	2.11

All manhours include checking out of job storage, handling, and job hauling.

Cable installation manhours include placing cable coils or spools on reel stands or racks or opening cable cartons, measuring and cutting cable, normal securing of cable in position, identifying or "polling-out" the conductors, and tagging for connections to branch circuits, panel boards, or equipment.

Service entrance or terminator manhours include stripping of cable sheath up to 36 inches, placing service entrance or terminator, and making connection.

Manhours exclude extensive identification, and scaffolding. See respective tables for these time frames.

For heights above 25 feet increase manhours 5% for each additional 5 feet or fraction thereof.

MINERAL INSULATED SHEATHED CABLE

Wire Size # 6 and Larger

Flat Slabs and Walls (Exposed)

MANHOURS PER FOOT OF CABLE AND FOR EACH OF FITTINGS

Cable Size	Number of Conductors	Cable Installation				Service Entrance or Terminator Installation			
		Manhours for Height to				Manhours for Height to			
		10′	15′	20′	25′	10′	15′	20′	25′
6	1	.083	.090	.093	.096	1.49	1.61	1.67	1.72
6	2	.088	.095	.099	.102	1.71	1.85	1.92	1.98
6	3	.099	.107	.111	.115	1.98	2.14	2.22	2.29
6	4	.110	.119	.124	.127	2.15	2.32	2.41	2.49
4	1	.088	.095	.099	.102	1.60	1.73	1.80	1.85
4	2	.094	.102	.106	.109	2.04	2.20	2.29	2.36
4	3	.110	.119	.124	.127	2.20	2.38	2.47	2.55
2	1	.094	.102	.106	.109	1.76	1.90	1.98	2.04
1	1	.099	.107	.111	.115	1.87	2.02	2.10	2.16
0	1	.105	.113	.118	.121	2.04	2.20	2.29	2.36
00	1	.110	.119	.124	.127	2.20	2.38	2.47	2.55
000	1	.121	.131	.136	.140	2.42	2.61	2.72	2.80
0000	1	.138	.149	.155	.160	2.64	2.85	2.97	3.05
250 MCM	1	.149	.160	.166	.171	3.03	3.27	3.40	3.51

All manhours include checking out of job storage, handling, and job hauling.

Cable installation manhours include placing cable coils or spools on reel stands or racks or opening cable cartons, measuring and cutting cable, normal securing of cable in position, identifying or "polling-out" the conductors, and tagging for connections to branch circuits, panel boards, or equipment.

Service entrance or terminator manhours include stripping of cable sheath up to 36 inches, placing service entrance or terminator, and making connection.

Manhours exclude extensive identification, and scaffolding. See respective tables for these time frames.

For heights above 25 feet increase manhours 5% for each additional 5 feet or fraction thereof.

MINERAL INSULATED SHEATHED CABLE
Wire Size #6 and Larger
Masonry Walls (Concealed)

MANHOURS PER FOOT OF CABLE AND FOR EACH OF FITTINGS

Cable Size	Number of Conductors	Cable Installation				Service Entrance or Terminator Installation			
		Manhours for Height to				Manhours for Height to			
		10'	15'	20'	25'	10'	15'	20'	25'
6	1	.071	.077	.080	.082	1.28	1.38	1.44	1.48
6	2	.076	.082	.085	.088	1.47	1.59	1.65	1.70
6	3	.085	.092	.095	.098	1.70	1.84	1.91	1.97
6	4	.095	.103	.107	.110	1.85	2.00	2.08	2.14
4	1	.076	.082	.085	.088	1.38	1.49	1.55	1.60
4	2	.080	.086	.090	.093	1.75	1.89	1.97	2.02
4	3	.095	.103	.107	.110	1.89	2.04	2.12	2.19
2	1	.080	.086	.090	.093	1.51	1.63	1.70	1.75
1	1	.085	.092	.095	.098	1.61	1.74	1.81	1.86
0	1	.090	.097	.101	.104	1.75	1.89	1.97	2.02
00	1	.095	.103	.107	.110	1.89	2.04	2.12	2.19
000	1	.104	.112	.117	.120	2.08	2.25	2.34	2.41
0000	1	.119	.129	.134	.138	2.27	2.45	2.55	2.63
250 MCM	1	.128	.138	.144	.148	2.61	2.82	2.93	3.20

All manhours include checking out of job storage, handling, and job hauling.

Cable installation manhours include placing cable coils or spools on reel stands or racks or opening cable cartons, measuring and cutting cable, normal securing of cable in position, identifying or "polling-out" the conductors, and tagging for connections to branch circuits, panel boards, or equipment.

Service entrance or terminator manhours include stripping of cable sheath up to 36 inches, placing service entrance or terminator, and making connection.

Manhours exclude extensive identification, and scaffolding. See respective tables for these time frames.

For heights above 25 feet increase manhours 5% for each additional 5 feet or fraction thereof.

MINERAL INSULATED SHEATHED CABLE

Wire Size #6 and Larger

Cable Trays or Channel (Exposed)

MANHOURS PER FOOT OF CABLE AND FOR EACH OF FITTINGS

Cable Size	Number of Conductors	Cable Installation				Service Entrance or Terminator Installation			
		Manhours for Height to				Manhours for Height to			
		10'	15'	20'	25'	10'	15'	20'	25'
6	1	.059	.064	.066	.068	1.04	1.12	1.17	1.20
6	2	.063	.068	.071	.073	1.20	1.30	1.35	1.39
6	3	.071	.077	.080	.082	1.39	1.50	1.56	1.61
6	4	.079	.085	.089	.091	1.51	1.63	1.69	1.75
4	1	.063	.068	.071	.073	1.12	1.21	1.26	1.30
4	2	.067	.072	.075	.078	1.43	1.54	1.61	1.65
4	3	.079	.085	.089	.091	1.54	1.66	1.73	1.78
2	1	.067	.072	.075	.078	1.23	1.33	1.38	1.42
1	1	.071	.077	.080	.082	1.31	1.41	1.47	1.52
0	1	.075	.081	.084	.087	1.43	1.54	1.61	1.65
00	1	.079	.085	.089	.091	1.54	1.66	1.73	1.78
000	1	.087	.094	.098	.101	1.69	1.83	1.90	1.96
0000	1	.091	.098	.102	.105	1.85	2.00	2.08	2.14
250 MCM	1	.099	.107	.111	.115	2.12	2.29	2.38	2.45

All manhours include checking out of job storage, handling, and job hauling.

Cable installation manhours include placing cable coils or spools on reel stands or racks or opening cable cartons, measuring and cutting cable, normal securing of cable in position, identifying or "polling-out" the conductors, and tagging for connections to branch circuits, panel boards, or equipment.

Service entrance or terminator manhours include stripping of cable sheath up to 36 inches, placing service entrance or terminator, and making connection.

Manhours exclude extensive identification, and scaffolding. See respective tables for these time frames.

For heights above 25 feet increase manhours 5% for each additional 5 feet or fraction thereof.

600-VOLT INTERLOCKED ARMORED CABLE
For Service and Feeders
Steel or Bronze Sheath—Varnished Cambric Insulated

MANHOURS PER 100 LINEAR FEET

Cable Size	Installation Manhours		
	2-Conductor	3-Conductor	4-Conductor
8	3.11	4.14	5.18
6	4.03	5.37	6.71
4	4.95	6.60	8.25
2	5.25	7.00	8.75
1	6.22	8.29	10.36
1/0	6.67	8.89	11.11
2/0	7.52	10.03	12.54
3/0	8.24	10.98	13.73
4/0	9.35	12.47	15.59
250 MCM	10.12	13.49	16.86
350 MCM	10.89	14.52	18.15
500 MCM	11.72	15.62	19.53
750 MCM	14.69	19.58	24.48

Manhours include checking out of job storage, handling, job hauling, placing wire coils or spools on reel stands, rigging with rollers, pulleys and rigging cables, pulling cable into troughs or channels or onto ladders with a pull-in cable and power winch, measuring and cutting cable, identifying or "polling-out" the conductors and tagging for connections.

Manhours exclude extensive identification, installation of cable fittings, cable splicing, and scaffolding. See respective tables for these time frames.

For rubber or rubber-like insulated cable, increase the manhours 10% to compensate for the additional cable weight.

For additional parallel runs to be installed at the same time apply the following percentages of the manhours.

For Second Parallel Run—85%
For Third Parallel Run—75%
For Fourth Parallel Run—70%
For Fifth Parallel Run—63%

5 KV INTERLOCKED ARMORED CABLE

For Service and Feeders

Steel or Bronze Sheath—Varnished Cambric Insulated

MANHOURS PER 100 LINEAR FEET

Cable Size	Installation Manhours	
	2-Conductor	3-Conductors
8	4.95	6.60
6	5.42	7.22
4	5.96	7.94
2	6.47	8.62
1	7.04	9.39
1/0	7.46	9.94
2/0	8.03	10.71
3/0	9.02	12.03
4/0	9.75	13.00
250 MCM	10.30	13.73
350 MCM	11.21	14.94
500 MCM	11.75	15.66
750 MCM	13.96	18.61

15 KV INTERLOCKED ARMORED CABLE

For Service and Feeders

Steel or Bronze Sheath—Varnished Cambric Insulated

MANHOURS PER 100 LINEAR FEET

Cable Size	Installation Manhours	
	2-Conductor	3-Conductor
8	–	–
6	7.40	9.87
4	7.92	10.56
2	8.88	11.84
1	9.26	12.34
1/0	9.82	13.09
2/0	10.18	13.57
3/0	10.56	14.08
4/0	10.96	14.61
250 MCM	11.19	14.92
350 MCM	11.72	15.62
500 MCM	13.05	17.40
750 MCM	–	–

Manhours include checking out of job storage, handling, job hauling, placing wire coils or spools on reel stands, rigging with rollers, pulleys and rigging cables, pulling cable into troughs or channels or onto ladders with a pull-in cable and power winch, measuring and cutting cable, identifying or "polling-out" the conductors, and tagging for connections.

Manhours exclude extensive identification, installation of cable fittings, cable splicing, and scaffolding. See respective tables for these time frames.

For rubber or rubber-like insulated cable, increase the manhours 10% to compensate for the additional cable weight.

For additional parallel runs to be installed at the same time apply the following percentages of the manhours.

For Second Parallel Run—85%
For Third Parallel Run—75%
For Fourth Parallel Run—70%
For Fifth Parallel Run—63%

NONWATERTIGHT AND WATERTIGHT FITTINGS FOR 600-VOLT INTERLOCKED ARMORED CABLE
For Service and Feeders

MANHOURS EACH

Cable Size	Installation Manhours					
	Nonwatertight Fittings			Watertight Fittings		
	2-Conductor	3-Conductor	4-Conductor	2-Conductor	3-Conductor	4-Conductor
8	1.38	1.65	1.93	1.65	1.93	2.20
6	1.65	2.20	2.75	2.20	2.48	3.03
4	2.20	2.48	3.03	2.48	2.75	3.30
2	2.75	3.03	3.58	3.03	3.30	3.85
1	3.03	3.30	4.13	3.30	3.58	4.40
1/0	3.30	3.58	4.40	3.58	3.85	4.95
2/0	3.58	3.85	4.95	3.85	4.13	5.50
3/0	4.13	4.40	5.50	4.40	4.95	6.05
4/0	5.50	6.05	7.90	6.05	6.33	8.25
250 MCM	6.05	6.60	8.25	6.60	6.88	9.08
350 MCM	6.60	7.15	9.08	7.15	7.43	9.90
500 MCM	7.15	7.70	11.00	7.70	8.25	12.10
750 MCM	8.80	9.90	–	9.35	10.45	–

Manhours include checking out of job storage, handling, job hauling, preparing cable ends, installing fitting, and insulating cable ends.

Manhours exclude installation of cable, identifying conductors, and scaffolding. See respective tables for these time frames.

COMPOUND FITTINGS AND CABLE SPLICING FOR 600-VOLT INTERLOCKED ARMORED CABLE

For Service and Feeders

MANHOURS EACH

Cable Size	Installation Manhours					
	Nonwatertight Fittings			Watertight Fittings		
	2-Conductor	3-Conductor	4-Conductor	2-Conductor	3-Conductor	4-Conductor
8	2.75	3.03	3.30	2.75	3.03	3.30
6	3.30	3.58	4.13	2.75	3.03	3.85
4	3.58	3.85	4.40	3.03	3.30	4.40
2	4.13	4.40	4.95	3.58	4.13	5.23
1	4.40	4.68	5.50	4.40	4.95	6.05
1/0	4.95	5.23	6.33	4.95	5.78	6.88
2/0	5.23	5.50	6.88	5.78	6.60	7.70
3/0	5.78	6.33	7.43	7.15	7.98	9.63
4/0	7.43	7.98	9.90	7.43	8.25	10.45
250 MCM	8.25	8.53	10.73	7.70	8.53	10.73
350 MCM	8.80	9.08	12.10	8.03	9.08	11.55
500 MCM	9.90	10.45	14.30	8.80	9.35	12.10
750 MCM	11.55	12.65	–	–	10.45	–

Fitting manhours include checking out of job storage, handling, job hauling, preparing cable ends, installing, fitting, and insulating cable ends.

Cable splicing manhours include preparing cable ends, making splice, and filling box or fitting with compound where required.

Manhours exclude installation of cable, identifying conductors, and scaffolding. See respective tables for these time frames.

Cable splicing manhours do not include installation of fitting or box. See respective tables for these time frames.

COMPOUND FITTINGS, POTHEADS, AND CABLE SPLICING FOR 5,000-VOLT INTERLOCKED ARMORED CABLE

For Service and Feeders

MANHOURS EACH

Cable Size	Installation Manhours					
	Nonwatertight Fittings			Watertight Fittings		
	2-Conductor	3-Conductor	4-Conductor	2-Conductor	3-Conductor	4-Conductor
8	4.40	4.95	11.00	13.20	3.30	3.85
6	4.40	4.95	11.00	13.20	3.30	3.85
4	4.95	5.50	11.00	13.20	4.40	4.40
2	4.95	5.50	11.00	13.20	4.40	4.40
1	6.05	6.60	11.00	13.20	4.68	5.23
1/0	6.60	7.15	11.00	13.20	6.05	7.15
2/0	7.15	7.70	15.40	17.60	6.60	7.70
3/0	7.70	8.25	15.40	17.60	8.25	8.80
4/0	9.63	10.45	15.40	17.60	8.25	9.35
250 MCM	10.73	11.55	15.40	17.60	8.80	9.90
350 MCM	13.20	14.30	17.60	19.80	9.08	10.18
500 MCM	14.30	15.40	17.60	19.80	9.35	10.45
750 MCM	15.95	17.60	19.80	22.00	9.63	11.00

Fitting manhours include checking out of job storage, handling, job hauling, preparing cable ends, installing, fitting, and insulating cable ends.

Cable splicing manhours include preparing cable ends, making splice, and filling box or fitting with compound where required.

Manhours exclude installation of cable, identifying conductors, and scaffolding. See respective tables for these time frames.

Cable splicing manhours do not include installation of fitting or box. See respective tables for these time frames.

COMPOUND FITTINGS, POTHEADS, AND CABLE SPLICING FOR 15,000-VOLT INTERLOCKED ARMORED CABLE

For Service and Feeders

MANHOURS EACH

Cable Size	Installation Manhours					
	Nonwatertight Fittings			Watertight Fittings		
	2-Conductor	3-Conductor	4-Conductor	2-Conductor	3-Conductor	4-Conductor
8	–	–	–	–	–	–
6	13.75	15.40	18.70	22.00	6.60	7.70
4	13.75	15.40	18.70	22.00	6.60	7.70
2	15.95	17.60	18.70	22.00	6.88	8.25
1	15.95	17.60	18.70	22.00	6.88	8.25
1/0	17.60	19.80	18.70	22.00	7.15	8.53
2/0	19.80	22.00	23.10	26.40	8.80	9.35
3/0	19.80	22.00	23.10	26.40	8.80	9.35
4/0	22.00	24.20	23.10	26.40	9.35	9.63
250 MCM	22.00	24.20	23.10	26.40	9.35	9.63
350 MCM	–	–	–	–	–	–
500 MCM	24.20	26.40	27.50	30.80	9.90	10.18
750 MCM	–	–	–	–	–	–

Fitting manhours include checking out of job storage, handling, job hauling, preparing cable ends, installing, fitting, and insulating cable ends.

Cable splicing manhours include preparing cable ends, making splice, and filling box or fitting with compound where required.

Manhours exclude installation of cable, identifying conductors, and scaffolding. See respective tables for these time frames.

Cable splicing manhours do not include installation of fitting or box. See respective tables for these time frames.

Section 3

WIRING DEVICES— SWITCHES, OUTLETS, AND RECEPTACLES

This section provides manhour units for the installation of switches, dimmers, receptacles, plates, and other devices for the control of lighting, power, and miscellaneous item circuits.

The units include time allowance, as may be required, for cleaning out boxes and the complete installation of the device or plate all in accordance with the notes appearing on the individual table pages.

The manhours are based on averages for the type of installation outlined. Consideration should be given to the variable productivity factors which may effect the overall time required for the installation of the various items. Some of these factors are outlined in the Introduction of this manual and we suggest that the estimator be familiar with these prior to the use and application of the following manhour units.

TUMBLER AND TUMBLER LOCK TYPE SWITCHES

MANHOURS EACH

Item Description	Amperes	Manhours Required for	
		Tumbler Switches	Lock Switches
Single-Pole	10	0.33	0.45
	15	0.33	0.45
	20	0.58	0.69
	30	0.77	0.85
Double-Pole	10	0.96	0.96
	15	0.96	0.96
	20	1.38	1.38
	30	1.60	1.60
Three-Way	10	0.63	0.63
	15	0.63	0.63
	20	0.85	0.85
	30	1.16	1.16
Four-Way	10	0.96	0.96
	15	0.96	0.96
	20	1.38	1.38
	30	1.60	1.60

Manhours include checking out of job storage, handling, job hauling, distributing, clearing box of debris, making copper conductor of normal gauge connections, and mounting device in box or fitting.

Manhours are based on installation of up to 50 switches on a single floor. If more than 50 switches are to be installed decrease manhours by the following percentages:

    51 to 100—Decrease 20%
    101 to 200—Decrease 30%
    201 or more—Decrease 35%

If aluminum conductors of a larger gauge for a given amperage are to be connected to the switches increase the manhours 10%.

Manhours exclude installation of conduit, boxes, conductors, and other devices. See respective tables for these time frames.

WEATHER PROOF, MARK TIME, AND THERMAL SWITCHES

MANHOURS EACH

Item Description	Manhours
Weather Proof Switches with Plates	
Single-Pole — 1-Gang — 10 Amp	0.58
Single-Pole — 1-Gang — 20 Amp	0.83
Single-Pole — 2-Gang — 10 Amp	0.96
Double-Pole — 1-Gang — 10 Amp	1.07
Double-Pole — 1-Gang — 20 Amp	1.24
Three-Way — 1-Gang — 10 Amp	0.83
Three-Way — 1-Gang — 20 Amp	0.96
Four-Way — 1-Gang — 10 Amp	1.07
Four-Way — 1-Gang — 20 Amp	1.24
Mark Time Switches	
Single-Pole — 10 Amp	0.58
Single-Pole — 20 Amp	0.83
Thermal Switches	
Single-Pole — Without Enclosure	0.58
Double-Pole — Without Enclosure	1.38
Single-Pole — With Enclosure	1.65
Double-Pole — With Enclosure	2.20

Manhours include checking out of job storage, handling, job hauling, distributing, clearing box of debris, making copper conductor of normal gauge connections, and mounting device in box or fitting.

Manhours are based on installation of up to 10 switches on a single floor. If more than 10 switches are to be installed decrease manhours by the following percentages:

11 to 50 — Decrease 20%
51 to 100 — Decrease 30%
101 or more — Decrease 35%

If aluminum conductors of a larger gauge for a given amperage are to be connected to the switches increase the manhours 10%.

Manhours exclude installation of conduit, boxes and plates except as noted, conductors and other devices. See respective tables for these time frames.

DOOR, MOMENTARY CONTACT, TUMBLER TYPE DIMMER, AND ROTARY DIMMER SWITCHES

MANHOURS EACH

Item Description	Manhours
Door Switch with Box	
Light On — Door Open	1.65
Light On — Door Close	1.65
Momentary Contact Switches	
Single Pole — 10 Amp	0.58
Double Throw — 20 Amp	0.83
Tumbler Type Dimmer Switches with Plates	
Hi-Lo Dimmer — 300 W	0.69
Hi-Lo Dimmer — 400 W	0.69
Hi-Lo Dimmer — 500 W	0.69
Rotary Dimmer Switches with Plates	
Incandescent Circuit Dimming — 300 W	0.96
Incandescent Circuit Dimming — 400 W	0.96
Incandescent Circuit Dimming — 500 W	0.96
Incandescent Circuit Dimming — 600 W	0.96
Incandescent Circuit Dimming — 900 W	0.96
Incandescent Circuit Dimming — 1000 W	0.96
Flourescent Circuit Dimming — 8-40 W	0.96

Manhours include checking out of job storage, handling, job hauling, distributing, clearing box of debris, making copper conductor of normal gauge connections, and mounting device in box or fitting.

Manhours are based on installation of up to 10 switches on a single floor. If more than 10 switches are to be installed decrease above manhours by the following percentages:

11 to 50 — Decrease 20%
51 to 100 — Decrease 30%
101 or more — Decrease 35%

If aluminum conductors of a larger gauge for a given amperage are to be connected to the switches increase the manhours 10%.

Manhours exclude installation of conduit, boxes and plates except as noted, conductors and other devices. See respective tables for these time frames.

SWITCH AND RECEPTACLE COMBINATIONS AND DUPLEX RECEPTACLES

MANHOURS EACH

Item Description	Manhours
Switch and Receptacle Combinations	
Single Pole Switch and 2-Wire Receptacle	1.38
Single Pole Switch and 3-Wire Grounding Receptacle	1.45
Two Single Pole Switches	1.38
One Single Pole and One 3-way Switch	1.45
Two 3-Way Switches	1.52
Duplex Receptacles	
Nongrounding — 125V — 2-Pole 2-Wire — 15 Amp	0.29
Nongrounding — 125V — 2-Pole 2-Circuit — 15 Amp	0.50
Grounding — 125V — 2-Pole 2-Wire — 15 Amp	0.33
Grounding — 125V — 2-Pole 2-Wire — 20 Amp	0.58
Grounding — 125V — 2-Pole 2-Circuit — 15 Amp	0.50
Grounding — 125V — 2-Pole 2-Circuit — 20 Amp	0.72
Grounding — 250V — 2-Pole 2-Wire — 15 Amp	0.44
Grounding — 277V — 2-Pole 2-Wire — 15 Amp	0.44

Manhours include checking out of job storage, handling, job hauling, distributing, clearing box of debris, making copper conductor of normal gauge connections, and mounting device in box or fitting.

Manhours are based on installation of up to 10 devices on a single floor. If more than 10 devices are to be installed decrease manhours by the following percentages:

11 to 50 — Decrease 20%
51 to 100 — Decrease 30%
101 or more — Decrease 35%

If aluminum conductors of a larger gauge for a given amperage are to be connected to the devices increase the manhours 10%.

Manhours exclude installation of conduit, boxes, plates, conductors and other devices. See respective tables for these time frames.

SINGLE NONGROUNDING RECEPTACLES

MANHOURS EACH

Item Description	Amperage	Number of Wires	Manhours
Nongrounding — 2-Pole — 125V	15	2	0.29
Nongrounding — 3-Pole — 125/250V	20	3	0.66
	30	3	0.88
	50	3	1.27
Nongrounding — 4-Pole — 120/208V	15	4	0.74
	20	4	0.83
	30	4	0.98
	50	4	1.43
	60	4	2.17
Nongrounding — 2-Pole — 250V	20	2	0.47
	30	2	0.80
Nongrounding — 3-Pole — 250V	15	3	0.55
	20	3	0.69
	30	3	0.91
	50	3	1.27

Manhours include checking out of job storage, handling, job hauling, distributing, clearing box of debris, making copper conductor of normal gauge connections, and mounting receptacle in box or fitting.

Manhours are based on installation of up to 10 receptacles on a single floor. If more than 10 receptacles are to be installed decrease manhours by the following percentages:

11 to 50 — Decrease 20%
51 to 100 — Decrease 30%
101 or more — Decrease 35%

If aluminum conductors of a larger gauge for a given amperage are to be connected to the receptacles increase the manhours 10%.

Manhours exclude installation of conduit, boxes, plates, conductors, and other devices. See respective tables for these time requirements.

SINGLE GROUNDING RECEPTACLES

MANHOURS EACH

Item Description	Amperage	Number of Wires	Manhours
Grounding – 2-Pole – 125V	15	2	0.33
	20	2	0.50
	30	2	0.85
	50	2	1.16
Grounding – 2-Pole – 125/250V	15	2	0.33
Grounding – 3-Pole – 125/250V	20	3	0.72
	30	3	0.96
	50	3	1.38
	60	3	1.93
Grounding – 2-Pole – 250V	15	2	0.33
	20	2	0.50
	30	2	0.85
	50	2	1.16
Grounding – 3-Pole – 250V	15	3	0.58
	20	3	0.72
	30	3	0.96
	50	3	1.38
	60	3	1.93

Manhours include checking out of job storage, handling, job hauling, distributing, clearing box of debris, making copper conductor of normal gauge connections, and mounting receptacle in box or fitting.

Manhours are based on installation of up to 10 receptacles on a single floor. If more than 10 receptacles are to be installed decrease manhours by the following percentages:

11 to 50 – Decrease 20%
51 to 100 – Decrease 30%
101 or more – Decrease 35%

If aluminum conductors of a larger gauge for a given amperage are to be connected to the receptacle increase the manhours 10%.

Manhours exclude installation of conduit, boxes, plates, conductors, and other devices. See respective tables for these time frames.

MISCELLANEOUS RECEPTACLES

MANHOURS EACH

Item Description	Manhours
Single Receptacles Without Plates	
Grounding — 2-Pole — 2-Wire — 277V — 15 Amp	0.33
Grounding — 2-Pole — 2-Wire — 277V — 20 Amp	0.50
Grounding — 2-Pole — 3-Wire — 277V — 30 Amp	0.96
Grounding — 2-Pole — 3-Wire — 277V — 50 Amp	1.38
Clock and Fan Receptacles With Plates	
Nongrounded Clock Receptacle — 2-Pole — 2-Wire	0.63
Grounded Clock Receptacle — 2-Pole — 2-Wire	0.69
Grounded Clock Receptacle — 3-Pole — 3-Wire	0.80
Nongrounded Fan Receptacle — 2-Pole — 2-Wire	1.07
Grounded Fan Receptacle — 2-Pole — 2-Wire	1.24
Pilot Light Receptacles	
Pilot Receptacle With Lamp — No Plate	0.77
Combination S.P. Switch and Pilot Receptacle — No Lamp	1.38
Combination S.P. Switch and Pilot Receptacle — With Neon Lamp	1.38
Red Jewell Switch Plate	0.11
Red Jewell Signal Receptacle Plate	0.13

Manhours include checking out of job storage, handling, job hauling, distributing, clearing box of debris, making copper conductor of normal gauge connections, and mounting receptacle in box or fitting.

Manhours are based on installation of up to 10 receptacles on a single floor. If more than 10 receptacles are to be installed decrease manhours by the following percentages:

    11 to 25 — Decrease 15%
    26 to 50 — Decrease 18%
    51 or more — Decrease 25%

Manhours exclude installation of conduit, boxes, plates except as noted, conductors, and other devices. See respective tables for these time frames.

SWITCH, RECEPTACLE, AND COMBINATION PLATES

MANHOURS EACH

Item Description	Gang Number	Manhours
Switch and Receptacle Plates		
	1	0.15
	2	0.29
Tumbler Switch Plates and	3	0.48
Single and Duplex Receptacle Plates	4	0.58
	5	0.96
	6	0.96
Switch and Receptacle Combination Plates		
Duplex Receptacle and One Switch	2	0.29
Duplex Receptacle and Two Switches	3	0.48
Duplex Receptacle and Three Switches	4	0.58
Single Receptacle and One Switch	2	0.29
Single Receptacle and Two Switches	3	0.48
Single Receptacle and Three Switches	4	0.58
Duplex Receptacle and Single Receptacle	2	0.29

Manhours include checking out of job storage, handling, job hauling, distributing, aligning device and plate, and installing plate.

Manhours are based on installing up to 50 plates on a single floor. If more or less than 50 plates are to be installed decrease or increase the manhours by the following percentages:

51 to 100 − Decrease 20%

101 to 200 − Decrease 30%

201 or more − Decrease 35%

0 to 50 − Increase 20%

Manhours exclude installation of conduit, boxes, conductors, switches, and receptacles. See respective tables for these time frames.

MISCELLANEOUS DEVICE PLATES

MANHOURS EACH

Item Description	Gang Number	Manhours
Telephone Plates	1	0.15
	2	0.29
	3	0.48
Blank Plates	1	0.15
	2	0.29
	3	0.48
	4	0.58
	5	0.96
	6	0.96
Power Receptacle Plates		
Receptacle Plate — 30-50 Amp	1	0.48
Receptacle Plate — 60 Amp	2	0.48
Weather Proof Plates		
Duplex Device — Verticle 1-Lid	1	0.15
Duplex Device — Verticle Lock	1	0.15
Duplex Device — Horizontal 1-Lid	1	0.15
Duplex Device — Horizontal 2-Lid	1	0.15
Duplex Device — Horizontal Lock	1	0.15
Duplex Combination — Horizontal 2-Lid	1	0.15
Single Device — Verticle 1-Lid	1	0.15
Signal Device — Horizontal 1-Lid	1	0.15
Blank Cover	1	0.12
Blank Cover	2	0.12
Special Plates		
Switch and Warning Light	—	0.22
Louvre Light	—	0.22
Oil Burner	—	0.83

Manhours include checking out of job storage, handling, job hauling, distributing, aligning device and plate, and installing plate.

Manhours are based on installing up to 10 plates on a single floor. If more than 10 plates are to be installed decrease the manhours by the following percentages:

11 to 25 — Decrease 15%
26 to 50 — Decrease 18%
51 or more — Decrease 25%

Manhours exclude installation of conduit, boxes, conductors, and devices. See respective tables for these time frames.

SURFACE METAL RACEWAY AND WIRE
One Piece—Types #200, #500, #700, and #1000

MANHOURS FOR 100 LINEAR FEET

Item Description	Manhours Required for Height to											
	Wood & Metal Frame Construction				Masonry Construction				Concrete Construction			
	10′	15′	20′	25′	10′	15′	20′	25′	10′	15′	20′	25′
Raceway												
# 200	7.50	8.10	8.42	8.68	9.68	10.45	10.87	11.20	9.15	9.88	10.28	10.55
# 500	7.50	8.10	8.42	8.68	9.68	10.45	10.87	11.20	9.15	9.88	10.28	10.55
# 700	9.00	9.72	10.11	10.41	11.61	12.54	13.04	13.43	10.98	11.85	12.33	12.70
#1000	12.00	12.96	13.48	13.88	15.48	16.72	17.39	17.91	14.64	15.81	16.44	16.94
Wiring												
#14	1.60	1.73	1.80	1.85	2.06	2.22	2.31	2.38	1.95	2.11	2.19	2.26
#12	1.90	2.05	2.13	2.20	2.45	2.65	2.75	2.83	2.99	3.23	3.36	3.46
#10	2.80	3.02	3.14	3.24	3.61	3.90	4.05	4.18	4.40	4.75	4.94	5.09
# 8	3.30	3.56	3.71	3.82	4.26	4.60	4.78	4.93	4.03	4.35	4.53	4.66
# 6	3.86	4.17	4.34	4.47	4.98	5.38	5.59	5.76	4.71	5.09	5.29	5.45

All manhours include checking out of job storage, handling, and job hauling of items as listed.

Raceway manhours include cutting, placing, and normal securing of raceway in position.

Wiring manhours include inserting of wire into raceway with the use of wire pulleys where required and connecting to power outlet or source.

Manhours exclude installation of raceway fitting, boxes, and scaffolding if required. See respective tables for these time frames.

SURFACE METAL RACEWAY AND MULTI-OUTLET ASSEMBLIES

Two-Piece—Types #2000, #2100, and #3000

MANHOURS FOR LISTED UNITS

Item Description	Unit	Manhours Required for		
		Wood or Metal Construction	Masonry Construction	Concrete Construction
Prewired Raceway Type #2000				
2-Conductor	C. Ft.	22.95	29.61	28.00
3-Conductor	C. Ft.	25.65	33.09	31.29
Elbows — Internal or External	each	0.38	0.49	0.46
Feed Connections	—	—	—	—
From existing outlet — 2-wire	each	1.35	1.74	1.65
From existing outlet — 3-wire	each	1.73	2.23	2.11
From concealed wiring — 2-wire	each	0.83	1.07	1.01
From concealed wiring — 3-wire	each	1.13	1.46	1.38
Unwired Raceway Types #2100 and #3000				
Install and Secure Base	C. Ft.	16.20	20.90	19.76
Attach Cover	C. Ft.	2.25	2.90	2.75
Elbows — Internal or External	each	0.45	0.58	0.55
Flat Elbows	each	0.53	0.68	.065
Tees	each	0.83	1.07	1.01

All manhours include checking out of job storage, handling, and job hauling of items as listed.

Raceway #2000 manhours are for 1-1/4″ x 3/4″ raceway with prewired devices and punched cover pieces, and include installation of base and two #12 conductors with prewired grounded or duplex receptacles, or three #12 conductors with prewired one side switched duplex receptacles in cover punched for devices and attaching the cover piece.

Raceway #2100 and #3000 manhours are for complete installation of 1-1/4″ x 7/8″ unwired devices and covers, but exclude installation of conductors and devices.

DESPARD DEVICES

MANHOURS EACH

Item Description	Manhours
Despard Tumbler Switches Excluding Plates or Straps	
1-Pole — 10 Amp	0.33
1-Pole — 15 Amp	0.33
1-Pole — 20 Amp	0.58
2-Pole — 10 Amp	0.96
2-Pole — 15 Amp	0.96
2-Pole — 20 Amp	1.38
3-Way — 10 Amp	0.63
3-Way — 15 Amp	0.63
3-Way — 20 Amp	0.85
4-Way — 10 Amp	0.96
4-Way — 15 Amp	0.96
4-Way — 20 Amp	1.38
Despard Receptacles Excluding Plates or Straps	
Single-Receptacle — 2-Pole	0.29
Single-Receptacle — 2-Pole — Grounded	0.33
Duplex-Receptacle — 2-Pole	0.29
Duplex-Receptacle — 2-Pole — 2-Circuit	0.50
Triplex-Receptable — 2-Pole	0.29
Triplex-Receptacle — 2-Pole — 2-Circuit	0.50
Despard Night and Pilot Lights and Mounting Straps	
Night Light	0.47
Pilot and Hood	0.47
Pilot and Jewell	0.47
Neon Pilot — 125 Volt	0.47
Mounting Strap	0.23

Manhours include checking out of job storage, handling, job hauling, distributing, clearing box of debris, making copper conductor of normal gauge connections, and mounting device in box or fitting.

Manhours are based on installation of up to 10 devices on a single floor. If more than 10 devices are to be installed decrease manhours by the following percentages:

    10 to 25 — Decrease 15%
    26 to 50 — Decrease 18%
    51 or more — Decrease 25%

Manhours exclude installation of conduit, boxes, plates, straps except as noted, conductors and other devices. See respective tables for these time frames.

DESPARD PLATES

MANHOURS EACH

Item Description	Manhours
1-Gang	
1-Verticle	0.15
1-Horizontal	0.15
2-Horizontal	0.18
3-Horizontal	0.20
2-Gang	
2-Verticle	0.29
2-Horizontal	0.29
3-Horizontal	0.33
4-Horizontal	0.35
6-Horizontal	0.38
3-Gang	
3-Vertical	0.48
3-Horizontal	0.48
6-Horizontal	0.53
9-Horizontal	0.60

Manhours include checking out of job storage, handling, job hauling, distributing, aligning device and plate, and installing plate.

Manhours are based on installing up to 10 plates on a single floor. If more than 10 plates are to be installed decrease the manhours by the following percentages:

    10 to 25 — Decrease 15%
    26 to 50 — Decrease 18%
    51 or more — Decrease 25%

Manhours exclude installation of conduit, boxes, conductors and devices. See respective tables for these time frames.

Section 4

SURFACE METAL RACEWAY AND BRANCH BUSWAY

These manhour tables are for the detail layout, fabrication, and installation of raceway, busway, and their respective accessory fittings.

In the manhour units for surface metal raceway, consideration has been given to all operations such as cutting, reaming, and actual fabrication of the metal raceway and fittings, the proper placement of raceway, the installation of all junction and device boxes, and the placement of wiring and receptacles.

The layout, fabrication, and placement of various lengths, fittings, and devices, of 50-ampere capacity have been given consideration in the manhour units for branch busway.

FITTINGS AND BOXES FOR SURFACE METAL RACEWAY

One Piece—Types #200, #500, #700, and #1000

MANHOURS EACH

Item Description	Manhours Required for Height to											
	Wood & Metal Frame Construction				Masonry Construction				Concrete Construction			
	10'	15'	20'	25'	10'	115'	20'	25'	10'	15'	20'	25'
Elbows – Intrl. & Extrl.	0.30	0.32	0.34	0.35	0.39	0.42	0.44	0.45	0.37	0.40	0.42	0.43
Flat Elbows	0.30	0.32	0.34	0.35	0.39	0.42	0.44	0.45	0.37	0.40	0.42	0.43
Tee	0.68	0.73	0.76	0.79	0.88	0.95	0.99	1.02	0.83	0.90	0.93	0.96
Offset Connectors	0.45	0.49	0.51	0.52	0.58	0.63	0.65	0.67	0.55	0.59	0.62	0.64
Kick Plates	0.45	0.49	0.51	0.52	0.58	0.63	0.65	0.67	0.55	0.59	0.62	0.64
Outlet Box	0.45	0.49	0.51	0.52	0.58	0.63	0.65	0.67	0.55	0.59	0.62	0.64
Extension Box	0.75	0.81	0.84	0.87	0.97	1.05	1.09	1.12	0.92	0.99	1.03	1.06
Fixture Box	1.13	1.22	1.27	1.31	1.46	1.58	1.64	1.69	1.38	1.49	1.55	1.60
Utility Box	0.68	0.73	0.76	0.79	0.88	0.95	0.99	1.02	0.83	0.90	0.93	0.96
Distribution Box	0.68	0.73	0.76	0.79	0.88	0.95	0.99	1.02	0.83	0.90	0.93	0.96
Switch & Recept. Box	0.68	0.73	0.76	0.79	0.88	0.95	0.99	1.02	0.83	0.90	0.93	0.96
Corner Box	0.75	0.81	0.84	0.87	0.97	1.05	1.09	1.12	0.92	0.99	1.03	1.06

Manhours include checking out of job storage, handling, job hauling, placing, and normal securing of fittings or boxes in position as may be required for surface metal raceway types number 200, 500, 700, and 1000.

Manhours exclude installation of raceway, wiring, and scaffolding if required. See respective tables for these time frames.

BRANCH BUSWAY
50-AMP PLUG-IN TYPE

MANHOURS FOR LISTED UNITS

Item Description	Unit	Manhours Required for			Roll-in Method
		Ceiling Height to			
		10′	18′	25′	20′-25′
Busway					
Straight Length — 2-Pole — 10 Ft.	C Lf.	15.60	17.55	19.50	6.50
Straight Length — 3-Pole — 10 Ft.	C Lf.	17.55	19.50	22.10	9.00
Straight Length — 4-Pole — 10 Ft.	C. Lf.	19.50	22.10	24.70	10.50
Straight Length — 2-Pole — 5 Ft.	C. Lf.	17.55	19.50	21.45	8.25
Straight Length — 3-Pole — 5 Ft.	C. Lf.	19.50	21.45	24.70	9.75
Straight Length — 4-Pole — 5 Ft.	C. Lf.	21.45	24.70	27.30	12.00
FIttings					
2-Pole Feed-in Box & Connections	each	1.80	1.80	2.55	2.55
3-Pole Feed-in Box & Connections	each	2.25	2.25	3.00	3.00
4-Pole Feed-in Box & Connections	each	2.70	2.70	3.45	3.45
2-Pole Elbow	each	0.23	0.30	0.30	–
3-Pole Elbow	each	0.30	0.38	0.38	–
4-Pole Elbow	each	0.45	0.45	0.45	–
2-Pole Plug	each	1.80	1.80	2.55	–
3-Pole Plug	each	2.25	2.25	3.00	–
4-Pole Plug	each	2.70	2.70	3.45	–
2-Pole End Cap	each	0.23	0.30	0.30	0.30
3-Pole End Cap	each	0.30	0.38	0.38	0.38
4-Pole End Cap	each	0.45	0.45	0.45	0.45
Hangers					
Surface Hangers	each	0.38	0.38	0.38	–
Rod Hangers — 3/8″	each	0.45	0.60	0.75	0.45
Strap Hangers	each	0.60	0.75	0.90	0.60
Roller Hangers	each	–	–	–	0.38
Messenger Cable	C. Lf.	–	–	–	2.60

Manhours include checking out of job storage, handling, job hauling, layout, fabrication, and installing of items as outlined.

Manhours exclude connection to power source, and scaffolding. See respective tables for these time frames.

Section 5

LIGHTING FIXTURES

This section includes labor manhour units for the installation of various types of lighting systems and lighting fixtures.

Throughout the tables consideration has been given to all necessary labor for operations such as unloading, handling, hauling, and installing that one might expect to encounter for this type of work. The manhours include time allowance for general foreman through apprentice.

Before applying the manhour units in this or any section of this manual, we caution the estimator to be thoroughly familiar with the Introduction of this manual.

COMMERCIAL INTERIOR INCANDESCENT FIXTURES SURFACE MOUNTED

1-Lamp Units

MANHOURS EACH

Item Description	Manhours Required to							
	Hang Only For Height to				Assemble and Hang For Height to			
	10′	15′	20′	25′	10′	15′	20′	25′
Metal Shade Type								
60 Watts	0.18	0.20	0.21	0.23	0.30	0.33	0.36	0.38
100 Watts	0.18	0.20	0.21	0.23	0.30	0.33	0.36	0.38
150 Watts	0.22	0.24	0.26	0.28	0.39	0.43	0.46	0.49
200 Watts	0.24	0.26	0.29	0.30	0.47	0.52	0.56	0.59
300 Watts	0.30	0.33	0.36	0.38	0.55	0.61	0.65	0.69
500 Watts	0.39	0.43	0.46	0.49	0.69	0.76	0.82	0.87
750 Watts	0.46	0.51	0.55	0.58	0.85	0.94	1.01	1.07
Glass Globe Type								
60 Watts	0.22	0.24	0.26	0.28	0.39	0.43	0.46	0.49
100 Watts	0.22	0.24	0.26	0.28	0.39	0.43	0.46	0.49
150 Watts	0.28	0.31	0.33	0.35	0.47	0.52	0.56	0.59
200 Watts	0.30	0.33	0.36	0.38	0.55	0.61	0.65	0.69
300 Watts	0.35	0.39	0.42	0.44	0.61	0.67	0.72	0.77
500 Watts	0.44	0.48	0.52	0.55	0.74	0.81	0.88	0.93
Plastic Globe Type								
60 Watts	0.19	0.21	0.23	0.24	0.30	0.33	0.36	0.38
100 Watts	0.19	0.21	0.23	0.24	0.30	0.33	0.36	0.38
150 Watts	0.22	0.24	0.26	0.28	0.39	0.43	0.46	0.49
200 Watts	0.26	0.29	0.31	0.33	0.47	0.52	0.56	0.59
300 Watts	0.30	0.33	0.36	0.38	0.55	0.61	0.65	0.69
500 Watts	0.39	0.43	0.46	0.49	0.69	0.76	0.82	0.87

All manhours include checking out of job storage, handling, job hauling, distributing, layout, and marking for installation.

Hang-Only manhours only include hanging of fixture.

Assemble-and-Hang manhours include complete assembling of fixture, installation, connecting the circuit power to the fixture, testing, and verifying fixture operation.

Manhours exclude installation of conduit, boxes, circuit wiring, switches, light bulbs, special hangers or supports not supplied with fixture, and ladders or scaffolding. See respective tables for these time frames.

For heights above 25 feet increase manhours 5% for each additional 5 feet or fraction thereof.

COMMERCIAL INTERIOR INCANDESCENT FIXTURES
SURFACE SURFACE
2-, 3-, and 4-Lamp Units

MANHOURS EACH

Item Description	Manhours Required to							
	Hang Only For Height to				Assemble and Hang For Height to			
	10'	15'	20'	25'	10'	15'	20'	25'
2 Metal Shades or Globes								
60 Watts	0.30	0.33	0.36	0.38	0.55	0.61	0.65	0.69
100 Watts	0.30	0.33	0.36	0.38	0.55	0.61	0.65	0.69
150 Watts	0.35	0.39	0.42	0.44	0.61	0.67	0.72	0.77
200 Watts	0.39	0.43	0.46	0.49	0.69	0.76	0.82	0.87
300 Watts	0.44	0.48	0.52	0.55	0.74	0.81	0.88	0.93
500 Watts	0.50	0.55	0.59	0.63	0.91	1.00	1.08	1.15
750 Watts	0.57	0.63	0.68	0.72	1.07	1.18	1.27	1.35
1000 Watts	0.72	0.79	0.86	0.91	1.30	1.43	1.54	1.64
3 Metal Shades or Gloves								
60 Watts	0.39	0.43	0.46	0.49	0.69	0.76	0.82	0.87
100 Watts	0.43	0.47	0.51	0.54	0.74	0.81	0.88	0.93
150 Watts	0.48	0.53	0.57	0.60	0.83	0.91	0.99	1.05
200 Watts	0.50	0.55	0.59	0.63	0.91	1.00	1.08	1.15
300 Watts	0.56	0.62	0.67	0.71	0.99	1.09	1.18	1.25
500 Watts	0.65	0.72	0.77	0.82	1.13	1.24	1.34	1.42
750 Watts	0.76	0.84	0.90	0.96	1.32	1.45	1.57	1.66
1000 Watts	0.89	0.98	1.06	1.12	1.54	1.69	1.83	1.94
4 Metal Shades or Globes								
60 Watts	0.50	0.55	0.59	0.63	0.91	1.00	1.08	1.15
100 Watts	0.56	0.62	0.67	0.71	0.96	1.06	1.14	1.21
150 Watts	0.59	0.65	0.70	0.74	1.05	1.16	1.25	1.32
200 Watts	0.65	0.72	0.77	0.82	1.13	1.24	1.34	1.42

All manhours include checking out of job storage, handling, job hauling, distributing, layout, and marking for installation.

Hang-Only manhours only include hanging of fixture.

Assemble-and-Hang manhours include complete assembling of fixture, installation, connecting the circuit power to the fixture, testing, and verifying fixture operation.

Manhours exclude installation of conduit, boxes, circuit wiring, switches, light bulbs, special hangers or supports not supplied with fixture, and ladders or scaffolding. See respective tables for these time frames.

For heights above 25 feet increase manhours 5% for each additional 5 feet or fraction thereof.

COMMERCIAL INTERIOR INCANDESCENT FIXTURES—PENDANT MOUNTED

1- and 4-Lamp Units

MANHOURS EACH

Item Description	Manhours Required to							
	Hang Only For Height to				Assemble and Hang For Height to			
	10'	15'	20'	25'	10'	15'	20'	25'
1 Metal Shade								
60 Watts	0.22	0.24	0.26	0.28	0.39	0.43	0.46	0.49
100 Watts	0.22	0.24	0.26	0.28	0.39	0.43	0.46	0.49
150 Watts	0.28	0.31	0.33	0.35	0.47	0.52	0.56	0.59
200 Watts	0.30	0.33	0.36	0.38	0.52	0.57	0.62	0.65
300 Watts	0.35	0.39	0.42	0.44	0.61	0.67	0.72	0.77
500 Watts	0.44	0.48	0.52	0.55	0.74	0.81	0.88	0.93
750 Watts	0.52	0.57	0.62	0.65	0.91	1.00	1.08	1.15
4 Metal Shades								
60 Watts	0.65	0.72	0.77	0.82	1.13	1.24	1.32	1.42
100 Watts	0.67	0.74	0.80	0.84	1.21	1.33	1.44	1.52
150 Watts	0.70	0.77	0.83	0.88	1.29	1.42	1.53	1.62
200 Watts	0.78	0.86	0.93	0.98	1.40	1.54	1.66	1.76

1-Lamp Units

MANHOURS EACH

Item Description	Manhours Required to							
	Hang Only For Height to				Assemble and Hang For Height to			
	10'	15'	20'	25'	10'	15'	20'	25'
Glass Globe Type								
60 Watts	0.24	0.26	0.29	0.30	0.47	0.52	0.56	0.59
100 Watts	0.24	0.26	0.29	0.30	0.52	0.57	0.62	0.65
150 Watts	0.30	0.33	0.36	0.38	0.61	0.67	0.72	0.77
200 Watts	0.33	0.36	0.39	0.42	0.69	0.76	0.82	0.87
300 Watts	0.39	0.43	0.46	0.49	0.85	0.94	1.01	1.07
500 Watts	0.46	0.51	0.55	0.58	0.97	1.07	1.15	1.22
Plastic Globe Type								
60 Watts	0.20	0.22	0.24	0.25	0.41	0.45	0.49	0.52
100 Watts	0.20	0.22	0.24	0.25	0.41	0.45	0.49	0.52
150 Watts	0.24	0.26	0.29	0.30	0.47	0.52	0.56	0.59
200 Watts	0.28	0.31	0.33	0.35	0.52	0.57	0.62	0.65
300 Watts	0.33	0.36	0.39	0.42	0.61	0.67	0.72	0.77
500 Watts	0.41	0.45	0.49	0.52	0.74	0.81	0.88	0.93

All manhours include checking out of job storage, handling, job hauling, distributing, layout, and marking for installation.

Hang-Only manhours only include hanging of fixture.

Assemble-and-Hang manhours include complete assembling of fixture, installation, connecting the circuit power to the fixture, testing, and verifying fixture operation.

Manhours exclude installation of conduit, boxes, circuit wiring, switches, light bulbs, special hangers or supports not supplied with fixture, and ladders or scaffolding. See respective tables for these time frames.

For heights above 25 feet increase manhours 5% for each additional 5 feet or fraction thereof.

COMMERCIAL INTERIOR INCANDESCENT
FIXTURES
PENDANT MOUNTED
2- and 3-Lamp Units

MANHOURS EACH

Item Description	Manhours Required to							
	Hang Only For Height to				Assemble and Hang For Height to			
	10′	15′	20′	25′	10′	15′	20′	25′
2 Metal Shades or Globes								
60 Watts	0.33	0.36	0.39	0.42	0.61	0.67	0.72	0.77
100 Watts	0.33	0.36	0.39	0.42	0.61	0.67	0.72	0.77
150 Watts	0.39	0.43	0.46	0.49	0.69	0.76	0.82	0.87
200 Watts	0.43	0.47	0.51	0.54	0.74	0.81	0.88	0.93
300 Watts	0.48	0.53	0.57	0.60	0.85	0.94	1.01	1.07
500 Watts	0.56	0.62	0.67	0.71	0.99	1.09	1.18	1.25
750 Watts	0.65	0.72	0.77	0.82	1.13	1.24	1.34	1.42
1000 Watts	0.76	0.84	0.90	0.96	1.32	1.45	1.57	1.66
3 Metal Shades or Globes								
60 Watts	0.46	0.51	0.55	0.58	0.85	0.94	1.01	1.07
100 Watts	0.50	0.55	0.59	0.63	0.91	1.00	1.08	1.15
150 Watts	0.56	0.62	0.67	0.71	0.99	1.09	1.18	1.25
200 Watts	0.59	0.65	0.70	0.74	1.07	1.18	1.27	1.35
300 Watts	0.65	0.72	0.77	0.82	1.16	1.28	1.38	1.46
500 Watts	0.72	0.79	0.86	0.91	1.29	1.42	1.53	1.62
750 Watts	0.83	0.91	0.99	1.05	1.51	1.66	1.79	1.90
1000 Watts	0.98	1.08	1.16	1.23	1.73	1.90	2.06	2.18

All manhours include checking out of job storage, handling, job hauling, distributing, layout, and marking for installation.

Hang-Only manhours only include hanging of fixture.

Assemble-and-Hang manhours include complete assembling of fixture, installation, connecting the circuit power to the fixture, testing, and verifying fixture operation.

Manhours exclude installation of conduit, boxes, circuit wiring, switches, light bulbs, special hangers or supports not supplied with fixture, and ladders or scaffolding. See respective tables for these time frames.

For heights above 25 feet increase manhours 5% for each additional 5 feet or fraction thereof.

COMMERCIAL INTERIOR INCANDESCENT FIXTURES
ROUND BODY RECESSED MOUNTED
1-Lamp Units

MANHOURS EACH

Item Description	Manhours Required to							
	Hang Only For Height to				Assemble and Hang For Height to			
	10′	15′	20′	25′	10′	15′	20′	25′
Prewired								
60 Watts	0.24	0.26	0.29	0.30	0.47	0.52	0.56	0.59
100 Watts	0.24	0.26	0.29	0.30	0.47	0.52	0.56	0.59
150 Watts	0.30	0.33	0.36	0.38	0.55	0.61	0.65	0.69
200 Watts	0.33	0.36	0.39	0.42	0.61	0.67	0.72	0.77
300 Watts	0.39	0.43	0.46	0.49	0.69	0.76	0.82	0.87
500 Watts	0.46	0.51	0.55	0.58	0.85	0.94	1.01	1.07
750 Watts	0.56	0.62	0.67	0.71	0.99	1.09	1.18	1.25
1000 Watts	0.67	0.74	0.80	0.84	1.21	1.33	1.44	1.52
1500 Watts	0.80	0.88	0.95	1.01	1.43	1.57	1.70	1.80
Unwired								
60 Watts	0.39	0.43	0.46	0.49	0.69	0.76	0.82	0.87
100 Watts	0.39	0.43	0.46	0.49	0.69	0.76	0.82	0.87
150 Watts	0.44	0.48	0.52	0.55	0.74	0.81	0.88	0.93
200 Watts	0.46	0.51	0.55	0.58	0.83	0.91	0.99	1.05
300 Watts	0.49	0.54	0.58	0.62	0.94	1.03	1.12	1.18
500 Watts	0.57	0.63	0.68	0.72	1.07	1.18	1.27	1.35
750 Watts	0.67	0.74	0.80	0.84	1.21	1.33	1.44	1.52
1000 Watts	0.80	0.88	0.95	1.01	1.43	1.57	1.70	1.80
1500 Watts	0.94	1.03	1.12	1.18	1.65	1.82	1.96	2.08

All manhours include checking out of job storage, handling, job hauling, distributing, layout, and marking for installation.

Hang-Only manhours only include hanging of fixture.

Assemble-and-Hang manhours include complete assembling of fixture, installation, connecting the circuit power to the fixture, testing, and verifying fixture operation.

Manhours exclude installation of conduit, boxes, circuit wiring, switches, light bulbs, special hangers or supports not supplied with fixture, and ladders or scaffolding. See respective tables for these time frames.

For heights above 25 feet increase manhours 5% for each additional 5 feet or fraction thereof.

COMMERCIAL INTERIOR INCANDESCENT FIXTURES
SQUARE BODY RECESSED MOUNTED
1- and 2-Lamp Units

MANHOURS EACH

Item Description	Hang Only For Height to				Assemble and Hang For Height to			
	10′	15′	20′	25′	10′	15′	20′	25′
Prewired — 1-Lamp								
60 Watts	0.28	0.31	0.33	0.35	0.55	0.61	0.65	0.69
100 Watts	0.28	0.31	0.33	0.35	0.55	0.61	0.65	0.69
150 Watts	0.35	0.39	0.42	0.44	0.61	0.67	0.72	0.77
200 Watts	0.39	0.43	0.46	0.49	0.69	0.76	0.82	0.87
300 Watts	0.44	0.48	0.52	0.55	0.74	0.81	0.88	0.93
500 Watts	0.50	0.55	0.59	0.63	0.91	1.00	1.08	1.15
750 Watts	0.57	0.63	0.68	0.72	1.07	1.18	1.27	1.35
1000 Watts	0.68	0.75	0.81	0.86	1.27	1.40	1.51	1.60
1500 Watts	0.83	0.91	0.99	1.05	1.51	1.66	1.79	1.90
Prewired — 2-Lamp								
60 Watts	0.33	0.36	0.39	0.42	0.61	0.67	0.72	0.77
100 Watts	0.33	0.36	0.39	0.42	0.61	0.67	0.72	0.77
150 Watts	0.39	0.43	0.46	0.49	0.69	0.76	0.82	0.87
200 Watts	0.41	0.45	0.49	0.52	0.74	0.81	0.88	0.93
300 Watts	0.46	0.51	0.55	0.58	0.85	0.95	1.01	1.07
500 Watts	0.56	0.62	0.67	0.71	0.99	1.09	1.18	1.25
750 Watts	0.65	0.72	0.77	0.82	1.16	1.28	1.38	1.46
1000 Watts	0.76	0.84	0.90	0.96	1.35	1.49	1.60	1.70

All manhours include checking out of job storage, handling, job hauling, distributing, layout, and marking for installation.

Hang-Only manhours only include hanging of fixture.

Assemble-and-Hang manhours include complete assembling of fixture, installation, connecting the circuit power to the fixture, testing, and verifying fixture operation.

Manhours exclude installation of conduit, boxes, circuit wiring, switches, light bulbs, special hangers or supports not supplied with fixture, and ladders or scaffolding. See respective tables for these time frames.

For heights above 25 feet increase manhours 5% for each additional 5 feet or fraction thereof.

COMMERCIAL INTERIOR INCANDESCENT FIXTURES
SQUARE BODY RECESSED MOUNTED
1- and 2-Lamp Units

MANHOURS EACH

Item Description	Manhours Required to							
	Hang Only For Height to				Assemble and Hang For Height to			
	10'	15'	20'	25'	10'	15'	20'	25'
Unwired − 1-Lamp								
60 Watts	0.43	0.47	0.51	0.54	0.74	0.81	0.88	0.93
100 Watts	0.43	0.47	0.51	0.54	0.74	0.81	0.88	0.93
150 Watts	0.48	0.53	0.57	0.60	0.83	0.91	0.99	1.05
200 Watts	0.50	0.55	0.59	0.63	0.91	1.00	1.08	1.15
300 Watts	0.56	0.62	0.67	0.71	0.99	1.09	1.18	1.25
500 Watts	0.63	0.69	0.75	0.79	1.16	1.28	1.38	1.46
750 Watts	0.72	0.79	0.86	0.91	1.32	1.45	1.57	1.66
1000 Watts	0.83	0.91	0.99	1.05	1.51	1.66	1.79	1.90
1500 Watts	0.98	1.08	1.16	1.23	1.73	1.90	2.06	2.18
Unwired − 2-Lamp								
60 Watts	0.46	0.51	0.55	0.58	0.85	0.94	1.01	1.07
100 Watts	0.46	0.51	0.55	0.58	0.85	0.94	1.01	1.07
150 Watts	0.50	0.55	0.59	0.63	0.91	1.00	1.08	1.15
200 Watts	0.56	0.62	0.67	0.71	0.99	1.09	1.18	1.25
300 Watts	0.59	0.65	0.70	0.74	1.10	1.21	1.31	1.39
500 Watts	0.67	0.74	0.80	0.84	1.21	1.33	1.44	1.52
750 Watts	0.76	0.84	0.90	0.96	1.38	1.52	1.64	1.74
1000 Watts	0.89	0.98	1.06	1.12	1.54	1.69	1.83	1.94

All manhours include checking out of job storage, handling, job hauling, distributing, layout, and marking for installation.

Hang-Only manhours only include hanging of fixture.

Assemble-and-Hang manhours include complete assembling of fixture, installation, connecting the circuit power to the fixture, testing, and verifying fixture operation.

Manhours exclude installation of conduit, boxes, circuit wiring, switches, light bulbs, special hangers or supports not supplied with fixture, and ladders or scaffolding. See respective tables for these time frames.

For heights above 25 feet increase manhours 5% for each additional 5 feet or fraction thereof.

COMMERCIAL INTERIOR INCANDESCENT FIXTURES
SQUARE BODY RECESSED MOUNTED
3- and 4-Lamp Units

MANHOURS EACH

Item Description	Manhours Required to							
	Hang Only For Height to				Assemble and Hang For Height to			
	10'	15'	20'	25'	10'	15'	20'	25'
Unwired − 3-Lamp								
60 Watts	0.50	0.55	0.59	0.63	0.91	1.00	1.08	1.15
100 Watts	0.56	0.62	0.67	0.71	0.99	1.09	1.18	1.25
150 Watts	0.59	0.65	0.70	0.74	1.10	1.21	1.31	1.39
200 Watts	0.65	0.72	0.77	0.82	1.18	1.30	1.40	1.49
Prewired − 3-Lamp								
60 Watts	0.39	0.43	0.46	0.49	0.69	0.76	0.82	0.87
100 Watts	0.43	0.47	0.51	0.54	0.74	0.81	0.88	0.93
150 Watts	0.46	·0.51	0.55	0.58	0.85	0.94	1.01	1.07
200 Watts	0.50	0.55	0.59	0.63	0.91	1.00	1.08	1.15
Unwired − 4-Lamp								
60 Watts	0.50	0.55	0.59	0.63	0.91	1.00	1.08	1.15
100 Watts	0.56	0.62	0.67	0.71	0.99	1.09	1.18	1.25
150 Watts	0.59	0.65	0.70	0.74	1.07	1.18	1.27	1.35
200 Watts	0.65	0.72	0.77	0.82	1.16	1.28	1.38	1.46
Prewired − 4-Lamp								
60 Watts	0.65	0.72	0.77	0.82	1.16	1.28	1.38	1.46
100 Watts	0.67	0.74	0.80	0.84	1.21	1.33	1.44	1.52
150 Watts	0.72	0.79	0.86	0.91	1.29	1.42	1.53	1.62
200 Watts	0.78	0.86	0.93	0.98	1.40	1.54	1.66	1.76

All manhours include checking out of job storage, handling, job hauling, distributing, layout, and marking for installation.

Hang-Only manhours only include hanging of fixture.

Assemble-and-Hang manhours include complete assembling of fixture, installation, connecting the circuit power to the fixture, testing, and verifying fixture operation.

Manhours exclude installation of conduit, boxes, circuit wiring, switches, light bulbs, special hangers or supports not supplied with fixture, and ladders or scaffolding. See respective tables for these time frames.

For heights above 25 feet increase manhours 5% for each additional 5 feet or fraction thereof.

INTERIOR LAMPHOLDER INCANDESCENT FIXTURES

Porcelain and Plastic 1-Lamp Units

MANHOURS EACH

Item Description	Manhours Required to							
	Hang Only For Height to				Assemble and Hang For Height to			
	10'	15'	20'	25'	10'	15'	20'	25'
Porcelain								
Keyless — 600W 250V	0.19	0.21	0.23	0.24	0.30	0.33	0.36	0.38
Pull Chain — 660W 250V	0.19	0.21	0.23	0.24	0.30	0.33	0.36	0.38
Pull Chain and Side — 660W 250V	0.19	0.21	0.23	0.24	0.30	0.33	0.36	0.38
Pull Chain and Outlet — 3 W Ground 660W 250V	0.22	0.24	0.26	0.28	0.39	0.43	0.46	0.49
Metalbase — 660W 250V	0.17	0.19	0.20	0.21	0.25	0.28	0.30	0.31
Plastic								
Keyless — 660W 250V	0.17	0.19	0.20	0.21	0.25	0.28	0.30	0.31
Pull Chain — 660W 250V	0.17	0.19	0.20	0.21	0.25	0.28	0.30	0.31
Pull Chain and Side — 660W 250V	0.17	0.19	0.20	0.21	0.25	0.28	0.30	0.31
Pull Chain and Outlet — 3W Ground 660W 250V	0.19	0.21	0.23	0.24	0.30	0.33	0.36	0.38

All manhours include checking out of job storage, handling, job hauling, distributing, layout, and marking for installation.

Hang-Only manhours only include hanging of fixture.

Assemble-and-Hang manhours include complete assembling of fixture, installation, connecting the circuit power to the fixture, testing, and verifying fixture operation.

Manhours exclude installation of conduit, boxes, circuit wiring, switches, light bulbs, special hangers or supports not supplied with fixture, and ladders or scaffolding. See respective tables for these time frames.

For heights above 25 feet increase manhours 5% for each additional 5 feet or fraction thereof.

COMMERCIAL AND INDUSTRIAL INCANDESCENT FIXTURES

Exit Light Units

MANHOURS EACH

Item Description	Hang Only For Height to				Assemble and Hang For Height to			
	10′	15′	20′	25′	10′	15′	20′	25′
Recess Mounted − Single Face	0.33	0.36	0.39	0.42	0.66	0.73	0.78	0.83
Surface Mounted − Single Face	0.26	0.29	0.31	0.33	0.55	0.61	0.65	0.69
Pandant Mounted − Single Face	0.30	0.33	0.36	0.38	0.61	0.67	0.72	0.77
Surface Mounted − Double Face	0.26	0.29	0.31	0.33	0.55	0.61	0.65	0.69
Pendant Mounted − Double Face	0.30	0.33	0.36	0.38	0.61	0.67	0.72	0.77
Wall Mounted − Triangular Face	0.26	0.29	0.31	0.33	0.55	0.61	0.65	0.69
Surface Mounted − Plexiglass Face	0.22	0.24	0.26	0.28	0.50	0.55	0.59	0.63
Surface Mounted − Glass Globe	0.33	0.36	0.39	0.42	0.66	0.73	0.78	0.83
Surface Mounted − Vaporproof	0.33	0.36	0.39	0.42	0.66	0.73	0.78	0.83

All manhours include checking out of job storage, handling, job hauling, distributing, layout, and marking for installation.

Hang-Only manhours only include hanging of fixture.

Assemble-and-Hang manhours include complete assembling of fixture, installation, connecting the circuit power to the fixture, testing, and verifying fixture operation.

Manhours exclude installation of conduit, boxes, circuit wiring, switches, light bulbs, special hangers or supports not supplied with fixture, and ladders or scaffolding. See respective tables for these time frames.

For heights above 25 feet increase manhours 5% for each additional 5 feet or fraction thereof.

INDUSTRIAL INTERIOR INCANDESCENT FIXTURES

Porcelain Enamel Reflector 1-Lamp Units

Surface Mounted

MANHOURS EACH

Item Description	Manhours Required to							
	Hang Only For Height to				Assemble and Hang For Height to			
	10'	15'	20'	25'	10'	15'	20'	25'
Separable Socket Type								
60 Watts	0.19	0.21	0.23	0.24	0.30	0.33	0.36	0.38
100 Watts	0.19	0.21	0.23	0.24	0.30	0.33	0.36	0.38
150 Watts	0.22	0.24	0.26	0.28	0.39	0.43	0.46	0.49
200 Watts	0.26	0.29	0.31	0.33	0.47	0.52	0.56	0.59
300 Watts	0.31	0.34	0.37	0.39	0.55	0.61	0.65	0.69
500 Watts	0.39	0.43	0.46	0.49	0.69	0.76	0.82	0.87
750 Watts	0.46	0.51	0.55	0.58	0.85	0.94	1.01	1.07
1000 Watts	0.57	0.63	0.68	0.72	1.05	1.16	1.25	1.32
1500 Watts	0.72	0.79	0.86	0.91	1.27	1.40	1.51	1.60
1-Piece Socket Type								
60 Watts	0.22	0.24	0.26	0.28	0.39	0.43	0.46	0.49
100 Watts	0.22	0.24	0.26	0.28	0.39	0.43	0.46	0.49
150 Watts	0.28	0.31	0.33	0.35	0.47	0.52	0.56	0.59
200 Watts	0.30	0.33	0.36	0.38	0.52	0.57	0.62	0.65
300 Watts	0.35	0.39	0.42	0.44	0.61	0.67	0.72	0.77
500 Watts	0.44	0.48	0.52	0.55	0.74	0.81	0.88	0.93
750 Watts	0.52	0.57	0.62	0.65	0.91	1.00	1.08	1.15
1000 Watts	0.65	0.72	0.77	0.82	1.10	1.21	1.31	1.39
1500 Watts	0.76	0.84	0.90	0.96	1.32	1.45	1.57	1.66

All manhours include checking out of job storage, handling, job hauling, distributing, layout, and marking for installation.

Hang-Only manhours only include hanging of fixture.

Assemble-and-Hang manhours include complete assembling of fixture, installation, connecting the circuit power to the fixture, testing, and verifying fixture operation.

Manhours exclude installation of conduit, boxes, circuit wiring, switches, light bulbs, special hangers or supports not supplied with fixture, and ladders or scaffolding. See respective tables for these time frames.

For heights above 25 feet increase manhours 5% for each additional 5 feet or fraction thereof.

INDUSTRIAL INTERIOR INCANDESCENT FIXTURES

Porcelain Enamel Reflector 1-Lamp Units
Surface Mounted

MANHOURS EACH

Item Description	Manhours Required to							
	Hang Only For Height to				Assemble and Hang For Height to			
	10'	15'	20'	25'	10'	15'	20'	25'
Glass Lens or Diffuser Type								
60 Watts	0.30	0.33	0.36	0.38	0.55	0.61	0.65	0.69
100 Watts	0.30	0.33	0.36	0.38	0.55	0.61	0.65	0.69
150 Watts	0.33	0.36	0.39	0.42	0.61	0.67	0.72	0.77
200 Watts	0.39	0.43	0.46	0.49	0.69	0.76	0.82	0.87
300 Watts	0.43	0.47	0.51	0.54	0.74	0.81	0.88	0.93
500 Watts	0.50	0.55	0.59	0.63	0.91	1.00	1.08	1.15
750 Watts	0.57	0.63	0.68	0.72	1.07	1.18	1.27	1.35
1000 Watts	0.68	0.75	0.81	0.86	1.27	1.40	1.51	1.60
1500 Watts	0.83	0.91	0.99	1.05	1.51	1.66	1.79	1.90
Vaportite Type								
60 Watts	0.33	0.36	0.39	0.42	0.61	0.67	0.72	0.77
100 Watts	0.39	0.43	0.46	0.49	0.69	0.76	0.82	0.87
150 Watts	0.43	0.47	0.51	0.54	0.74	0.81	0.88	0.93
200 Watts	0.48	0.53	0.57	0.60	0.80	0.88	0.95	1.01
300 Watts	0.52	0.57	0.62	0.65	0.91	1.00	1.08	1.15
500 Watts	0.59	0.65	0.70	0.74	1.05	1.16	1.25	1.32
750 Watts	0.67	0.74	0.80	0.84	1.21	1.33	1.44	1.52

All manhours include checking out of job storage, handling, job hauling, distributing, layout, and marking for installation.

Hang-Only manhours only include hanging of fixture.

Assemble-and-Hang manhours include complete assembling of fixture, installation, connecting the circuit power to the fixture, testing, and verifying fixture operation.

Manhours exclude installation of conduit, boxes, circuit wiring, switches, light bulbs, special hangers or supports not supplied with fixture, and ladders or scaffolding. See respective tables for these time frames.

For heights above 25 feet increase manhours 5% for each additional 5 feet or fraction thereof.

INDUSTRIAL INTERIOR INCANDESCENT FIXTURES

Vaportite Lampholder 1-Lamp Units

MANHOURS EACH

Item Description	Manhours Required to							
	Hang Only For Height to				Assemble and Hang For Height to			
	10'	15'	20'	25'	10'	15'	20'	25'
Surface Mounted Glass Globe								
60 Watts	0.19	0.21	0.23	0.24	0.30	0.33	0.36	0.38
100 Watts	0.22	0.24	0.26	0.28	0.39	0.43	0.46	0.49
150 Watts	0.28	0.31	0.33	0.35	0.44	0.48	0.52	0.55
200 Watts	0.31	0.34	0.37	0.39	0.50	0.55	0.59	0.63
300 Watts	0.37	0.41	0.44	0.47	0.61	0.67	0.72	0.77
500 Watts	0.44	0.48	0.52	0.55	0.74	0.81	0.88	0.93
750 Watts	0.52	0.57	0.62	0.65	0.91	1.00	1.08	1.15
Surface Mounted Globe and Wire Guard								
60 Watts	0.22	0.24	0.26	0.28	0.39	0.43	0.46	0.49
100 Watts	0.24	0.26	0.29	0.30	0.47	0.52	0.56	0.59
150 Watts	0.30	0.33	0.36	0.38	0.52	0.57	0.62	0.65
200 Watts	0.35	0.39	0.42	0.44	0.61	0.67	0.72	0.77
300 Watts	0.41	0.45	0.49	0.52	0.69	0.76	0.82	0.87
Pendant Mounted Globe and Wire Guard								
60 Watts	0.30	0.33	0.36	0.38	0.52	0.57	0.62	0.65
100 Watts	0.33	0.36	0.39	0.42	0.61	0.67	0.72	0.77
150 Watts	0.39	0.43	0.46	0.49	0.69	0.76	0.82	0.87
200 Watts	0.44	0.48	0.52	0.55	0.77	0.85	0.91	0.97
300 Watts	0.50	0.55	0.59	0.63	0.88	0.97	1.05	1.11

All manhours include checking out of job storage, handling, job hauling, distributing, layout, and marking for installation.

Hang-Only manhours only include hanging of fixture.

Assemble-and-Hang manhours include complete assembling of fixture, installation, connecting the circuit power to the fixture, testing, and verifying fixture operation.

Manhours exclude installation of conduit, boxes, circuit wiring, switches, light bulbs, special hangers or supports not supplied with fixture, and ladders or scaffolding. See respective tables for these time frames.

For heights above 25 feet increase manhours 5% for each additional 5 feet or fraction thereof.

INDUSTRIAL INTERIOR INCANDESCENT FIXTURES

Explosion Proof Enclosed Lampholder 1-Lamp Units

MANHOURS EACH

Item Description	Manhours Required to							
	Hang Only For Height to				Assemble and Hang For Height to			
	10'	15'	20'	25'	10'	15'	20'	25'
Class II Explosion Proof Surface Mounted								
60 Watts	0.33	0.36	0.39	0.42	0.61	0.67	0.72	0.77
100 Watts	0.35	0.39	0.42	0.44	0.66	0.73	0.78	0.83
150 Watts	0.41	0.45	0.49	0.52	0.74	0.81	0.88	0.93
200 Watts	0.46	0.51	0.55	0.58	0.80	0.88	0.95	1.01
300 Watts	0.52	0.57	0.62	0.65	0.91	1.00	1.08	1.15
Class I Explosion Proof Pendant Mounted								
60 Watts	0.43	0.47	0.51	0.54	0.74	0.81	0.88	0.93
100 Watts	0.45	0.50	0.53	0.57	0.80	0.88	0.95	1.01
150 Watts	0.50	0.55	0.59	0.63	0.91	1.00	1.08	1.15
200 Watts	0.56	–	–	–	0.99	1.09	1.18	1.25
300 Watts	0.61	–	–	–	1.14	1.25	1.35	1.44
Class I Explosion Proof Surface Mounted								
60 Watts	0.43	0.47	0.51	0.54	0.74	0.81	0.88	0.93
100 Watts	0.45	0.50	0.53	0.57	0.80	0.88	0.95	1.01
150 Watts	0.50	0.55	0.59	0.63	0.91	1.00	1.08	1.15
200 Watts	0.56	0.62	0.67	0.71	0.99	1.09	1.18	1.25
300 Watts	0.61	0.67	0.72	0.77	1.10	1.21	1.31	1.39
Class I Explosion Proof Pendant Mounted								
60 Watts	0.50	0.55	0.59	0.63	0.91	1.00	1.08	1.15
100 Watts	0.52	0.57	0.62	0.65	0.96	1.06	1.14	1.21
150 Watts	0.61	0.67	0.72	0.77	1.02	1.12	1.21	1.28
200 Watts	0.63	0.69	0.75	0.79	1.10	1.21	1.31	1.39
300 Watts	0.68	0.75	0.81	0.86	1.21	1.33	1.44	1.52

All manhours include checking out of job storage, handling, job hauling, distributing, layout, and marking for installation.

Hang-Only manhours only include hanging of fixture.

Assemble-and-Hang manhours include complete assembling of fixture, installation, connecting the circuit power to the fixture, testing, and verifying fixture operation.

Manhours exclude installation of conduit, boxes, circuit wiring, switches, light bulbs, special hangers or supports not supplied with fixture, and ladders or scaffolding. See respective tables for these time frames.

For heights above 25 feet increase manhours 5% for each additional 5 feet or fraction thereof.

INDUSTRIAL INTERIOR INCANDESCENT FIXTURES

Pendant Mounted Porcelain Enamel Reflector

1-Lamp Units

MANHOURS EACH

Item Description	Manhours Required to							
	Hang Only For Height to				Assemble and Hang For Height to			
	10'	15'	20'	25'	10'	15'	20'	25'
Separable Socket								
60 Watts	0.26	0.29	0.31	0.33	0.47	0.52	0.56	0.59
100 Watts	0.26	0.29	0.31	0.33	0.47	0.52	0.56	0.59
150 Watts	0.31	0.34	0.37	0.39	0.52	0.57	0.62	0.65
200 Watts	0.33	0.36	0.39	0.42	0.61	0.67	0.72	0.77
300 Watts	0.39	0.43	0.46	0.49	0.69	0.76	0.82	0.87
500 Watts	0.46	0.51	0.55	0.58	0.85	0.94	1.01	1.07
750 Watts	0.56	0.62	0.67	0.71	0.99	1.09	1.18	1.25
1000 Watts	0.67	0.74	0.80	0.84	1.21	1.33	1.44	1.52
1500 Watts	0.80	0.88	0.95	1.01	1.43	1,57	1.70	1.80
1-Piece Socket								
60 Watts	0.30	0.33	0.36	0.38	0.55	0.61	0.65	0.69
100 Watts	0.30	0.33	0.36	0.38	0.55	0.61	0.65	0.69
150 Watts	0.33	0.36	0.39	0.42	0.61	0.67	0.72	0.77
200 Watts	0.39	0.43	0.46	0.49	0.69	0.76	0.82	0.87
300 Watts	0.43	0.47	0.51	0.54	0.74	0.81	0.88	0.93
500 Watts	0.50	0.55	0.59	0.63	0.91	1.00	1.08	1.15
750 Watts	0.57	0.63	0.68	0.72	1.07	1.18	1.27	1.35
1000 Watts	0.68	0.75	0.81	0.86	1.27	1.40	1.51	1.60
1500 Watts	0.83	0.91	0.99	1.05	1.51	1.66	1.79	1.90

All manhours include checking out of job storage, handling, job hauling, distributing, layout, and marking for installation.

Hang-Only manhours only include hanging of fixture.

Assemble-and-Hang manhours include complete assembling of fixture, installation, connecting the circuit power to the fixture, testing, and verifying fixture operation.

Manhours exclude installation of conduit, boxes, circuit wiring, switches, light bulbs, special hangers or supports not supplied with fixture, and ladders or scaffolding. See respective tables for these time frames.

For heights above 25 feet increase manhours 5% for each additional 5 feet or fraction thereof.

INDUSTRIAL INTERIOR INCANDESCENT FIXTURES

Pendant Mounted Porcelain Enamel Reflector

1-Lamp Units

MANHOURS EACH

Item Description	Manhours Required to							
	Hang Only For Height to				Assemble and Hang For Height to			
	10′	15′	20′	25′	10′	15′	20′	25′
Glass Lens or Diffuser								
60 Watts	0.39	0.43	0.46	0.49	0.69	0.76	0.82	0.87
100 Watts	0.39	0.43	0.46	0.49	0.69	0.76	0.82	0.87
150 Watts	0.44	0.48	0.52	0.55	0.74	0.81	0.88	0.93
200 Watts	0.46	0.51	0.55	0.58	0.83	0.91	0.99	1.05
300 Watts	0.52	0.57	0.62	0.65	0.94	1.03	1.12	1.18
500 Watts	0.57	0.63	0.68	0.72	1.07	1.18	1.27	1.35
750 Watts	0.67	0.74	0.80	0.74	1.21	1.33	1.44	1.52
1000 Watts	0.80	0.88	0.95	1.01	1.43	1.57	1.70	1.80
1500 Watts	0.94	1.03	1.12	1.18	1.65	1.82	1.96	2.08
Vaportite								
60 Watts	0.39	0.43	0.46	0.49	0.69	0.76	0.82	0.87
100 Watts	0.44	0.48	0.52	0.55	0.74	0.81	0.88	0.93
150 Watts	0.48	0.53	0.57	0.60	0.80	0.88	0.95	1.01
200 Watts	0.54	0.59	0.64	0.68	0.88	0.97	1.05	1.11
300 Watts	0.57	0.63	0.68	0.72	0.99	1.09	1.18	1.25
500 Watts	0.65	0.72	0.77	0.82	1.10	1.21	1.31	1.39
750 Watts	0.72	0.79	0.86	0.91	1.27	1.40	1.51	1.60

All manhours include checking out of job storage, handling, job hauling, distributing, layout, and marking for installation.

Hang-Only manhours only include hanging of fixture.

Assemble-and-Hang manhours include complete assembling of fixture, installation, connecting the circuit power to the fixture, testing, and verifying fixture operation.

Manhours exclude installation of conduit, boxes, circuit wiring, switches, light bulbs, special hangers or supports not supplied with fixture, and ladders or scaffolding. See respective tables for these time frames.

For heights above 25 feet increase manhours 5% for each additional 5 feet or fraction thereof.

INDUSTRIAL INTERIOR INCANDESCENT FIXTURES

High Bay Rigid Aluminum Reflector Lampholder
and
Explosion-Proof porcelain Enamel Reflector
1-Lamp Units

MANHOURS EACH

Item Description	Manhours Required to							
	Hang Only For Height to				Assemble and Hang For Height to			
	10'	15'	20'	25'	10'	15'	20'	25'
High Bay Aluminum Reflector								
200 Watts	0.28	0.31	0.33	0.35	0.39	0.43	0.46	0.49
300 Watts	0.33	0.36	0.39	0.42	0.47	0.52	0.56	0.59
500 Watts	0.44	0.48	0.52	0.55	0.61	0.67	0.72	0.77
750 Watts	0.55	0.61	0.65	0.69	0.74	0.81	0.88	0.93
1000 Watts	0.72	0.79	0.86	0.91	0.99	1.09	1.18	1.25
1500 Watts	0.88	0.97	1.05	1.11	1.21	1.33	1.44	1.52
Explosion-Proof Porcelain reflector								
Class II — 60 Watts	0.43	0.47	0.51	0.54	0.74	0.81	0.88	0.93
Class II — 100 Watts	0.44	0.48	0.52	0.55	0.69	0.76	0.82	0.77
Class II — 150 watts	0.50	0.55	0.59	0.63	0.88	0.97	1.05	1.11
Class II — 200 Watts	0.56	0.62	0.67	0.71	0.99	1.09	1.18	1.25
Class II — 300 Watts	0.61	0.67	0.72	0.77	1.10	1.21	1.31	1.39
Class II — 500 Watts	0.68	0.75	0.81	0.86	1.21	1.33	1.44	1.52
Class II — 750 Watts	0.78	0.86	0.93	0.98	1.35	1.49	1.60	1.70

All manhours include checking out of job storage, handling, job hauling, distributing, layout, and marking for installation.

Hang-Only manhours only include hanging of fixture.

Assemble-and-Hang manhours include complete assembling of fixture, installation, connecting the circuit power to the fixture, testing, and verifying fixture operation.

Manhours exclude installation of conduit, boxes, circuit wiring, switches, light bulbs, special hangers or supports not supplied with fixture, and ladders or scaffolding. See respective tables for these time frames.

For heights above 25 feet increase manhours 5% for each additional 5 feet or fraction thereof.

INTERIOR INCANDESCENT VAPORTITE LIGHT FIXTURES AND ACCESSORIES

MANHOURS EACH

Item Description	Manhours for Height to			
	10'	15'	20'	25'
Pendant Mounted Vaportiee Lampholder with Glass Globe				
60 Watts	0.33	0.36	0.39	0.42
100 Watts	0.36	0.40	0.43	0.45
150 Watts	0.44	0.48	0.52	0.55
200 Watts	0.50	0.55	0.59	0.63
300 Watts	0.58	0.64	0.69	0.73
500 Watts	0.67	0.74	0.80	0.84
750 Watts	0.79	0.87	0.94	0.99
Accessories				
Threaded Neck Lampholder Steel Wire Guard for				
60 Watts	0.14	0.15	0.17	0.18
100 Watts	0.14	0.15	0.17	0.18
150 Watts	0.14	0.15	0.17	0.18
200 Watts	0.18	0.20	0.21	0.23
300 Watts	0.18	0.20	0.21	0.23
Reflector Wire Cage Locking Guard for				
60 Watts	0.18	0.20	0.21	0.23
100 Watts	0.18	0.20	0.21	0.23
150 Watts	0.22	0.24	0.26	0.28
200 Watts	0.22	0.24	0.26	0.28
300 Watts	0.22	0.24	0.26	0.28
500 Watts	0.29	0.32	0.34	0.37
750 Watts	0.36	0.40	0.43	0.45

Manhours include checking out of job storage, handling, job hauling, distributing, layout and marking for installation, assembling of fixture, installation, connectiong the circuit power to the fixture, testing, and verifying fixture operation.

Manhours exclude installation of conduit, boxes, circuit wiring, switches, light bulbs, special hangers or supports not supplied with fixture, and ladders or scaffolding. See respective tables for these time frames.

For heights above 25 feet increase manhours 5% for each additional 5 feet or fraction thereof.

INTERIOR MERCURY VAPOR FIXTURES
Separable Socket Rigid 1- and 2-Lampholder Units

MANHOURS EACH

Item Description	Manhours Required to							
	Hang Only For Height to				Assemble and Hang For Height to			
	10'	15'	20'	25'	10'	15'	20'	25'
1-Lamp Units								
250 Watts	0.83	0.91	0.99	1.05	1.51	1.66	1.79	1.90
300 Watts	0.83	0.91	0.99	1.05	1.51	1.66	1.79	1.90
400 Watts	1.00	1.10	1.19	1.26	1.82	2.00	2.16	2.29
500 Watts	1.00	1.10	1.19	1.26	1.82	2.00	2.16	2.89
750 Watts	1.30	1.43	1.54	1.64	2.31	2.54	2.74	2.91
1000 Watts	1.76	1.84	1.98	2.10	3.03	3.33	3.60	3.92
1500 Watts	2.11	2.32	2.51	2.66	3.80	4.18	4.51	4.79
2-Lamp Units								
250 Watts	1.30	1.43	1.54	1.64	2.31	2.54	2.74	2.91
300 Watts	1.46	1.61	1.73	1.84	2.64	2.90	3.14	3.32
400 Watts	1.67	1.84	1.98	2.10	3.03	3.33	3.60	3.82
500 Watts	2.11	2.32	2.51	2.66	3.74	4.11	4.44	4.71
750 Watts	2.52	2.77	2.99	3.17	4.57	5.03	5.43	5.75
1000 Watts	2.96	3.26	3.52	3.73	5.34	5.87	6.34	6.72
1500 Watts	3.33	3.66	3.96	4.19	6.05	6.66	7.19	7.62

1-Piece Socket Rigid 1- and 2-Lampholder Units

Item Description	10'	15'	20'	25'	10'	15'	20'	25'
1-Lamp Units								
250 Watts	0.94	1.03	1.12	1.18	1.71	1.88	2.03	2.15
300 Watts	0.94	1.03	1.12	1.18	1.71	1.88	2.03	2.15
400 Watts	1.11	1.22	1.32	1.40	1.98	2.17	2.34	2.48
500 Watts	1.11	1.22	1.32	1.40	1.98	2.17	2.34	2.48
750 Watts	1.44	1.58	1.71	1.81	2.53	2.78	3.01	3.19
1000 Watts	1.81	1.99	2.15	2.28	3.36	3.70	3.99	4.23
1500 Watts	2.33	2.56	2.77	2.93	4.24	4.66	5.04	5.34
2-Lamp Units								
250 Watts	1.44	1.58	1.71	1.81	2.53	2.78	3.01	2.19
300 Watts	1.67	1.84	1.98	2.10	3.03	3.33	3.60	3.82
400 Watts	1.92	2.11	2.28	2.42	3.52	3.87	4.18	4.43
500 Watts	2.33	2.56	2.77	2.93	4.24	4.66	5.04	5.34
750 Watts	2.78	3.06	3.30	3.50	5.00	5.50	5.94	6.30
1000 Watts	3.60	3.96	4.28	4.53	5.78	6.36	6.87	7.28
1500 Watts	3.66	4.03	4.35	4.61	6.66	7.33	7.91	8.39

All manhours include checking out of job storage, handling, job hauling, distributing, layout, and marking for installation.

Hang-Only manhours only include hanging of fixture.

Assemble-and-Hang manhours include complete assembling of fixture, installation, connecting the circuit power to the fixture, testing, and verifying fixture operation.

Manhours exclude installation of conduit, boxes, circuit wiring, switches, light bulbs, special hangers or supports not supplied with fixture, and ladders or scaffolding. See respective tables for these time frames.

For heights above 25 feet increase manhours 5% for each additional 5 feet or fraction thereof.

INTERIOR MERCURY VAPOR FIXTURES
RIGID VAPORTITE 1-LAMPHOLDER WITH GLASS COVER

MANHOURS EACH

Item Description	Manhours Required to							
	Hang Only For Height to				Assemble and Hang For Height to			
	10'	15'	20'	25'	10'	15'	20'	25'
1-Lamp Units								
250 Watts	1.00	1.10	1.19	1.26	1.82	2.00	2.16	2.28
300 Watts	1.09	1.20	1.29	1.37	1.93	2.17	2.29	2.43
400 Watts	1.18	1.30	1.40	1.49	2.09	2.30	2.48	2.63
500 Watts	1.28	1.41	1.52	1.61	2.31	2.54	2.74	2.91
750 Watts	1.42	1.56	1.69	1.79	2.59	2.85	3.08	3.26
1000 Watts	1.81	1.99	2.15	2.28	3.36	3.70	3.99	4.23
1500 Watts	2.33	2.56	2.77	2.93	4.24	4.66	5.04	5.34

INTERIOR MERCURY VAPOR FIXTURES
RIGID EXPLOSION PROOF 1-LAMPHOLDER UNITS

	10'	15'	20'	25'	10'	15'	20'	25'
Class II – Explosion Proof								
250 Watts	1.26	1.39	1.50	1.59	2.31	2.54	2.74	2.91
300 Watts	1.33	1.46	1.58	1.67	2.42	2.66	2.87	3.05
400 Watts	1.42	1.56	1.69	1.79	2.53	2.78	3.01	3.19
500 Watts	1.52	1.67	1.81	1.91	2.75	3.03	3.27	3.46
750 Watts	1.67	1.84	1.98	2.10	3.03	3.33	3.60	3.82
1000 Watts	2.11	2.32	2.51	2.66	3.80	4.18	4.51	4.79
1500 Watts	2.52	2.77	2.99	3.17	4.57	5.03	5.43	5.75
Class I – Explosion Proof								
250 Watts	1.67	1.84	1.98	2.10	3.03	3.33	3.60	3.82
300 Watts	1.74	1.94	2.07	2.19	3.14	3.45	3.73	3.95
400 Watts	1.81	1.99	2.15	2.28	3.36	3.70	3.99	4.23
500 Watts	1.92	2.11	2.28	2.42	3.52	3.87	4.18	4.43
750 Watts	2.11	2.32	2.51	2.66	3.80	4.18	4.51	4.79
1000 Watts	2.52	2.77	2.99	3.17	4.57	5.03	5.43	5.75
1500 Watts	2.96	3.26	3.52	3.73	5.34	5.87	6.34	6.72

All manhours include checking out of job storage, handling, job hauling, distributing, layout, and marking for installation.

Hang-Only manhours only include hanging of fixture.

Assemble-and-Hang manhours include complete assembling of fixture, installation, connecting the circuit power to the fixture, testing, and verifying fixture operation.

Manhours exclude installation of conduit, boxes, circuit wiring, switches, light bulbs, special hangers or supports not supplied with fixture, and ladders or scaffolding. See respective tables for these time frames.

For heights above 25 feet increase manhours 5% for each additional 5 feet or fraction thereof.

FLUORESCENT LIGHTING FIXTURES
INTERIOR SURFACE MOUNTED
1-Tube Fixtures

MANHOURS EACH

Item Description	Tube Length (Inches)	Manhours Required to							
		Hang Only For Height to				Assemble and Hang For Height to			
		10'	15'	20'	25'	10'	15'	20'	25'
Commercial Grade Single Fixtures	18	0.57	0.63	0.68	0.72	1.32	1.45	1.57	1.66
	24	0.67	0.74	0.80	0.84	1.54	1.69	1.83	1.94
	36	0.74	0.81	0.88	0.93	1.73	1.90	2.06	2.18
	48	0.83	0.91	0.99	1.05	1.93	2.12	2.29	2.43
	72	1.00	1.10	1.19	1.26	2.31	2.54	2.74	2.91
	96	1.18	1.30	1.40	1.49	2.70	2.97	3.21	3.40
Continuous Row Fixtures	48	0.67	0.74	0.80	0.84	1.54	1.69	1.83	1.94
	72	0.74	0.81	0.88	0.93	1.73	1.90	2.06	2.18
	96	0.91	1.00	1.08	1.15	2.12	2.33	2.52	2.67
Industrial Grade Single Fixtures	18	0.56	0.62	0.67	0.71	1.27	1.40	1.51	1.60
	24	0.63	0.69	0.75	0.79	1.43	1.57	1.70	1.80
	36	0.72	0.79	0.86	0.91	1.65	1.89	1.96	2.08
	48	0.80	0.88	0.95	1.01	1.82	2.00	2.16	2.29
	72	0.96	1.06	1.14	1.21	2.20	2.42	2.61	2.77
	96	1.13	1.24	1.34	1.42	2.59	2.85	3.08	3.26
Continuous Row Fixtures	48	0.63	0.69	0.75	0.79	1.43	1.57	1.70	1.80
	72	0.72	0.79	0.86	0.91	1.65	1.89	1.96	2.08
	96	0.83	0.91	0.99	1.05	1.93	2.12	2.29	2.43

All manhours include checking out of job storage, handling, job hauling, distributing, layout, and marking for installation.

Hang-Only manhours only include hanging of fixture.

Assemble-and-Hang manhours include complete assembling of fixture, installation, connecting circuit power to the fixture, testing, and verifying fixture operation.

Manhours exclude installation of conduit, boxes, circuit wiring, switches, tubes, special hangers or supports not supplied with fixture, change out of or separate ballast, and ladder or scaffold set-up. See respective tables for these time frames.

For heights above 25 feet increase manhours 5% for each additional 5 feet or fraction thereof.

FLUORESCENT LIGHTING FIXTURES
INTERIOR SURFACE MOUNTED
2-Tube Fixtures

MANHOURS EACH

Item Description	Tube Length (Inches)	Manhours Required to							
		Hang Only For Height to				Assemble and Hang For Height to			
		10′	15′	20′	25′	10′	15′	20′	25′
Commercial Grade Single Fixtures	18	0.67	0.74	0.80	0.84	1.54	1.69	1.83	1.94
	24	0.76	0.84	0.90	0.96	1.73	1.90	2.06	2.18
	36	0.83	0.91	0.99	1.05	1.93	2.12	2.29	2.43
	48	0.91	1.00	1.08	1.15	2.12	2.33	2.52	2.67
	72	1.09	1.20	1.29	1.37	2.53	2.78	3.01	3.19
	96	1.26	1.39	1.50	1.59	2.89	3.18	3.43	3.64
Continuous Row Fixtures	48	0.76	0.84	0.90	0.96	1.73	1.90	2.06	2.18
	72	0.83	0.91	0.99	1.05	1.93	2.12	2.29	2.43
	96	1.00	1.10	1.19	1.26	2.31	2.54	2.74	2.91
Industrial Grade Single Fixtures	18	0.63	0.69	0.75	0.79	1.43	1.57	1.70	1.80
	24	0.72	0.79	0.86	0.91	1.65	1.82	1.96	2.08
	36	0.80	0.88	0.95	1.01	1.82	2.00	2.16	2.29
	48	0.89	0.98	1.06	1.12	2.04	2.24	2.42	2.57
	72	1.05	1.16	1.25	1.32	2.42	2.66	2.87	3.05
	96	1.22	1.34	1.45	1.54	2.81	3.09	3.34	3.54
Continuous Row Fixtures	48	0.72	0.79	0.86	0.91	1.65	1.82	1.96	2.08
	72	0.80	0.88	0.95	1.01	1.82	2.00	2.16	2.29
	96	0.91	1.00	1.08	1.15	2.12	2.33	2.52	2.67

All manhours include checking out of job storage, handling, job hauling, distributing, layout, and marking for installation.

Hang-Only manhours only include hanging of fixture.

Assemble-and-Hang manhours include complete assembling of fixture, installation, connecting circuit power to the fixture, testing, and verifying fixture operation.

Manhours exclude installation of conduit, boxes, circuit wiring, switches, tubes, special hangers or supports not supplied with fixture, change out of or separate ballast, and ladder or scaffold set-up. See respective tables for these time frames.

For heights above 25 feet increase manhours 5% for each additional 5 feet or fraction thereof.

FLUORESCENT LIGHTING FIXTURES
INTERIOR SURFACE MOUNTED

3-Tube Fixtures

MANHOURS EACH

Item Description	Tube Length (Inches)	Manhours Required to							
		Hang Only For Height to				Assemble and Hang For Height to			
		10'	15'	20'	25'	10'	15'	20'	25'
Commercial Grade Single Fixtures	18	0.76	0.84	0.90	0.96	1.73	1.90	2.06	2.18
	24	0.83	0.91	0.99	1.05	1.93	2.12	2.29	2.43
	36	0.91	1.00	1.08	1.15	2.12	2.33	2.52	2.67
	48	1.00	1.10	1.19	1.26	2.31	2.54	2.74	2.91
	72	1.18	1.30	1.40	1.49	2.70	2.97	3.21	3.40
	96	1.26	1.39	1.50	1.59	3.08	3.39	3.66	3.88
Continuous Row Fixtures	48	0.83	0.91	0.99	1.05	1.93	2.12	2.29	2.43
	72	0.91	1.00	1.08	1.15	2.12	2.33	2.52	2.67
	96	1.09	1.20	1.29	1.37	2.53	2.78	3.01	3.19
Industrial Grade Single Fixtures	18	0.72	0.79	0.86	0.91	1.65	1.82	1.96	2.08
	24	0.80	0.88	0.95	1.01	1.82	2.00	2.16	2.29
	36	0.89	0.98	1.06	1.12	2.04	2.24	2.42	2.57
	48	1.43	1.57	1.70	1.80	2.20	2.42	2.61	2.77
	72	1.71	1.88	2.03	2.15	2.61	2.87	3.10	3.29
	96	1.93	2.12	2.29	2.43	3.00	3.30	3.56	3.78
Continuous Row Fixtures	48	0.80	0.88	0.95	1.01	1.82	2.00	2.16	2.29
	72	0.89	0.98	1.06	1.12	2.04	2.24	2.42	2.57
	96	1.05	1.16	1.25	1.32	2.42	2.66	2.87	3.05

All manhours include checking out of job storage, handling, job hauling, distributing, layout, and marking for installation.

Hang-Only manhours only include hanging of fixture.

Assemble-and-Hang manhours include complete assembling of fixture, installation, connecting circuit power to the fixture, testing, and verifying fixture operation.

Manhours exclude installation of conduit, boxes, circuit wiring, switches, tubes, special hangers or supports not supplied with fixture, change out of or separate ballast, and ladder or scaffold set-up. See respective tables for these time frames.

For heights above 25 feet increase manhours 5% for each additional 5 feet or fraction thereof.

FLUORESCENT LIGHTING FIXTURES INTERIOR SURFACE MOUNTED

4-Tube Fixtures

MANHOURS EACH

Item Description	Tube Length (Inches)	Manhours Required to							
		Hang Only For Height to				Assemble and Hang For Height to			
		10'	15'	20'	25'	10'	15'	20'	25'
Commercial Grade Single Fixtures	18	0.83	0.91	0.99	1.05	1.93	2.12	2.29	2.43
	24	0.91	1.00	1.08	1.15	2.12	2.33	2.52	2.67
	36	1.00	1.10	1.19	1.26	2.31	2.54	2.74	2.91
	48	1.09	1.20	1.29	1.37	2.53	2.78	3.01	3.19
	72	1.26	1.39	1.50	1.59	2.89	3.18	3.43	3.64
	96	1.42	1.56	1.69	1.79	3.27	3.60	3.88	4.12
Continuous Row Fixtures	48	0.91	1.00	1.08	1.15	2.12	2.33	2.52	2.67
	72	1.00	1.10	1.19	1.26	2.31	2.54	2.74	2.91
	96	1.18	1.30	1.40	1.49	2.70	2.97	3.21	3.40
Industrial Grade Single Fixtures	18	0.80	0.88	0.95	1.01	1.82	2.00	2.16	2.29
	24	0.89	0.98	1.06	1.12	2.04	2.24	2.42	2.67
	36	0.96	1.06	1.14	1.21	2.20	2.42	2.61	2.77
	48	0.81	0.89	0.96	1.02	2.42	2.66	2.87	3.05
	72	1.18	1.30	1.40	1.49	2.81	3.09	3.34	3.54
	96	1.39	1.53	1.65	1.75	3.19	3.51	3.79	4.02
Continuous Row Fixtures	48	0.89	0.98	1.06	1.12	2.04	2.24	2.42	2.67
	72	0.96	1.06	1.14	1.21	2.20	2.42	2.61	2.77
	96	1.15	1.27	1.37	1.45	2.61	2.87	3.10	3.29

All manhours include checking out of job storage, handling, job hauling, distributing, layout, and marking for installation.

Hang-Only manhours only include hanging of fixture.

Assemble-and-Hang manhours include complete assembling of fixture, installation, connecting circuit power to the fixture, testing, and verifying fixture operation.

Manhours exclude installation of conduit, boxes, circuit wiring, switches, tubes, special hangers or supports not supplied with fixture, change out of or separate ballast, and ladder or scaffold set-up. See respective tables for these time frames.

For heights above 25 feet increase manhours 5% for each additional 5 feet or fraction thereof.

FLUORESCENT LIGHTING FIXTURES INTERIOR SURFACE MOUNTED

6-Tube Fixtures

MANHOURS EACH

Item Description	Tube Length (Inches)	Manhours Required to							
		Hang Only For Height to				Assemble and Hang For Height to			
		10′	15′	20′	25′	10′	15′	20′	25′
Commercial Grade Single Fixtures	36	1.18	1.30	1.40	1.49	2.70	2.97	3.21	3.40
	48	1.26	1.39	1.50	1.59	2.89	3.18	3.43	3.64
	72	1.42	1.56	1.69	1.79	3.27	3.60	3.88	4.12
	96	1.57	1.73	1.87	1.98	3.66	4.03	4.35	4.61
Continuous Row Fixtures	48	1.09	1.20	1.29	1.37	2.53	2.78	3.01	3.18
	72	1.18	1.30	1.40	1.49	2.70	2.97	3.21	3.40
	96	1.33	1.46	1.58	1.67	3.08	3.39	3.66	3.88
Industrial Grade Single Fixtures	36	1.15	1.27	1.37	1.45	2.61	2.87	3.10	3.29
	48	1.22	1.34	1.45	1.54	2.81	3.09	3.34	3.54
	72	1.38	1.52	1.64	1.74	3.19	3.51	3.79	4.02
	96	1.55	1.71	1.84	1.95	3.58	3.94	4.25	4.51
Continuous Row Fixtures	48	1.05	1.16	1.25	1.32	2.42	2.66	2.87	3.05
	72	1.15	1.27	1.37	1.45	2.61	2.87	3.10	3.39
	96	1.30	1.43	1.54	1.64	3.00	3.30	3.56	3.78

All manhours include checking out of job storage, handling, job hauling, distributing, layout, and marking for installation.

Hang-Only manhours only include hanging of fixture.

Assemble-and-Hang manhours include complete assembling of fixture, installation, connecting circuit power to the fixture, testing, and verifying fixture operation.

Manhours exclude installation of conduit, boxes, circuit wiring, switches, tubes, special hangers or supports not supplied with fixture, change out of or separate ballast, and ladder or scaffold set-up. See respective tables for these time frames.

For heights above 25 feet increase manhours 5% for each additional 5 feet or fraction thereof.

FLUORESCENT LIGHTING FIXTURES
INTERIOR SURFACE MOUNTED
8-Tube Fixtures

MANHOURS EACH

Item Description	Tube Length (Inches)	Manhours Required to							
		Hang Only For Height to				Assemble and Hang For Height to			
		10′	15′	20′	25′	10′	15′	20′	25′
Commercial Grade Single Fixtures	48	1.42	1.56	1.69	1.79	3.27	3.60	3.88	4.12
	72	1.57	1.73	1.87	1.98	3.66	4.03	4.35	4.61
	96	1.79	1.96	2.11	2.23	4.04	4.44	4.80	5.09
Continuous Row Fixtures	48	1.26	1.39	1.50	1.59	2.89	3.18	3.43	3.64
	72	1.33	1.46	1.58	1.67	3.08	3.39	3.66	3.88
	96	1.50	1.65	1.78	1.89	3.47	3.82	4.12	4.37
Industrial Grade Single Fixtures	48	1.39	1.53	1.65	1.75	3.19	3.51	3.79	4.02
	72	1.54	1.69	1.83	1.94	3.58	3.94	4.25	4.51
	96	1.70	1.87	2.02	2.14	3.96	4.36	4.70	4.99
Continuous Row Fixtures	48	1.20	1.32	1.43	1.51	2.81	3.09	3.34	3.54
	72	1.30	1.43	1.54	1.64	3.00	3.30	3.56	3.78
	96	1.46	1.61	1.73	1.84	3.36	3.70	3.99	4.23

All manhours include checking out of job storage, handling, job hauling, distributing, layout, and marking for installation.

Hang-Only manhours only include hanging of fixture.

Assemble-and-Hang manhours include complete assembling of fixture, installation, connecting circuit power to the fixture, testing, and verifying fixture operation.

Manhours exclude installation of conduit, boxes, circuit wiring, switches, tubes, special hangers or supports not supplied with fixture, change out of or separate ballast, and ladder or scaffold set-up. See respective tables for these time frames.

For heights above 25 feet increase manhours 5% for each additional 5 feet or fraction thereof.

FLUORESCENT LIGHTING FIXTURES
INTERIOR CEILING SUSPENDED

1-Tube Fixtures

MANHOURS EACH

Item Description	Tube Length (Inches)	Manhours Required to							
		Hang Only For Height to				Assemble and Hang For Height to			
		10'	15'	20'	25'	10'	15'	20'	25'
Commercial Grade Single Fixtures	18	0.74	0.81	0.88	0.93	1.76	1.94	2.09	2.22
	24	0.83	0.91	0.99	1.05	1.93	2.12	2.29	2.43
	36	0.91	1.00	1.08	1.15	2.12	2.33	2.52	2.67
	48	1.00	1.10	1.19	1.26	2.31	2.54	2.74	2.91
	72	1.18	1.30	1.40	1.49	2.70	2.97	3.21	3.40
	96	1.33	1.46	1.58	1.67	3.08	3.39	3.66	3.88
Continuous Row Fixtures	48	0.83	0.91	0.99	1.05	1.93	2.12	2.29	2.43
	72	0.91	1.00	1.08	1.15	2.12	2.33	2.52	2.67
	96	1.09	1.20	1.29	1.37	2.50	2.75	2.97	3.15
Industrial Grade Single Fixtures	18	0.67	0.74	0.80	0.84	1.54	1.69	1.83	1.94
	24	0.74	0.81	0.88	0.93	1.73	1.90	2.06	2.18
	36	0.83	0.91	0.99	1.05	1.93	2.12	2.29	2.43
	48	0.91	1.00	1.08	1.15	2.12	2.33	2.52	2.67
	72	1.09	1.20	1.29	1.37	2.50	2.75	2.97	3.15
	96	1.27	1.40	1.51	1.60	2.89	3.18	3.43	3.64
Continuous Row Fixtures	48	0.74	0.81	0.88	0.93	1.73	1.90	2.06	2.18
	72	0.83	0.91	0.99	1.05	1.93	2.12	2.29	2.43
	96	1.00	1.10	1.19	1.26	2.31	2.54	2.74	2.91

All manhours include checking out of job storage, handling, job hauling, distributing, layout, and marking for installation.

Hang-Only manhours only include hanging of fixture.

Assemble-and-Hang manhours include complete assembling of fixture, installation, connecting circuit power to the fixture, testing, and verifying fixture operation.

Manhours exclude installation of conduit, boxes, circuit wiring, switches, tubes, special hangers or supports not supplied with fixture, change out of or separate ballast, and ladder or scaffold set-up. See respective tables for these time frames.

For heights above 25 feet increase manhours 5% for each additional 5 feet or fraction thereof.

FLUORESCENT LIGHTING FIXTURES
INTERIOR CEILING SUSPENDED
2-Tube Fixtures

MANHOURS EACH

Item Description	Tube Length (Inches)	Manhours Required to							
		Hang Only For Height to				Assemble and Hang For Height to			
		10′	15′	20′	25′	10′	15′	20′	25′
Commercial Grade Single Fixtures	18	0.83	0.91	0.99	1.05	1.93	2.12	2.29	2.43
	24	0.91	1.00	1.08	1.15	2.12	2.33	2.52	2.67
	36	1.00	1.10	1.19	1.26	2.31	2.54	2.74	2.91
	48	1.09	1.20	1.29	1.37	2.53	2.78	3.01	3.19
	72	1.26	1.39	1.50	1.59	2.89	3.18	3.43	3.64
	96	1.42	1.56	1.69	1.79	3.27	3.60	3.88	4.12
Continuous Row Fixtures	48	0.91	1.00	1.08	1.15	2.12	2.33	2.52	2.67
	72	1.00	1.10	1.19	1.26	2.31	2.54	2.74	2.91
	96	1.18	1.30	1.40	1.49	2.70	2.97	3.21	3.40
Industrial Grade Single Fixtures	18	0.76	—	—	—	1.73	—	—	—
	24	0.83	0.91	0.99	1.05	1.93	2.12	2.29	2.43
	36	0.91	1.00	1.08	1.15	2.12	2.33	2.52	2.67
	48	1.00	1.10	1.19	1.26	2.31	2.54	2.74	2.91
	72	1.18	1.30	1.40	1.49	2.70	2.97	3.21	3.40
	96	1.33	1.46	1.58	1.67	3.08	3.39	3.66	3.88
Continuous Row Fixtures	48	0.83	0.91	0.99	1.05	1.93	2.12	2.79	2.43
	72	0.91	1.00	1.08	1.15	2.12	2.33	2.52	2.67
	96	1.09	1.20	1.29	1.37	2.53	2.78	3.01	3.19

All manhours include checking out of job storage, handling, job hauling, distributing, layout, and marking for installation.

Hang-Only manhours only include hanging of fixture.

Assemble-and-Hang manhours include complete assembling of fixture, installation, connecting circuit power to the fixture, testing, and verifying fixture operation.

Manhours exclude installation of conduit, boxes, circuit wiring, switches, tubes, special hangers or supports not supplied with fixture, change out of or separate ballast, and ladder or scaffold set-up. See respective tables for these time frames.

For heights above 25 feet increase manhours 5% for each additional 5 feet or fraction thereof.

FLUORESCENT LIGHTING FIXTURES
INTERIOR CEILING SUSPENDED
3-Tube Fixtures

MANHOURS EACH

Item Description	Tube Length (Inches)	Manhours Required to							
		Hang Only For Height to				Assemble and Hang For Height to			
		10'	15'	20'	25'	10'	15'	20'	25'
Commercial Grade Single Fixtures	18	0.91	1.00	1.08	1.15	2.12	2.33	2.52	2.67
	24	1.00	1.10	1.19	1.26	2.31	2.54	2.74	2.91
	36	1.09	1.20	1.29	1.37	2.53	2.78	3.01	3.19
	48	1.18	1.30	1.40	1.49	2.70	2.97	3.21	3.40
	72	1.33	1.46	1.58	1.67	3.08	3.39	3.66	3.88
	96	1.50	1.65	1.78	1.89	3.47	3.82	4.12	4.37
Continuous Row Fixtures	48	1.00	1.10	1.19	1.26	2.31	2.54	2.74	2.91
	72	1.09	1.20	1.29	1.37	2.53	2.78	3.01	3.19
	96	1.26	1.39	1.50	1.59	2.89	3.18	3.43	3.64
Industrial Grade Single Fixtures	18	0.83	0.91	0.99	1.05	1.93	2.12	2.29	2.43
	24	0.91	1.00	1.08	1.15	2.12	2.33	2.52	2.67
	36	1.00	1.10	1.19	1.26	2.31	2.54	2.74	2.91
	48	1.09	1.20	1.29	1.37	2.53	2.78	3.01	3.19
	72	1.25	1.38	1.49	1.57	2.89	3.18	3.43	3.64
	96	1.42	1.56	1.69	1.79	3.27	3.60	3.88	4.12
Continuous Row Fixtures	48	0.91	1.00	1.08	1.15	2.12	2.33	2.52	2.67
	72	1.00	1.10	1.19	1.26	2.31	2.54	2.74	2.91
	96	1.18	1.30	1.40	1.49	2.70	2.97	3.21	3.40

All manhours include checking out of job storage, handling, job hauling, distributing, layout, and marking for installation.

Hang-Only manhours only include hanging of fixture.

Assemble-and-Hang manhours include complete assembling of fixture, installation, connecting circuit power to the fixture, testing, and verifying fixture operation.

Manhours exclude installation of conduit, boxes, circuit wiring, switches, tubes, special hangers or supports not supplied with fixture, change out of or separate ballast, and ladder or scaffold set-up. See respective tables for these time frames.

For heights above 25 feet increase manhours 5% for each additional 5 feet or fraction thereof.

FLUORESCENT LIGHTING FIXTURES
INTERIOR CEILING SUSPENDED
4-Tube Fixtures

MANHOURS EACH

Item Description	Tube Length (Inches)	Manhours Required to							
		Hang Only For Height to				Assemble and Hang For Height to			
		10'	15'	20'	25'	10'	15'	20'	25'
Commercial Grade Single Fixtures	18	1.00	1.10	1.19	1.26	2.31	2.54	2.74	2.91
	24	1.09	1.20	1.29	1.37	2.53	2.78	3.01	3.19
	36	1.18	1.30	1.40	1.49	2.70	2.97	3.21	3.40
	48	1.26	1.39	1.50	1.59	2.89	3.18	3.43	3.64
	72	1.42	1.56	1.69	1.79	3.27	3.60	3.88	4.12
	96	1.57	1.73	1.87	1.98	3.66	4.03	4.35	4.61
Continuous Row Fixtures	48	1.09	1.20	1.29	1.37	2.53	2.78	3.01	3.19
	72	1.18	1.30	1.40	1.49	2.70	2.97	3.21	3.40
	96	1.33	1.46	1.58	1.67	3.08	3.39	3.66	3.88
Industrial Grade Single Fixtures	18	0.91	1.00	1.08	1.15	2.12	2.33	2.52	2.67
	24	1.00	1.10	1.19	1.26	2.31	2.54	2.74	2.91
	36	1.09	1.20	1.29	1.37	2.53	2.78	3.01	3.19
	48	1.18	1.30	1.40	1.49	2.70	2.97	3.21	3.40
	72	1.33	1.46	1.58	1.67	3.08	3.39	3.66	3.88
	96	1.50	1.65	1.78	1.89	3.47	3.82	4.12	4.37
Continuous Row Fixtures	48	1.00	1.10	1.19	1.26	2.31	2.54	2.74	2.91
	72	1.09	1.20	1.29	1.37	2.53	2.78	3.01	3.19
	96	1.26	1.39	1.50	1.59	2.89	3.18	3.43	3.64

All manhours include checking out of job storage, handling, job hauling, distributing, layout, and marking for installation.

Hang-Only manhours only include hanging of fixture.

Assemble-and-Hang manhours include complete assembling of fixture, installation, connecting circuit power to the fixture, testing, and verifying fixture operation.

Manhours exclude installation of conduit, boxes, circuit wiring, switches, tubes, special hangers or supports not supplied with fixture, change out of or separate ballast, and ladder or scaffold set-up. See respective tables for these time frames.

For heights above 25 feet increase manhours 5% for each additional 5 feet or fraction thereof.

FLUORESCENT LIGHTING FIXTURES
INTERIOR CEILING SUSPENDED

6-Tube Fixtures

MANHOURS EACH

Item Description	Tube Length (Inches)	Manhours Required to							
		Hang Only For Height to				Assemble and Hang For Height to			
		10′	15′	20′	25′	10′	15′	20′	25′
Commercial Grade Single Fixtures	36	1.33	1.46	1.58	1.67	3.08	3.39	3.66	3.88
	48	1.42	1.56	1.69	1.79	3.27	3.60	3.88	4.12
	72	1.57	1.73	1.87	1.98	3.66	4.03	4.35	4.61
	96	1.76	1.94	2.09	2.22	4.04	4.44	4.80	5.09
Continuous Row Fixtures	48	1.26	1.39	1.50	1.59	2.89	3.18	3.42	3.64
	72	1.33	1.46	1.58	1.67	3.08	3.39	3.66	3.88
	96	1.50	1.65	1.78	1.89	3.47	3.82	4.12	4.37
Industrial Grade Single Fixtures	36	1.09	1.20	1.29	1.37	2.53	2.78	3.01	3.19
	48	1.18	1.30	1.40	1.49	2.70	2.97	3.21	3.40
	72	1.33	1.46	1.58	1.67	3.08	3.09	3.66	3.88
	96	1.50	1.65	1.78	1.89	3.47	3.82	4.12	4.37
Continuous Row Fixtures	48	1.04	1.14	1.24	1.31	1.76	1.94	2.09	2.22
	72	1.09	1.20	1.29	1.37	2.53	2.78	3.01	3.19
	96	1.26	1.39	1.50	1.59	2.89	3.18	3.43	3.64

All manhours include checking out of job storage, handling, job hauling, distributing, layout, and marking for installation.

Hang-Only manhours only include hanging of fixture.

Assemble-and-Hang manhours include complete assembling of fixture, installation, connecting circuit power to the fixture, testing, and verifying fixture operation.

Manhours exclude installation of conduit, boxes, circuit wiring, switches, tubes, special hangers or supports not supplied with fixture, change out of or separate ballast, and ladder or scaffold set-up. See respective tables for these time frames.

For heights above 25 feet increase manhours 5% for each additional 5 feet or fraction thereof.

FLUORESCENT LIGHTING FIXTURES
INTERIOR CEILING SUSPENDED
8-Tube Fixtures

MANHOURS EACH

Item Description	Tube Length (Inches)	Manhours Required to							
		Hang Only For Height to				Assemble and Hang For Height to			
		10′	15′	20′	25′	10′	15′	20′	25′
Commercial Grade Single Fixtures	48	1.57	1.73	1.87	1.98	3.66	4.03	4.35	4.61
	72	1.76	1.94	2.09	2.22	4.04	4.44	4.80	5.09
	96	1.91	2.10	2.26	2.39	4.43	4.87	5.26	5.58
Continuous Row Fixtures	48	1.42	1.56	1.69	1.79	3.27	3.60	3.88	4.12
	72	1.50	1.65	1.78	1.89	3.47	3.82	4.12	4.37
	96	1.67	1.84	1.98	2.10	3.85	4.24	4.57	4.85
Industrial Grade Single Fixtures	48	1.50	1.65	1.78	1.89	3.47	3.82	4.12	4.37
	72	1.67	1.84	1.98	2.10	3.85	4.24	4.57	4.85
	96	1.83	2.01	2.17	2.30	4.24	4.66	5.04	5.34
Continuous Row Fixtures	48	1.33	1.46	1.58	1.67	3.08	3.39	3.66	3.88
	72	1.42	1.56	1.69	1.79	3.27	3.60	3.88	4.12
	96	1.57	1.73	1.87	1.98	3.66	4.03	4.35	4.61

All manhours include checking out of job storage, handling, job hauling, distributing, layout, and marking for installation.

Hang-Only manhours only include hanging of fixture.

Assemble-and-Hang manhours include complete assembling of fixture, installation, connecting circuit power to the fixture, testing, and verifying fixture operation.

Manhours exclude installation of conduit, boxes, circuit wiring, switches, tubes, special hangers or supports not supplied with fixture, change out of or separate ballast, and ladder or scaffold set-up. See respective tables for these time frames.

For heights above 25 feet increase manhours 5% for each additional 5 feet or fraction thereof.

FLUORESCENT LIGHTING FIXTURES
INTERIOR RECESSED MOUNTED
1- and 2-Tube Fixtures

MANHOURS EACH

Item Description	Tube Length (Inches)	Manhours Required to							
		Hang Only For Height to				Assemble and Hang For Height to			
		10'	15'	20'	25'	10'	15'	20'	25'
Commercial Grade Single Fixtures 1-Tube	18	0.83	0.91	0.99	1.05	1.93	2.12	2.29	3.43
	24	0.91	1.00	1.08	1.15	2.12	2.33	2.52	2.67
	36	1.00	1.10	1.19	1.26	2.31	2.54	2.74	2.91
	48	1.09	1.20	1.29	1.37	2.50	2.75	2.97	3.15
	72	1.26	1.39	1.50	1.59	2.89	3.18	3.42	3.64
	96	1.42	1.56	1.69	1.79	3.27	3.60	3.88	4.12
Continuous Row Fixtures 1-Tube	48	0.91	1.00	1.08	1.15	2.12	2.33	2.52	2.67
	72	1.00	1.10	1.19	1.26	2.31	2.54	2.74	2.91
	96	1.17	1.29	1.39	1.47	2.70	2.97	3.21	3.40
Commercial Grade Single Fixtures 2-Tube	18	0.91	1.00	1.08	1.15	1.57	1.73	1.87	1.98
	24	1.00	1.10	1.19	1.26	2.31	2.54	2.74	2.91
	36	1.09	1.20	1.29	1.37	2.53	2.78	3.01	3.19
	48	1.18	1.30	1.40	1.49	2.70	2.97	3.21	3.40
	72	1.33	1.46	1.58	1.67	3.08	3.39	3.66	3.88
	96	1.50	1.65	1.78	1.89	3.47	3.82	4.12	4.37
Continuous Row Fixtures 2-Tube	48	1.00	1.10	1.19	1.26	2.31	2.54	2.74	2.91
	72	1.09	1.20	1.29	1.37	2.53	2.78	3.01	3.19
	96	1.26	1.39	1.50	1.59	2.89	3.18	3.42	3.64

All manhours include checking out of job storage, handling, job hauling, distributing, layout, and marking for installation.

Hang-Only manhours only include hanging of fixture.

Assemble-and-Hang manhours include complete assembling of fixture, installation, connecting circuit power to the fixture, testing, and verifying fixture operation.

Manhours exclude installation of conduit, boxes, circuit wiring, switches, tubes, special hangers or supports not supplied with fixture, change out of or separate ballast, and ladder or scaffold set-up. See respective tables for these time frames.

For heights above 25 feet increase manhours 5% for each additional 5 feet or fraction thereof.

FLUORESCENT LIGHTING FIXTURES
INTERIOR RECESSED MOUNTED
3- and 4-Tube Fixtures

MANHOURS EACH

Item Description	Tube Length (Inches)	Manhours Required to							
		Hang Only For Height to				Assemble and Hang For Height to			
		10'	15'	20'	25'	10'	15'	20'	25'
Commercial Grade Single Fixtures 3-Tube	18	1.00	1.10	1.19	1.26	1.76	1.94	2.09	2.22
	24	1.09	1.20	1.29	1.37	2.53	2.78	3.01	3.19
	36	1.18	1.30	1.40	1.49	2.70	2.97	3.21	3.40
	48	1.26	1.39	1.50	1.59	2.89	3.18	3.43	3.64
	72	1.42	1.56	1.69	1.79	3.27	3.60	3.88	4.12
	96	1.57	1.73	1.87	1.98	3.68	4.05	4.37	4.63
Continuous Row Fixtures 3-Tube	48	1.09	1.20	1.29	1.37	2.53	2.78	3.01	3.19
	72	1.18	1.30	1.40	1.49	2.70	2.97	3.21	3.40
	96	1.33	1.46	1.58	1.67	3.08	3.39	3.66	3.88
Commercial Grade Single Fixtures 4-Tube	18	1.09	1.20	1.29	1.37	2.53	2.78	3.01	3.19
	24	1.18	1.30	1.40	1.49	2.70	2.97	3.21	3.40
	36	1.26	1.39	1.50	1.59	2.89	3.18	3.43	3.64
	48	1.33	1.46	1.58	1.67	3.08	3.39	3.66	3.88
	72	1.50	1.65	1.78	1.89	3.47	3.82	4.12	4.37
	96	1.67	1.84	1.98	2.10	3.85	4.24	4.57	4.85
Continuous Row Fixtures 4-Tube	48	1.18	1.30	1.40	1.49	2.70	2.97	3.21	3.40
	72	1.26	1.39	1.50	1.59	2.89	3.18	3.43	3.64
	96	1.42	1.56	1.69	1.79	3.27	3.60	3.88	4.12

All manhours include checking out of job storage, handling, job hauling, distributing, layout, and marking for installation.

Hang-Only manhours only include hanging of fixture.

Assemble-and-Hang manhours include complete assembling of fixture, installation, connecting circuit power to the fixture, testing, and verifying fixture operation.

Manhours exclude installation of conduit, boxes, circuit wiring, switches, tubes, special hangers or supports not supplied with fixture, change out of or separate ballast, and ladder or scaffold set-up. See respective tables for these time frames.

For heights above 25 feet increase manhours 5% for each additional 5 feet or fraction thereof.

FLUORESCENT LIGHTING FIXTURES
INTERIOR RECESSED MOUNTED
6- and 8-Tube Fixtures

MANHOURS EACH

Item Description	Tube Length (Inches)	Manhours Required to							
		Hang Only For Height to				Assemble and Hang For Height to			
		10'	15'	20'	25'	10'	15'	20'	25'
Commercial Grade Single Fixtures 6-Tube	36	1.42	1.56	1.69	1.79	3.27	3.60	3.88	4.12
	48	1.50	1.65	1.78	1.89	3.63	3.99	4.31	4.57
	72	1.67	1.84	1.98	2.10	3.85	4.24	4.57	4.85
	96	1.83	2.01	2.17	2.30	4.24	4.66	5.04	5.34
Continuous Row Fixtures 6-Tube	48	1.33	1.46	1.58	1.67	3.08	3.39	3.66	3.88
	72	1.42	1.56	1.69	1.79	3.27	3.60	3.88	4.12
	96	1.57	1.73	1.87	1.98	3.66	4.03	4.35	4.61
Commercial Grade Single Fixtures 8-Tube	36	–	–	–	–	–	–	–	–
	48	1.67	1.84	1.98	2.10	3.85	4.24	4.57	4.85
	72	1.83	2.01	2.17	2.30	4.24	4.66	5.04	5.34
	96	2.00	2.20	2.37	2.51	4.62	5.08	5.49	5.82
Continuous Row Fixtures 8-Tube	48	1.50	1.65	1.78	1.89	3.47	3.82	4.12	4.37
	72	1.57	1.73	1.87	1.98	3.67	4.04	4.36	4.62
	96	1.79	1.96	2.11	2.23	4.04	4.44	4.80	5.09

All manhours include checking out of job storage, handling, job hauling, distributing, layout, and marking for installation.

Hang-Only manhours only include hanging of fixture.

Assemble-and-Hang manhours include complete assembling of fixture, installation, connecting circuit power to the fixture, testing, and verifying fixture operation.

Manhours exclude installation of conduit, boxes, circuit wiring, switches, tubes, special hangers or supports not supplied with fixture, change out of or separate ballast, and ladder or scaffold set-up. See respective tables for these time frames.

For heights above 25 feet increase manhours 5% for each additional 5 feet or fraction thereof.

FLUORESCENT LIGHTING FIXTURES
INTERIOR WALL OR COVE MOUNTED
1- and 2-Tube Fixtures

MANHOURS EACH

Item Description	Tube Length (Inches)	Manhours Required to							
		Hang Only For Height to				Assemble and Hang For Height to			
		10'	15'	20'	25'	10'	15'	20'	25'
Commercial Grade Single Fixtures 1-Tube	18	0.50	0.55	0.59	0.63	1.16	1.28	1.38	1.46
	24	0.57	0.63	0.68	0.72	1.32	1.45	1.57	1.66
	36	0.67	0.74	0.80	0.84	1.54	1.69	1.83	1.94
	48	0.74	0.81	0.88	0.93	1.73	1.90	2.06	2.18
	72	0.91	1.00	1.08	1.15	2.12	2.33	2.52	2.67
	96	1.09	1.20	1.29	1.37	2.50	2.75	2.97	3.15
Continuous Row Fixtures 1-Tube	48	0.57	0.63	0.68	0.72	1.32	1.45	1.57	1.66
	72	0.67	0.74	0.80	0.84	1.54	1.69	1.83	1.94
	96	0.83	0.91	0.99	1.05	1.93	2.12	2.29	2.43
Commercial Grade Single Fixtures 2-Tube	18	0.57	0.63	0.68	0.72	1.32	1.45	1.57	1.66
	24	0.67	0.74	0.80	0.84	1.54	1.69	1.83	1.94
	36	0.76	0.84	0.90	0.96	1.73	1.90	2.06	2.18
	48	0.83	0.91	0.99	1.05	1.93	2.12	2.29	2.43
	72	1.00	1.10	1.19	1.26	2.31	2.54	2.74	2.91
	96	1.18	1.30	1.40	1.49	2.70	2.97	3.21	3.40
Continuous Row Fixtures 2-Tube	48	0.67	0.74	0.80	0.84	1.54	1.69	1.83	1.94
	72	0.76	0.84	0.90	0.96	1.73	1.90	2.06	2.18
	96	0.91	1.00	1.08	1.15	2.12	2.33	2.52	2.67

All manhours include checking out of job storage, handling, job hauling, distributing, layout, and marking for installation.

Hang-Only manhours only include hanging of fixture.

Assemble-and-Hang manhours include complete assembling of fixture, installation, connecting circuit power to the fixture, testing, and verifying fixture operation.

Manhours exclude installation of conduit, boxes, circuit wiring, switches, tubes, special hangers or supports not supplied with fixture, change out of or separate ballast, and ladder or scaffold set-up. See respective tables for these time frames.

For heights above 25 feet increase manhours 5% for each additional 5 feet or fraction thereof.

FLUORESCENT LIGHTING FIXTURES
INTERIOR WALL OR COVE MOUNTED

3- and 4-Tube Fixtures

MANHOURS EACH

Item Description	Tube Length (Inches)	Manhours Required to							
		Hang Only For Height to				Assemble and Hang For Height to			
		10'	15'	20'	25'	10'	15'	20'	25'
Commercial Grade Single Fixtures 3-Tube	18	0.67	0.74	0.80	0.84	1.54	1.69	1.83	1.94
	24	0.76	0.84	0.90	0.96	1.73	1.90	2.06	2.18
	36	0.83	0.91	0.99	1.05	1.93	2.12	2.29	2.43
	48	0.91	1.00	1.08	1.15	2.12	2.33	2.52	2.67
	72	1.09	1.20	1.29	1.37	2.53	2.78	3.01	3.19
	96	1.26	1.39	1.50	1.59	2.89	3.18	3.43	3.64
Continuous Row Fixtures 3-Tube	48	0.76	0.84	0.90	0.96	1.73	1.90	2.06	2.18
	72	0.83	0.91	0.99	1.05	1.93	2.12	2.29	2.43
	96	1.00	1.10	1.19	1.26	2.31	2.54	2.74	2.91
Commercial Grade Single Fixtures 4-Tube	18	0.76	0.84	0.90	0.96	1.73	1.90	2.06	2.18
	24	0.83	0.91	0.90	1.05	1.93	2.12	2.29	2.43
	36	0.91	1.00	1.08	1.15	2.12	2.33	2.52	2.67
	48	1.00	1.10	1.19	1.26	2.31	2.54	2.74	2.91
	72	1.18	1.30	1.40	1.49	2.70	2.97	3.21	3.40
	96	1.33	1.46	1.58	1.67	3.08	3.39	3.66	3.88
Continuous Row Fixtures 4-Tube	48	0.83	0.91	0.99	1.05	1.93	2.12	2.29	2.43
	72	0.91	1.00	1.08	1.15	2.12	2.33	2.52	2.67
	96	1.09	1.20	1.29	1.37	2.53	2.78	3.01	3.19

All manhours include checking out of job storage, handling, job hauling, distributing, layout, and marking for installation.

Hang-Only manhours only include hanging of fixture.

Assemble-and-Hang manhours include complete assembling of fixture, installation, connecting circuit power to the fixture, testing, and verifying fixture operation.

Manhours exclude installation of conduit, boxes, circuit wiring, switches, tubes, special hangers or supports not supplied with fixture, change out of or separate ballast, and ladder or scaffold set-up. See respective tables for these time frames.

For heights above 25 feet increase manhours 5% for each additional 5 feet or fraction thereof.

FLUORESCENT LIGHTING FIXTURES
INTERIOR WALL AND COVE MOUNTED
6- and 8-Tube Fixtures

MANHOURS EACH

Item Description	Tube Length (Inches)	Manhours Required to							
		Hang Only For Height to				Assemble and Hang For Height to			
		10′	15′	20′	25′	10′	15′	20′	25′
Commercial Grade Single Fixtures 6-Tube	36	1.09	1.20	1.29	1.37	2.53	2.78	3.01	3.19
	48	1.18	1.30	1.40	1.49	2.70	2.97	3.21	3.40
	72	1.33	1.46	1.58	1.67	3.08	3.39	3.66	3.88
	96	1.50	1.65	1.78	1.89	3.47	3.82	4.12	4.37
Continuous Row Fixtures 6-Tube	48	1.00	1.10	1.19	1.26	2.31	2.54	2.74	2.91
	72	1.09	1.20	1.29	1.37	2.53	2.78	3.01	3.19
	96	1.26	1.39	1.50	1.59	2.89	3.18	3.43	3.64
Commercial Grade Single Fixtures 8-Tube	36	–	–	–	–	–	–	–	–
	48	1.33	1.46	1.58	1.67	3.08	3.33	3.60	3.82
	72	1.50	1.65	1.78	1.89	3.47	3.82	4.12	4.37
	96	1.67	1.84	1.98	2.10	3.85	4.24	4.57	4.85
Continuous Row Fixtures 8-Tube	48	1.18	1.30	1.40	1.49	2.70	2.97	3.21	3.40
	72	1.26	1.39	1.50	1.59	2.89	3.18	3.43	3.64
	96	1.42	1.56	1.69	1.79	3.27	3.60	3.88	4.12

All manhours include checking out of job storage, handling, job hauling, distributing, layout, and marking for installation.

Hang-Only manhours only include hanging of fixture.

Assemble-and-Hang manhours include complete assembling of fixture, installation, connecting circuit power to the fixture, testing, and verifying fixture operation.

Manhours exclude installation of conduit, boxes, circuit wiring, switches, tubes, special hangers or supports not supplied with fixture, change out of or separate ballast, and ladder or scaffold set-up. See respective tables for these time frames.

For heights above 25 feet increase manhours 5% for each additional 5 feet or fraction thereof.

FLUORESCENT LIGHTING FIXTURES
INTERIOR FLUSH MOUNTED AIR HANDLING
1- and 2-Tube Fixtures

MANHOURS EACH

Item Description	Tube Length (Inches)	Manhours Required to							
		Hang Only For Height to				Assemble and Hang For Height to			
		10'	15'	20'	25'	10'	15'	20'	25'
Commercial Grade Single Fixtures 1-Tube	18	1.00	1.10	1.19	1.26	2.31	2.54	2.74	2.91
	24	1.09	1.20	1.29	1.37	2.50	2.75	2.97	3.15
	36	1.17	1.29	1.39	1.47	2.70	2.97	3.21	3.40
	48	1.26	1.39	1.50	1.59	2.89	3.18	3.43	3.64
	72	1.42	1.56	1.69	1.79	3.27	3.60	3.88	4.12
	96	1.59	1.75	1.89	2.00	3.66	4.03	4.35	4.61
Continuous Row Fixtures 1-Tube	48	1.09	1.20	1.29	1.37	2.50	2.75	2.97	3.15
	72	1.17	1.29	1.39	1.47	2.70	2.97	3.21	3.40
	96	1.33	1.46	1.58	1.67	3.08	3.39	3.66	3.88
Commercial Grade Single Fixtures 2-Tube	18	1.09	1.20	1.29	1.37	2.53	–	–	–
	24	1.18	–	–	–	2.70	2.97	3.21	3.40
	36	1.26	1.39	1.50	1.59	2.89	3.18	3.43	3.64
	48	1.33	1.46	1.58	1.67	3.08	3.39	3.66	3.88
	72	1.50	1.65	1.78	1.89	3.47	3.82	4.12	4.37
	96	1.67	1.84	1.98	2.10	3.85	4.24	4.57	4.85
Continuous Row Fixtures 2-Tube	48	1.18	1.30	1.40	1.49	2.70	2.97	3.21	2.40
	72	1.26	1.39	1.50	1.59	2.89	3.18	3.43	3.64
	96	1.42	1.56	1.69	1.79	3.27	3.60	3.88	4.12

All manhours include checking out of job storage, handling, job hauling, distributing, layout, and marking for installation.

Hang-Only manhours only include hanging of fixture.

Assemble-and-Hang manhours include complete assembling of fixture, installation, connecting circuit power to the fixture, testing, and verifying fixture operation.

Manhours exclude installation of conduit, boxes, circuit wiring, switches, tubes, special hangers or supports not supplied with fixture, change out of or separate ballast, and ladder or scaffold set-up. See respective tables for these time frames.

For heights above 25 feet increase manhours 5% for each additional 5 feet or fraction thereof.

FLUORESCENT LIGHTING FIXTURES
INTERIOR FLUSH MOUNTED AIR HANDLING

3- and 4-Tube Fixtures

MANHOURS EACH

Item Description	Tube Length (Inches)	Manhours Required to							
		Hang Only For Height to				Assemble and Hang For Height to			
		10'	15'	20'	25'	10'	15'	20'	25'
Commercial Grade Single Fixtures 3-Tube	18	1.18	1.30	1.40	1.49	2.70	2.97	3.21	3.40
	24	1.26	1.39	1.50	1.59	2.89	3.18	3.43	3.64
	36	1.33	1.46	1.58	1.67	3.08	3.39	3.66	3.88
	48	1.42	1.56	1.69	1.79	3.27	3.60	3.88	4.12
	72	1.57	1.73	1.87	1.98	3.66	4.03	4.35	4.61
	96	1.76	–	–	–	4.04	4.44	4.80	5.09
Continuous Row Fixtures 3-Tube	48	1.26	1.39	1.50	1.59	2.89	3.18	3.43	3.64
	72	1.33	1.46	1.58	1.67	3.08	3.39	3.66	3.88
	96	1.50	1.65	1.78	1.89	3.47	3.82	4.12	4.37
Commercial Grade Single Fixtures 4-Tube	18	1.26	1.39	1.50	1.59	2.89	3.18	3.43	3.64
	24	1.33	1.46	1.58	1.67	3.08	3.39	3.66	3.88
	36	1.42	1.56	1.69	1.79	3.27	3.60	3.88	4.12
	48	1.50	1.65	1.78	1.89	3.47	3.82	4.12	4.37
	72	1.67	1.84	1.98	2.10	3.85	4.24	4.57	4.85
	96	1.83	–	–	–	4.23	4.65	5.03	5.33
Continuous Row Fixtures 4-Tube	48	1.33	1.46	1.58	1.67	3.08	3.39	3.66	3.88
	72	1.42	1.56	1.69	1.79	3.27	3.60	3.88	4.12
	96	1.57	1.73	1.87	1.98	3.66	4.03	4.35	4.61

All manhours include checking out of job storage, handling, job hauling, distributing, layout, and marking for installation.

Hang-Only manhours only include hanging of fixture.

Assemble-and-Hang manhours include complete assembling of fixture, installation, connecting circuit power to the fixture, testing, and verifying fixture operation.

Manhours exclude installation of conduit, boxes, circuit wiring, switches, tubes, special hangers or supports not supplied with fixture, change out of or separate ballast, and ladder or scaffold set-up. See respective tables for these time frames.

For heights above 25 feet increase manhours 5% for each additional 5 feet or fraction thereof.

FLUORESCENT LIGHTING FIXTURES
INTERIOR FLUSH MOUNTED
6- and 8-Tube Fixtures

MANHOURS EACH

Item Description	Tube Length (Inches)	Manhours Required to							
		Hang Only For Height to				Assemble and Hang For Height to			
		10'	15'	20'	25'	10'	15'	20'	25'
Commercial Grade Single Fixtures 6-Tube	36	1.57	1.73	1.87	1.98	3.66	4.03	4.35	4.61
	48	1.67	1.84	1.98	2.10	3.85	4.24	4.57	4.85
	72	1.83	2.01	2.17	2.30	4.24	4.66	5.04	5.34
	96	2.00	2.20	2.37	2.51	4.62	5.08	5.49	5.82
Continuous Row Fixtures	48	1.50	1.65	1.78	1.89	3.47	3.82	4.12	4.37
	72	1.57	1.73	1.87	1.98	3.66	4.03	4.35	4.61
	96	1.79	1.96	2.11	2.23	4.04	4.44	4.80	5.09
Commercial Grade Single Fixtures 8-Tube	36	–	–	–	–	–	–	–	–
	48	1.83	2.01	2.17	2.30	4.24	4.66	5.04	5.34
	72	2.00	2.20	2.37	2.51	4.62	5.08	5.49	5.82
	96	2.16	2.37	2.55	2.70	5.00	5.50	5.94	6.30
Continuous Row Fixtures 8-Tube	48	1.67	1.84	1.98	2.10	3.85	4.24	4.57	4.85
	72	1.76	1.94	2.09	2.22	4.04	4.44	4.80	5.09
	96	1.91	2.10	2.26	2.39	4.43	4.87	5.26	5.58

All manhours include checking out of job storage, handling, job hauling, distributing, layout, and marking for installation.

Hang-Only manhours only include hanging of fixture.

Assemble-and-Hang manhours include complete assembling of fixture, installation, connecting circuit power to the fixture, testing, and verifying fixture operation.

Manhours exclude installation of conduit, boxes, circuit wiring, switches, tubes, special hangers or supports not supplied with fixture, change out of or separate ballast, and ladder or scaffold set-up. See respective tables for these time frames.

For heights above 25 feet increase manhours 5% for each additional 5 feet or fraction thereof.

FLUORESCENT LIGHTING FIXTURES
INTERIOR VAPORTITE

1- and 2-Tube Fixtures

MANHOURS EACH

Item Description	Tube Length (Inches)	Manhours Required to							
		Hang Only For Height to				Assemble and Hang For Height to			
		10'	15'	20'	25'	10'	15'	20'	25'
Industrial Grade Single Fixtures 1-Tube	18	1.09	1.20	1.29	1.37	2.50	2.75	2.97	3.15
	24	1.19	1.30	1.40	1.49	2.70	2.97	3.21	3.40
	36	1.26	1.39	1.50	1.59	2.89	3.18	3.43	3.64
	48	1.33	1.46	1.58	1.67	3.08	3.39	3.66	3.88
	72	1.50	1.65	1.78	1.89	3.30	3.63	3.92	4.16
	96	1.67	1.84	1.98	2.10	3.85	4.24	4.57	4.85
Continuous Row Fixtures 1-Tube	48	1.18	1.30	1.40	1.49	2.70	2.97	3.21	3.40
	72	1.26	1.39	1.50	1.59	2.89	3.18	3.43	3.64
	96	1.42	1.56	1.69	1.79	3.27	3.60	3.88	4.12
Industrial Grade Single Fixtures 2-Tube	18	1.18	1.30	1.40	1.49	2.70	2.97	3.21	3.40
	24	1.26	1.39	1.50	1.59	2.89	3.18	3.43	3.64
	36	1.33	1.46	1.58	1.67	3.08	3.39	3.66	3.88
	48	1.42	1.56	1.69	1.79	3.27	3.60	3.88	4.17
	72	1.57	1.73	1.87	1.98	3.66	4.03	4.35	4.61
	96	1.76	—	—	—	4.04	4.44	4.80	5.09
Continuous Row Fixtures 2-Tube	48	1.26	1.39	1.50	1.59	2.89	3.18	3.43	3.64
	72	1.33	1.46	1.58	1.67	3.08	3.39	3.66	3.88
	96	1.50	1.65	1.78	1.89	3.47	3.82	4.12	4.37

All manhours include checking out of job storage, handling, job hauling, distributing, layout, and marking for installation.

Hang-Only manhours only include hanging of fixture.

Assemble-and-Hang manhours include complete assembling of fixture, installation, connecting circuit power to the fixture, testing, and verifying fixture operation.

Manhours exclude installation of conduit, boxes, circuit wiring, switches, tubes, special hangers or supports not supplied with fixture, change out of or separate ballast, and ladder or scaffold set-up. See respective tables for these time frames.

For heights above 25 feet increase manhours 5% for each additional 5 feet or fraction thereof.

FLUORESCENT LIGHTING FIXTURES
INTERIOR VAPORTITE
3- and 4-Tube Fixtures

MANHOURS EACH

Item Description	Tube Length (Inches)	Manhours Required to							
		Hang Only For Height to				Assemble and Hang For Height to			
		10′	15′	20′	25′	10′	15′	20′	25′
Industrial Grade Single Fixtures 3-Tube	18	1.26	1.39	1.50	1.59	2.89	3.18	3.43	3.64
	24	1.33	1.46	1.58	1.67	3.08	3.39	3.66	3.88
	36	1.42	1.56	1.69	1.79	3.27	3.60	3.88	4.12
	48	1.50	1.65	1.78	1.89	3.47	3.82	4.12	4.37
	72	1.67	1.84	1.98	2.10	3.85	4.24	4.57	4.85
	96	1.83	2.01	2.17	2.30	4.24	4.66	5.04	5.34
Continuous Row Fixtures 3-Tube	48	1.33	1.46	1.58	1.67	3.08	3.39	3.66	3.88
	72	1.42	1.56	1.69	1.79	3.27	3.60	3.88	4.12
	96	1.57	1.73	1.87	1.98	3.66	4.03	4.35	4.61
Industrial Grade Single Fixtures 4-Tube	18	1.33	1.46	1.58	1.67	3.08	3.39	3.66	3.88
	24	1.42	1.56	1.69	1.79	3.27	3.60	3.88	4.12
	36	1.50	1.65	1.78	1.89	3.47	3.82	4.12	4.37
	48	1.61	1.77	1.91	2.03	3.66	4.03	4.35	4.61
	73	1.76	1.94	2.09	2.22	4.04	4.44	4.80	5.09
	96	1.91	2.10	2.26	2.39	4.43	4.87	5.26	5.58
Continuous Row Fixtures 4-Tube	48	1.42	1.56	1.69	1.79	3.27	3.60	3.88	4.12
	72	1.50	1.65	1.78	1.89	3.47	3.82	4.12	4.37
	96	1.67	1.84	1.98	2.10	3.85	4.24	4.57	4.85

All manhours include checking out of job storage, handling, job hauling, distributing, layout, and marking for installation.

Hang-Only manhours only include hanging of fixture.

Assemble-and-Hang manhours include complete assembling of fixture, installation, connecting circuit power to the fixture, testing, and verifying fixture operation.

Manhours exclude installation of conduit, boxes, circuit wiring, switches, tubes, special hangers or supports not supplied with fixture, change out of or separate ballast, and ladder or scaffold set-up. See respective tables for these time frames.

For heights above 25 feet increase manhours 5% for each additional 5 feet or fraction thereof.

FLUORESCENT LIGHTING FIXTURES
INTERIOR VAPORTITE
6- and 8-Tube Fixtures

MANHOURS EACH

Item Description	Tube Length (Inches)	Manhours Required to							
		Hang Only For Height to				Assemble and Hang For Height to			
		10'	15'	20'	25'	10'	15'	20'	25'
Industrial Grade Single Fixtures 6-Tube	36	1.50	1.65	1.78	1.89	3.47	3.82	4.12	4.37
	48	1.57	1.73	1.87	1.98	3.66	4.03	4.35	4.61
	72	1.76	1.94	2.09	2.22	4.04	4.44	4.80	5.09
	96	1.91	2.10	2.26	2.39	4.43	4.87	5.26	5.58
Continuous Row Fixtures 6-Tube	48	1.42	1.56	1.69	1.79	3.27	3.60	3.88	4.12
	72	1.50	1.65	1.78	1.89	3.47	3.82	4.12	4.37
	96	1.67	1.84	1.98	2.10	3.85	4.24	4.57	4.85
Industrial Grade Single Fixtures 8-Tube	36	–	–	–	–	–	–	–	–
	48	1.91	2.10	2.26	2.39	4.43	4.87	5.26	5.58
	72	2.09	2.30	2.48	2.63	4.81	5.29	5.71	6.06
	96	2.24	2.46	2.65	2.80	5.20	5.72	6.18	6.55
Continuous Row Fixtures 8-Tube	48	1.76	1.94	2.09	2.22	4.04	4.44	4.80	5.09
	72	1.83	2.01	2.17	2.30	4.24	4.66	5.04	5.34
	96	2.00	2.20	2.37	2.51	4.62	5.08	5.49	5.82

All manhours include checking out of job storage, handling, job hauling, distributing, layout, and marking for installation.

Hang-Only manhours only include hanging of fixture.

Assemble-and-Hang manhours include complete assembling of fixture, installation, connecting circuit power to the fixture, testing, and verifying fixture operation.

Manhours exclude installation of conduit, boxes, circuit wiring, switches, tubes, special hangers or supports not supplied with fixture, change out of or separate ballast, and ladder or scaffold set-up. See respective tables for these time frames.

For heights above 25 feet increase manhours 5% for each additional 5 feet or fraction thereof.

FLUORESCENT LIGHTING FIXTURES
INTERIOR CLASS I EXPLOSION PROOF
1- and 2-Tube Fixtures

MANHOURS EACH

Item Description	Tube Length (Inches)	Manhours Required to							
		Hang Only For Height to				Assemble and Hang For Height to			
		10'	15'	20'	25'	10'	15'	20'	25'
Industrial Grade Single Fixtures 1-Tube	18	1.26	1.39	1.50	1.59	2.89	3.18	3.43	3.64
	24	1.33	1.46	1.58	1.67	3.08	3.39	3.66	3.88
	36	1.42	1.56	1.69	1.79	3.27	3.60	3.88	4.12
	48	1.50	1.65	1.78	1.89	3.47	3.82	4.12	4.37
	72	1.67	1.84	1.98	2.10	3.85	4.24	4.57	4.85
	96	1.83	2.01	2.17	2.30	4?24	4.66	5.04	5.34
Industrial Grade Single Fixtures 2-Tube	18	1.33	1.46	1.58	1.67	3.08	3.39	3.66	3.88
	24	1.42	1.56	1.69	1.79	3.27	3.60	3.88	4.12
	36	1.50	1.65	1.78	1.89	3.47	3.82	4.12	4.37
	48	1.57	1.73	1.87	1.98	3.66	4.03	4.35	4.61
	72	1.79	1.96	2.11	2.23	4.04	4.44	4.80	5.09
	96	1.91	2.10	2.26	2.39	4.43	4.87	5.26	5.58

All manhours include checking out of job storage, handling, job hauling, distributing, layout, and marking for installation.

Hang-Only manhours only include hanging of fixture.

Assemble-and-Hang manhours include complete assembling of fixture, installation, connecting circuit power to the fixture, testing, and verifying fixture operation.

Manhours exclude installation of conduit, boxes, circuit wiring, switches, tubes, special hangers or supports not supplied with fixture, change out of or separate ballast, and ladder or scaffold set-up. See respective tables for these time frames.

For heights above 25 feet increase manhours 5% for each additional 5 feet or fraction thereof.

FLUORESCENT LIGHTING FIXTURES
INTERIOR CLASS II EXPLOSION PROOF
3- and 4-Tube Fixtures

MANHOURS EACH

Item Description	Tube Length (Inches)	Manhours Required to							
		Hang Only For Height to				Assemble and Hang For Height to			
		10′	15′	20′	25′	10′	15′	20′	25′
Industrial Grade Single Fixtures 3-Tube	18	1.42	1.56	1.69	1.79	3.27	3.60	3.88	4.12
	24	1.50	1.65	1.78	1.89	3.47	3.82	4.12	4.37
	36	1.57	1.73	1.87	1.98	3.66	4.03	4.35	4.61
	48	1.67	1.84	1.98	2.10	3.85	4.24	4.57	4.85
	72	1.83	2.01	2.17	2.30	4.24	4.66	5.04	5.34
	96	2.00	2.20	2.37	2.51	4.62	5.08	5.49	5.82
Industrial Grade Single Fixtures 4-Tube	18	1.50	1.65	1.78	1.89	3.47	3.82	4.12	4.37
	24	1.57	1.73	1.87	1.98	3.66	4.03	4.35	4.61
	36	1.67	1.84	1.98	2.10	3.85	4.24	4.57	4.85
	48	1.79	1.96	2.11	2.23	4.04	4.44	4.80	5.09
	72	1.91	2.10	2.26	2.39	4.43	4.87	5.26	5.58
	96	2.09	2.30	2.48	2.63	4.81	5.29	5.71	6.06

6- and 8 Tube Fixtures

MANHOURS EACH

Item Description	Tube Length (Inches)	Manhours Required to							
		Hang Only For Height to				Assemble and Hang For Height to			
		10′	15′	20′	25′	10′	15′	20′	25′
Industrial Grade Single Fixtures 6-Tube	36	1.83	2.01	2.17	2.30	4.24	4.66	5.04	5.34
	48	1.91	2.10	2.26	2.39	4.43	4.87	5.26	5.58
	72	2.09	2.30	2.48	2.63	4.81	5.29	5.71	6.06
	96	2.24	2.46	2.65	2.80	5.20	5.72	6.18	6.55
Industrial Grade Single Fixtures 8-Tube	48	2.09	2.30	2.48	2.63	4.81	5.29	5.71	6.06
	72	2.24	2.46	2.65	2.80	5.20	5.72	6.18	6.55
	96	2.42	2.66	2.87	3.05	5.58	6.14	6.63	7.03

All manhours include checking out of job storage, handling, job hauling, distributing, layout, and marking for installation.

Hang-Only manhours only include hanging of fixture.

Assemble-and-Hang manhours include complete assembling of fixture, installation, connecting circuit power to the fixture, testing, and verifying fixture operation.

Manhours exclude installation of conduit, boxes, circuit wiring, switches, tubes, special hangers or supports not supplied with fixture, change out of or separate ballast, and ladder or scaffold set-up. See respective tables for these time frames.

For heights above 25 feet increase manhours 5% for each additional 5 feet or fraction thereof.

FLUORESCENT LIGHTING FIXTURES
INTERIOR CLASS I EXPLOSION PROOF
1- and 2-Tube Fixtures

MANHOURS EACH

Item Description	Tube Length (Inches)	Manhours Required to							
		Hang Only For Height to				Assemble and Hang For Height to			
		10′	15′	20′	25′	10′	15′	20′	25′
Industrial Grade Single Fixtures 1-Tube	18	1.42	1.56	1.69	1.79	3.27	3.60	3.88	4.12
	24	1.50	1.65	1.78	1.89	3.47	3.82	4.12	4.37
	36	1.57	1.73	1.87	1.98	3.66	4.03	4.35	4.61
	48	1.67	1.84	1.98	2.10	3.85	4.24	4.57	4.85
	72	1.83	2.01	2.17	2.30	4.24	4.66	5.04	5.34
	96	2.00	2.20	2.37	2.51	4.62	5.08	5.49	5.82
Industrial Grade Single Fixtures 2-Tube	18	1.50	1.65	1.78	1.89	3.47	3.82	4.12	4.37
	24	1.57	1.73	1.87	1.98	3.66	4.03	4.35	4.61
	36	1.67	1.84	1.98	2.10	3.85	4.24	4.57	4.85
	48	1.79	1.96	2.11	2.23	4.04	4.44	4.80	5.09
	72	1.91	2.10	2.26	2.39	4.43	4.87	5.26	5.58
	96	2.09	2.30	2.48	2.63	4.81	5.29	5.71	6.06

All manhours include checking out of job storage, handling, job hauling, distributing, layout, and marking for installation.

Hang-Only manhours only include hanging of fixture.

Assemble-and-Hang manhours include complete assembling of fixture, installation, connecting circuit power to the fixture, testing, and verifying fixture operation.

Manhours exclude installation of conduit, boxes, circuit wiring, switches, tubes, special hangers or supports not supplied with fixture, change out of or separate ballast, and ladder or scaffold set-up. See respective tables for these time frames.

For heights above 25 feet increase manhours 5% for each additional 5 feet or fraction thereof.

FLUORESCENT LIGHTING FIXTURES
INTERIOR CLASS I EXPLOSION PROOF

3- and 4-Tube Fixtures

MANHOURS EACH

Item Description	Tube Length (Inches)	Manhours Required to							
		Hang Only For Height to				Assemble and Hang For Height to			
		10'	15'	20'	25'	10'	15'	20'	25'
Industrial Grade Single Fixtures 3-Tube	18	1.57	1.73	1.87	1.98	3.66	4.03	4.35	4.61
	24	1.67	1.84	1.98	2.10	3.85	4.24	4.57	4.85
	36	1.79	1.96	2.11	2.23	4.04	4.44	4.80	5.09
	48	1.83	2.01	2.17	2.30	4.24	4.66	5.04	5.34
	72	2.00	2.20	2.37	2.51	4.62	5.08	5.49	5.82
	96	2.16	2.37	2.55	2.70	5.00	5.50	5.94	6.30
Industrial Grade Single Fixtures 4-Tube	18	1.67	1.84	1.98	2.10	3.85	4.24	4.57	4.85
	24	1.76	1.94	2.09	2.22	4.04	4.44	4.80	5.09
	36	1.83	2.01	2.17	2.30	4.24	4.66	5.04	5.34
	48	1.91	2.10	2.26	2.39	4.43	4.87	5.26	5.58
	72	2.09	2.30	2.48	2.63	4.81	5.29	5.71	6.06
	96	2.24	2.46	2.65	2.80	5.20	5.72	6.18	6.55

6- and 8-Tube Fixtures

MANHOURS EACH

Item Description	Tube Length (Inches)	Manhours Required to							
		Hang Only For Height to				Assemble and Hang For Height to			
		10'	15'	20'	25'	10'	15'	20'	25'
Industrial Grade Single Fixtures 6-Tube	36	2.00	2.20	2.37	2.51	4.62	5.08	5.49	5.82
	48	2.09	2.30	2.48	2.63	4.81	5.29	5.71	6.06
	72	2.24	2.46	2.65	2.80	5.20	5.72	6.18	6.55
	96	2.42	2.66	2.87	3.05	5.58	6.14	6.63	7.03
Industrial Grade Single Fixtures 8-Tube	48	2.24	2.36	2.65	2.80	5.20	6.72	6.18	6.55
	72	2.42	2.66	2.87	3.05	5.58	6.14	6.63	7.03
	96	2.57	2.82	3.04	3.22	5.97	6.57	7.09	7.52

All manhours include checking out of job storage, handling, job hauling, distributing, layout, and marking for installation.

Hang-Only manhours only include hanging of fixture.

Assemble-and-Hang manhours include complete assembling of fixture, installation, connecting circuit power to the fixture, testing, and verifying fixture operation.

Manhours exclude installation of conduit, boxes, circuit wiring, switches, tubes, special hangers or supports not supplied with fixture, change out of or separate ballast, and ladder or scaffold set-up. See respective tables for these time frames.

For heights above 25 feet increase manhours 5% for each additional 5 feet or fraction thereof.

FLUORESCENT LUMINOUS CEILING SYSTEMS

Conditions 1, 2, and 3

MANHOURS PER SQUARE FOOT

Item Description	Manhours Required for Diffusers Made of		
	Glass	Plastic	Metal
Condition 1			
2' x 2' Diffusers – 2' Fixture Spacing	0.22	0.21	0.21
2' x 4' Diffusers – 4' Fixture Spacing	0.15	0.14	0.15
3' x 3' Diffusers – 3' Fixture Spacing	0.16	0.16	0.16
4' x 4' Diffusers – 4' Fixture Spacing	0.11	0.11	0.10
3' Strip – 3' Fixture Spacing	–	0.13	–
Condition 2			
2' x 2' Diffusers – 2' Fixture Spacing	0.25	0.24	0.24
2' x 4' Diffusers – 4' Fixture Spacing	0.18	0.17	0.17
3' x 3' Diffusers – 3' Fixture Spacing	0.19	0.18	0.18
4' x 4' Diffusers – 4' Fixture Spacing	0.14	0.13	0.14
3' Strip – 3' Fixture Spacing	–	0.16	–
Condition 3			
2' x 2' Diffusers – 2' Fixture Spacing	0.24	0.23	0.23
2' x 4' Diffusers – 4' Fixture Spacing	0.16	0.15	0.16
3' x 3' Diffusers – 3' Fixture Spacing	0.17	0.17	0.17
4' x 4' Diffusers – 4' Fixture Spacing	0.12	0.12	0.12
3' Strip – 3' Fixture Spacing	–	0.14	–
Condition 4			
2' x 2' Diffusers – 2' Lamp Spacing	0.19	0.18	0.18
2' x 4' Diffusers – 4' Lamp Spacing	0.17	0.16	0.16
3' x 3' Diffusers – 3' Lamp Spacing	0.15	0.15	0.15
4' x 4' Diffusers – 4' Lamp Spacing	0.13	0.12	0.12
3' Strip – 3' Lamp Spacing	–	0.13	–
Condition 5			
2' x 2' Diffusers – 2' Lamp Spacing	0.20	0.19	0.19
2' x 4' Diffusers – 4' Lamp Spacing	0.17	0.17	0.17
3' x 3' Diffusers – 3' Lamp Spacing	0.16	0.16	0.16
4' x 4' Diffusers – 4' Lamp Spacing	0.14	0.13	0.13
3' Strip – 3' Lamp Spacing	–	0.14	–

Condition 1–Fixtures fastened to and diffusing ceiling suspended from structure or existing ceiling.

Condition 2–Fixtures and diffusing ceiling fastened to channels suspended from structure or existing ceiling.

Condition 3–Fixtures suspended from structure or existing ceiling with diffusing ceiling hung from fixture or existing ceiling.

Condition 4–Wiring Channels with intergral lamp sockets attached to existing ceiling or structure and diffusing ceiling suspended from channel.

Condition 5–Wiring channels with intergral lamp sockets suspended from existing ceiling or structure and diffusing ceiling suspended from channel.

Manhours include checking out of job storage, handling, job hauling, distributing, layout, and complete installation of suspension system, fluorescent strip fixtures, inner wire connecting of fixtures, tubes, diffusers, and testing and verifying system operation.

Manhours are average for heights to 16 feet for complete wall to wall or rectangular area systems and do not include time allowance for curved or offset sections.

Manhours exclude make-up connection of home runs and scaffolding. See respective tables for these time frames.

AUDITORIUM LIGHTING

MANHOURS FOR UNITS LISTED

Item Description	Unit	Manhours
Footlight Troughs		
Surface Type Trough	l.f.	0.66
Disappearing Type Trough	l.f.	0.95
Sockets — One Circuit	each	0.39
Sockets — Alternate Circuits	each	0.50
Asbestos Wire	c.l.f.	1.10
Color Lenses	each	0.25
Border and Proscenium Lighting		
Border Troughs (Excluding Batten and Overhead Rigging)	l.f.	0.83
Proscenium Strip	l.f.	0.94
Sockets	each	0.44
Reflectors	each	0.28
Asbestos Wire	c.l.f.	1.10
Stage Cable	c.l.f.	5.50
Color Lenses	each	0.25
Stage Pocket Lighting		
1-Gang — Floor	each	5.50
2-Gang — Floor	each	6.60
3-Gang — Floor	each	7.70
1-Gang — Wall	each	6.60
2-Gang — Wall	each	7.70
3-Gang — Wall	each	9.35
Miscellaneous Lighting		
Aisle Light Outlet	each	1.10
Aisle Light Fixture — Mounted on Seat	each	1.10
Balcony Flood Outlet	each	1.50
Balcony Flood — 200 Watt Unit	each	1.10

Manhours include checking out of job storage, handling, job hauling, distributing, layout, and installing items as listed.

Manhours exclude installation of conduit, boxes, circuit wiring, and switches. See respective tables for these time frames.

FLUORESCENT LIGHTING FIXTURES

Fluorescent Tube Installation

MANHOURS EACH

Tube Length (Inches)	Manhours for Height to			
	10'	15'	20'	25'
6	0.07	0.08	0.09	0.10
9	0.07	0.08	0.09	0.10
12	0.08	0.09	0.10	0.11
15	0.10	0.11	0.12	0.13
18	0.10	0.11	0.12	0.13
21	0.14	0.15	0.17	0.18
24	0.14	0.15	0.17	0.18
33	0.19	0.21	0.23	0.24
36	0.20	0.22	0.24	0.25
42	0.25	0.28	0.30	0.31
48	0.25	0.28	0.30	0.31
60	0.32	0.35	0.38	0.40
64	0.33	0.36	0.39	0.42
72	0.40	0.44	0.48	0.50
84	0.51	0.56	0.61	0.64
96	0.58	0.64	0.69	0.73

Manhours include checking out of job storage, handling, job hauling, distributing, and installing of fluorescent tubes as listed.

Manhours exclude installation of conduit, boxes, fixtures, switches, wiring, ballast, and ladder or scaffolding. See respective tables for these time frames.

For heights above 25 feet increase manhours 3% for each additional 5 feet or fraction thereof.

FLUORESCENT LIGHTING FIXTURES
Ballast and Filter Installation

MANHOURS EACH

Item Description	Manhours for Height to			
	10'	15'	20'	25'
Ballast				
Indoor Type – Less than 2 Pounds	0.70	0.87	0.94	0.99
Indoor Type – 2 to 5 Pounds	0.90	0.99	1.07	1.13
Indoor Type – Over 5 Pounds	1.00	1.10	1.19	1.26
Vaportite Type – Less than 2 Pounds	0.90	0.99	1.07	1.13
Vaportite Type – 2 to 5 Pounds	1.00	1.10	1.19	1.26
Vaportite Type – Over 5 Pounds	1.12	1.23	1.33	1.41
Filters for Suppression of Radio Interference				
Indoor Type	0.79	0.87	0.94	0.99
Vaportite Type	0.90	0.99	1.07	1.13

Manhours include checking out of job storage, handling, job hauling, distributing, and installing of ballast or filter as outlined.

Manhours exclude installation of conduit, boxes, circuit wiring, fixtures, switches, fixture connection, tubes, and ladder or scaffolding. See respective tables for these time frames.

For heights above 25 feet increase manhours 3% for each additional 5 feet or fraction thereof.

INCANDESCENT LIGHTING FIXTURES

Incandescent Bulb Installation

MANHOURS EACH

Incandescent Bulb Wattage	Manhours for Height to							
	Open Units				Enclosed Units			
	10'	15'	20'	25'	10'	15'	20'	25'
25	0.09	0.10	0.10	0.11	0.15	0.17	0.18	0.19
40	0.09	0.10	0.10	0.11	0.15	0.17	0.18	0.19
60	0.09	0.10	0.10	0.11	0.15	0.17	0.18	0.19
75	0.09	0.10	0.10	0.11	0.15	0.17	0.18	0.19
100	0.09	0.10	0.10	0.11	0.15	0.17	0.18	0.19
150	0.10	0.11	0.12	0.13	0.17	0.19	0.20	0.21
200	0.10	0.11	0.12	0.13	0.17	0.19	0.20	0.21
300	0.10	0.11	0.12	0.13	0.17	0.19	0.20	0.21
500	0.12	0.13	0.14	0.15	0.20	0.22	0.24	0.25
750	0.12	0.13	0.14	0.15	0.20	0.22	0.24	0.25
1000	0.12	0.13	0.14	0.15	0.20	0.22	0.24	0.25

Manhours include checking out of job storage, handling, job hauling, distributing, and installing incadescent light bulbs as outlined.

Manhours exclude installation of conduit, boxes, fixtures, switches, wiring, and ladder or scaffolding. See respective tables for these time frames.

For heights above 25 feet increase manhours 3% for each additional 5 feet or fraction thereof.

OUTDOOR FLOOR LIGHTING

MANHOURS EACH

Item Description	Manhours Required to					
	Install Only For Height to			Assemble and Install For Height to		
	15'	20'	30'	15'	20'	30'
Open Floods						
100 Watts	1.10	1.65	2.75	1.65	2.00	3.20
150 Watts	1.12	1.68	2.80	1.68	2.10	3.30
200 Watts	1.15	1.72	2.95	1.72	2.25	3.35
300 Watts	1.65	2.00	3.50	2.00	2.80	4.20
500 Watts	1.70	2.10	3.85	2.10	2.95	4.40
750 Watts	2.10	4.00	5.50	3.30	3.05	6.50
1000 Watts	2.20	4.20	6.00	3.50	5.40	6.90
1500 Watts	2.25	4.35	6.60	3.70	5.50	7 40
Louvers or Visors	0.55	0.83	1.10	0.55	0.83	1.00
Steel Pole and Single Flood	–	–	–	–	–	11.00
Steel Pole and Double Flood	–	–	–	–	–	14.30
Enclosed Floods						
200/250 Watts	2.48	3.00	4.95	3.30	4.40	6.60
300/500Watts	3.30	4.12	6.50	4.40	5.50	8.00
750/1000 Watts	5.25	6.70	10.50	5.90	8.00	12.50
1500 Watts	5.50	6.89	11.00	6.60	8.25	13.20
Louvers or Visors	0.55	0.83	1.10	0.55	0.83	1.10

All manhours include checking out of job storage, handling, job hauling, distributing, layout, and marking for installation.

Install-Only manhours only include installing of fixture.

Assemble-and-Install manhours include complete assembling of fixture, installation, connecting circuit power to the fixture, testing, and verifying fixture operation.

Manhours exclude installation of conduit, boxes, circuit wiring, switches, bulbs, special hangers or supports not supplied with fixture, and ladder or scaffold set-up. See respective tables for these time frames.

Section 6

UNDERFLOOR DUCT

This section presents time requirements in the form of man-hour units for operations as may be involved in the total installation of underfloor fiber and steel duct systems.

In the manhour units for underfloor ducts consideration has been given to all operations such as hauling, handling, unloading, laying out, measuring, cutting, deburring, and installing ducts, fittings, and devices as may be required.

The manhours are averages of many projects installed under varied conditions and therefore may need adjusting to fit the needs of a specific project all in accordance with the Introduction of this manual.

FIBER DUCTS

MANHOURS FOR UNITS LISTED

Item Description	Unit	Manhours
Open Bottom Fiber Duct		
Duct	c.l.f.	3.13
Sealing Duct with Compound	c.l.f.	0.81
Floor Elbows	each	0.60
Wall Elbows — Low Tension	each	0.78
Wall Elbows — High Tension	each	0.96
Junction Boxes — Single System 4-Way	each	1.38
Junction Boxes — Double System 4-Way	each	1.98
Junction Boxes — Triple System 4-Way	each	2.40
Crossunder	each	1.20
End Marker	each	0.30
Concrete Inserts to 1-1/4″	each	1.10
Concrete Inserts 1-1/4″ to 1-1/2″	each	1.10
Closed Bottom Fiber Duct		
Duct	c.l.f.	6.69
Sealing Duct with Compound	c.l.f.	0.44
Floor Elbows	each	0.78
Wall Elbows	each	1.02
Coupling and Support — Single	each	0.42
Coupling and Support — Double	each	0.78
Coupling and Support — Triple	each	0.78
Junction Boxes — Single System — 4-Way	each	1.80
Junction Boxes — Double System — 4-Way	each	2.40
Junction Boxes — Triple System — 4-Way	each	2.82
Junction Box Closure Plate	each	0.42
Cover Plate — Single or Double Box	each	0.42
Crossunder	each	1.02
Prelokaylet Unit	each	0.42
Flush Flange Unit	each	0.78
Carpet Flange Assembly	each	0.78
Duct Receptacle Unit	each	0.42
Telephone, Double Opening Unit	each	0.42
End Marker and Conduit Adapter	each	0.60
Concrete Insert to 1-1/2″	each	1.11
Concrete Insert 1-1/2″ to 3″	each	1.30

Manhours include checking out of job storage, handling, job hauling, layout, measuring, cutting of duct, installing of duct or fittings, and checking alignment and elevations.

Manhours are for duct installed on concrete pads or rough floor furnished by others.

Manhours exclude wire or cable pull-in. See respective tables for these time frames.

STANDARD STEEL DUCTS AND FITTINGS
Sizes 3-Inch and 6-Inch

MANHOURS FOR UNITS LISTED

Item Description	Unit	Slab on Grade		Elevated Slab Height to							
				10′		15′		20′		25′	
		3″	6″	3″	6″	3″	6″	3″	6″	3″	6″
Blank Duct	lin. ft.	0.05	0.08	0.06	0.09	0.07	0.11	0.08	0.12	0.09	0.13
Insert Duct	lin. ft.	0.07	0.10	0.08	0.12	0.09	0.13	0.10	0.15	0.11	0.16
Duct Coupling	each	0.10	0.12	0.12	0.14	0.13	0.16	0.15	0.17	0.16	0.19
Elbow – Steel Wall	each	0.48	0.51	0.55	0.59	0.63	0.67	0.70	0.74	0.75	0.80
Elbow – Cast Wall	each	0.50	0.54	0.58	0.62	0.66	0.71	0.73	0.79	0.79	0.85
90° Flat Elbow – Steel	each	0.48	0.51	0.55	0.59	0.63	0.67	0.70	0.74	0.75	0.80
90° Flat Elbow – Cast	each	0.50	0.54	0.58	0.62	0.66	0.71	0.73	0.79	0.79	0.85
45° Flat Elbow – Steel	each	0.48	0.51	0.55	0.59	0.63	0.67	0.70	0.74	0.75	0.80
45° Flat Elbow – Cast	each	0.50	0.54	0.58	0.62	0.66	0.71	0.73	0.79	0.79	0.85
Offset – Steel	each	0.69	0.74	0.79	0.85	0.91	0.98	1.00	1.08	1.08	1.16
Offset – Cast	each	0.75	0.80	0.86	0.92	0.99	1.06	1.09	1.16	1.18	1.26
Tee – Steel	each	0.63	0.67	0.72	0.77	0.83	0.89	0.92	0.97	0.99	1.05
One Level Cross – Steel	each	0.69	0.74	0.79	0.85	0.91	0.98	1.00	1.08	1.08	1.16
Two Level Cross – Steel	each	0.75	0.80	0.86	0.92	0.99	1.06	1.09	1.16	1.18	1.26
End Cap	each	0.08	0.09	0.09	0.10	0.11	0.12	0.12	0.13	0.13	0.14
Female Adapter	each	0.15	0.16	0.17	0.18	0.20	0.21	0.22	0.23	0.24	0.25
Male Adapter	each	0.15	0.16	0.17	0.18	0.20	0.21	0.22	0.23	0.24	0.25
Expansion Joint	each	0.75	0.80	0.86	0.92	0.99	1.06	1.09	1.16	1.18	1.26
Connector – 3″ x 6″	each	0.08	–	0.09	–	0.11	–	0.12	–	0.13	–
Reducer – 6″ x 3″	each	–	0.09	–	0.10	–	0.12	–	0.13	–	0.14
Panel Connector	each	0.15	0.16	0.17	0.18	0.20	0.21	0.22	0.23	0.24	0.25
Duct Support – Single	each	0.13	0.14	0.15	0.16	0.17	0.19	0.19	0.20	0.20	0.22
Duct Support – Double	each	0.16	0.17	0.18	0.20	0.21	0.22	0.23	0.25	0.25	0.27
Duct Support – Triple	each	0.19	0.20	0.22	0.23	0.25	0.26	0.28	0.29	0.30	0.31
Combination Support	each	0.19	0.20	0.22	0.23	0.25	0.26	0.28	0.29	0.30	0.31

Manhours include checking out of job storage, handling, job hauling, layout, measuring, cutting of duct, installing of duct or fittings, and checking alignment and elevations.

Manhours are based on installing up to 1,000 linear feet. If more than 1,000 linear feet are to be installed, decrease the manhours by the following percentages:

Runs from 1,000 to   5,000 Linear Feet–Decrease 10%
Runs from 5,000 to 10,000 Linear Feet–Decrease 22%
Runs greater than 10,000 Feet   –Decrease 35%

Manhours exclude welding of duct, connecting of conduits, and wire or cable pull-in. See respective tables for these time frames.

STANDARD STEEL DUCT JUNCTION BOXES AND ACCESSORIES

MANHOURS EACH

Item Description	Slabs on Grade	Manhours Required for			
		Elevated Slab Height to			
		10'	15'	20'	25'
1-Level Junction Boxes					
8" x 8"	0.94	1.03	1.11	1.15	1.17
10" x 10"	0.99	1.09	1.17	1.21	1.24
12" x 12"	1.04	1.14	1.22	1.27	1.30
14" x 14"	1.09	1.20	1.28	1.33	1.36
16" x 16"	1.13	1.24	1.33	1.38	1.41
18" x 18"	1.18	1.30	1.39	1.44	1.47
20" x 20"	1.24	1.36	1.46	1.52	1.55
24" x 24"	1.29	1.42	1.52	1.58	1.61
30" x 30"	1.35	1.49	1.59	1.65	1.69
2-Level Junction Boxes					
8" x 8"	1.13	1.24	1.33	1.38	1.41
10" x 10"	1.18	1.30	1.39	1.44	1.47
12" x 12"	1.24	1.36	1.46	1.52	1.55
14" x 14"	1.29	1.42	1.52	1.58	1.61
16" x 16"	1.35	1.49	1.59	1.65	1.69
18" x 18"	1.41	1.55	1.66	1.73	1.76
20" x 20"	1.48	1.63	1.74	1.81	1.85
24" x 24"	1.54	1.69	1.81	1.89	1.92
30" x 30"	1.60	1.76	1.88	1.96	2.00
Box Accessories					
Closure Cap	0.06	0.07	0.08	0.09	0.10
Access Ring	0.06	0.07	0.08	0.09	0.10
Conduit Adapter	0.09	0.10	0.12	0.13	0.14
Outlet Reducer	0.06	0.07	0.08	0.09	0.10
Corner Adapter	0.09	0.10	0.12	0.13	0.14
Cap and Marker	0.19	0.22	0.25	0.28	0.30

Manhours include checking out of job storage, handling, job hauling, and installing junction boxes and accessories as outlined.

Manhours exclude installation of ducts and duct fittings. See respective table for this time frame.

Manhours are based on 5 to 7 junction boxes for each 1,000 linear feet of duct and should be adjusted if more or less boxes are to be installed.

STEEL TRENCH DUCT

MANHOURS PER LINEAR FOOT

Duct Width (Inches)	Manhours Required for				
	Slab on Grade	Elevated Slab Height to			
		10'	15'	20'	25'
6	0.38	0.44	0.50	0.55	0.60
9	0.48	0.55	0.63	0.70	0.75
12	0.58	0.67	0.77	0.84	0.91
18	0.67	0.77	0.89	0.97	1.05
24	0.77	0.89	1.02	1.12	1.21
30	0.86	0.99	1.14	1.25	1.35

Manhours include checking out of job storage, handling, job hauling, layout, measuring, installing, and checking alignment and elevations.

Manhours are based on installing up to 100 linear feet. If more than 100 linear feet are to be installed, decrease the manhours by the following percentages:

Runs from 100 to 250 Linear Feet—Decrease 10%
Runs from 250 to 400 Linear Feet—Decrease 22%
Runs greater than 400 Linear Feet—Decrease 35%

Manhours exclude welding of duct, connecting of conduits and wire or cable pull-in. See respective tables for these time frames.

STEEL TRENCH DUCT FITTINGS

MANHOURS EACH

Item Description	Duct Width (Inches)	Slab on Grade	Manhours Required for Elevated Slab Height to			
			10'	15'	20'	25'
90° Elbows	6	3.84	4.15	4.31	4.36	4.38
	9	4.80	5.18	5.39	5.45	5.47
	12	5.76	6.22	6.47	6.53	6.57
	18	6.72	7.26	7.55	7.62	7.66
	24	7.68	8.29	8.63	8.71	8.76
	30	8.64	9.33	9 70	9.80	9.85
Crosses	6	9.60	10.37	10.78	10.89	10.94
	9	10.56	11.40	11.86	11.98	12.04
	12	11.42	12.33	12.83	12.96	13.02
	18	12.48	13.48	14.02	14.16	14.23
	24	13.44	14.52	15.10	15.25	15.32
	30	14.40	15.55	16.17	16.34	16.42
Tees	6	6.72	7.26	7.55	7.62	7.66
	9	7.68	8.29	8.63	8.71	8.76
	12	8.64	9.33	9.70	9.80	9.85
	18	9.60	10.37	10.78	10.89	10.94
	24	11.28	12.18	12.67	12.80	12.86
	30	12.48	13.48	14.02	14.16	14.23
End Closures	6	0.96	1.04	1.08	1.09	1.10
	9	1.46	1.58	1.64	1.66	1.67
	12	1.92	2.07	2.16	2.18	2.19
	18	2.24	2.61	2.72	2.75	2.76
	24	2.88	3.11	3.23	3.27	3.28
	30	3.84	4.15	4.31	4.36	4.38

Manhours include checking out of job storage, handling, job hauling, and installing fittings as outlined.

Manhours are based on installing average number of fittings with 100 linear feet of duct. If more than 100 linear feet of duct is to be installed, decrease the fitting manhours by the following percentages:

Runs from 100 to 250 Linear Feet—Decrease 8%
Runs from 250 to 400 Linear Feet—Decrease 20%
Runs greater than 400 Linear Feet—Decrease 30%

Manhours exclude preparation for and installation of duct. See respective table for this time frame.

STEEL TRENCH DUCT FITTINGS

MANHOURS EACH

Item Description	Duct Width (Inches)	Manhours Required for				
		Slab on Grade	Elevated Slab Height to			
			10'	15'	20'	25'
Wall Elbows	6	2.88	3.11	3.23	3.27	3.28
	9	3.84	4.15	4.31	4.36	4.38
	12	4.80	5.18	5.39	5.45	5.47
	18	5.76	6.22	6.47	6.53	6.57
	24	6.72	7.26	7.55	7.62	7.66
	30	7.68	8.29	8.63	8.71	8.76
Panel Risers	6	3.84	4.15	4.31	4.36	4.38
	9	4.80	5.18	5.39	5.45	5.47
	12	5.76	6.22	6.47	6.53	6.57
	18	6.72	7.26	7.55	7.62	7.66
	24	7.68	8.29	8.63	8.71	8.76
	30	8.64	9.33	9.70	9.80	9.85
Box Connectors	6	1.92	2.07	2.16	2.18	2.19
	9	2.88	3.11	3.23	3.27	3.28
	12	3.84	4.15	4.31	4.36	4.38
	18	4.80	5.18	5.39	5.45	5.47
	24	5.76	6.22	6.47	6.53	6.57
	30	6.72	7.26	7.55	7.62	7.66
Cover Plates	6	0.03	0.04	0.04	0.05	0.05
	9	0.04	0.05	0.05	0.06	0.06
	12	0.05	0.06	0.07	0.07	0.08
	18	0.06	0.07	0.08	0.09	0.09
	24	0.07	0.08	0.09	0.10	0.11
	30	0.08	0.09	0.11	0.12	0.13
Marker Screw Assembly	—	0.18	0.21	0.24	0.26	0.28

Manhours include checking out of job storage, handling, job hauling, and installing fittings as outlined.

Manhours are based on installing average number of fittings with 100 linear feet of duct. If more than 100 linear feet of duct is to be installed, decrease the fitting manhours by the following percentages:

Runs from 100 to 250 Linear Feet—Decrease 8%
Runs from 250 to 400 Linear Feet—Decrease 20%
Runs greater than 400 Linear Feet—Decrease 30%

Manhours exclude preparation for and installation of duct. See respective tables for this time frame.

STEEL INDUSTRIAL DUCT

Size 4 Inches by 4 Inches

MANHOURS FOR UNITS LISTED

Item Description	Unit	Slab on Grade	Elevated Slab Height to			
			10'	15'	20'	25'
Blank Duct	lin. ft.	0.06	0.07	0.08	0.09	0.10
Insert Duct	lin. ft.	0.07	0.08	0.09	0.10	0.11
Duct Coupling	each	0.56	0.64	0.74	0.81	0.88
90° Elbow	each	0.38	0.44	0.50	0.55	0.60
End Cap	each	0.09	0.10	0.12	0.13	0.14
Female Adapter — 1-Conduit	each	0.19	0.22	0.25	0.28	0.30
Female Adapter — 2-Conduit	each	0.23	0.26	0.30	0.33	0.36
Male Adapter — 1-Conduit	each	0.19	0.22	0.25	0.28	0.30
Male Adapter — 2-Conduit	each	0.23	0.26	0.30	0.33	0.36
Panel Connector	each	0.15	0.17	0.20	0.22	0.24
Expansion Joint	each	0.56	0.64	0.74	0.81	0.88
Junction Box — Single	each	0.94	1.08	1.24	1.37	1.48
Junction Box — Double	each	1.13	1.30	1.49	1.64	1.78
Junction Box — Plug	each	0.08	0.09	0.11	0.12	0.13
Junction Box — Connector	each	0.09	0.10	0.12	0.13	0.14
Insert Cap and Market	each	0.19	0.22	0.25	0.28	0.30
Duct Support — Single	each	0.09	0.10	0.12	0.13	0.14
Duct Support — Double	each	0.11	0.13	0.15	0.16	0.17

Manhours include checking out of job storage, handling, job hauling, layout, measuring, cutting of duct, installing of duct or fittings, and checking alignment and elevations.

Manhours are based on installing up to 100 linear feet of duct with average number of fittings. If more than 100 linear feet of duct are to be installed, decrease the manhours by the following percentages:

	Decrease	
	Duct	Fittings
Runs from 100 to 250 Linear Feet—	10%	8%
Runs from 250 to 400 Linear Feet—	22%	20%
Runs greater than 400 Linear Feet—	35%	30%

Manhours exclude welding of duct, connecting of conduits, and wire or cable pull-in. See respective tables for these time frames.

DUCT WIRING DEVICES AND ACCESSORIES

MANHOURS EACH

Item Description	Size (Inches)	Manhours Required for				
		Slab on Grade	Elevated Slab Height to			
			10'	15'	20'	25'
Junction Box Covers						
	6	0.12	0.14	0.16	0.17	0.19
	8	0.13	0.15	0.18	0.19	0.20
	10	0.14	0.16	0.19	0.20	0.22
	12	0.16	0.18	0.21	0.23	0.25
Set in Tile	14	0.17	0.20	0.22	0.25	0.27
	16	0.18	0.21	0.24	0.26	0.28
	18	0.19	0.22	0.25	0.28	0.30
	20	0.21	0.24	0.28	0.31	0.33
	24	0.22	0.25	0.29	0.32	0.35
Junction Box Covers						
	6	0.14	0.16	0.19	0.20	0.22
	8	0.16	0.18	0.21	0.23	0.25
	10	0.17	0.20	0.22	0.25	0.27
	12	0.18	0.21	0.24	0.26	0.28
Set in Terrazzo	14	0.19	0.22	0.25	0.28	0.30
	16	0.21	0.24	0.28	0.31	0.33
	18	0.22	0.25	0.29	0.32	0.35
	20	0.23	0.26	0.30	0.33	0.36
	24	0.24	0.28	0.32	0.35	0.38
Insert and Insert Accessories						
Locate and Open Insert	–	1.08	1.24	1.43	1.57	1.70
Afterset Insert	–	0.90	1.04	1.19	1.31	1.41
Insert Extension	–	0.45	0.52	0.60	0.65	0.71
Insert Bushing	–	0.45	0.45	0.60	0.65	0.71
Insert Chase Nipple	–	0.45	0.45	0.60	0.65	0.71
Insert Standpipe	–	0.45	0.45	0.60	0.65	0.71
Insert Slotted Cover	–	0.23	0.26	0.30	0.33	0.36

Manhours include checking out of job storage, handling, job hauling, and installing up to 50-duct accessories as outlined.

If more than 50 of the items are to be installed on a single floor decrease the manhours by the following percentages:

51 to 100—Decrease 15%
101 to 200—Decrease 25%
201 or more—Decrease 32%

Manhours exclude installation of ducts, duct fittings, other devices and wire pull-in. See respective tables for these time frames.

LOW AND HIGH POTENTIAL DEVICES

MANHOURS EACH

Item Description	Manhours Required for				
	Slab on Grade	Elevated Slab Height to			
		10′	15′	20′	25′
Low Potential Devices					
Square Outlet with Rubber Bushing	0.47	0.54	0.59	0.64	0.67
Post Outlet with Double Bushing	0.47	0.54	0.59	0.64	0.67
Phone Outlet with Terminal Block	0.56	0.64	0.71	0.77	0.80
Cable Outlet	0.47	0.54	0.59	0.64	0.67
Phone and Power Outlet Without Receptacle	0.23	0.26	0.30	0.33	0.36
Outlet Abandoning Assembly	0.56	0.64	0.71	0.77	0.80
High Potential Receptacles					
Single Duplex — Grounding 125V 2P2W 15 Amp	0.69	0.79	0.87	0.94	0.98
Double Duplex — Grounding 125V 2P2W 15 Amp	0.69	0.79	0.87	0.94	0.98
Single Duplex — Grounding 125V 2P2W 20 Amp	0.91	1.05	1.16	1.25	1.30
Double Duplex — Grounding 125V 2P2W 20 Amp	0.91	1.05	1.16	1.25	1.30
Non-Grounding 125/250V 3P3W 20 Amp	1.13	1.30	1.43	1.54	1.60
Non-Grounding 125/250V 3P3W 30 Amp	1.34	1.54	1.69	1.83	1.90
Non-Grounding 125/250V 3P3W 50 Amp	1.78	2.05	2.26	2.44	2.54
Grounding 250V 2P2W 20 Amp	0.91	1.05	1.16	1.25	1.30
Grounding 125/250V 3P3W 30 Amp	1.44	1.66	1.83	1.98	2.06
Grounding 125/250V 3P3W 50 Amp	1.91	2.20	2.42	2.61	2.71
Grounding 125/250V 3P3W 60 Amp	2.53	2.91	3.20	3.46	3.60
High Tension Service Fitting — No Recept.	0.47	0.54	0.59	0.64	0.67

Manhours include checking out of job storage, handling, job hauling and installing up to 50 devices as outlined.

If more than 50 of the items are to be installed on a single floor decrease the manhours by the following percentages:

51 to 100—Decrease 15%
101 to 200—Decrease 25%
201 or more—Decrease 32%

Manhours exclude installation of ducts, duct fittings, other devices and wire pull-in. See respective tables for these time frames.

Section 7

BUS DUCT

This section includes manhour tables for the installation of plug-in type bus duct from 250-amp to 4,000-amp rating and feeder type bus duct from 800-amp to 5,000-amp rating.

Separate tables are provided for both copper and aluminum bus duct as well as time requirements for the installation of fittings, 250- and 600-volt fusible and nonfusible switches and plug-in molded case circuit breakers.

The manhour units for bus duct are intended to suffice for the installation of either ventilated or nonventilated duct. However, if the weight per linear foot of duct varies significantly from those listed in the tables, the manhours may need an adjustment.

Supports, fasteners and hangers are included under the section entitled "Anchors, Fasteners, Hangers and Supports". Surface metal raceway and branch busway will be found under the section bearing that title.

3-PHASE, 3-POLE BUS DUCT
COPPER PLUG-IN TYPE
For Installation Height to 10 Feet

MANHOURS PER LINEAR FOOT

Approx. Weight Per Foot (Pounds)	Amperes	Manhours			
		Vertical Surface	Horizontal Suspended	Concrete Ceilings	Open Truss or Bar Joist
10	225	0.42	0.44	0.48	0.46
12	400	0.47	0.49	0.54	0.52
14	600	0.54	0.56	0.62	0.59
16	800	0.59	0.61	0.67	0.65
18	1000	0.65	0.68	0.74	0.71
20	1350	0.73	0.76	0.84	0.80
24	1600	0.81	0.84	0.93	0.89
28	2000	1.08	1.12	1.24	1.19
32	2500	1.20	1.25	1.37	1.32
40	3000	1.34	1.39	1.53	1.47
54	4000	1.69	1.76	1.93	1.86

For Installation Height to 15 Feet

MANHOURS PER LINEAR FOOT

Approx. Weight Per Foot (Pounds)	Amperes	Manhours			
		Vertical Surface	Horizontal Suspended	Concrete Ceilings	Open Truss or Bar Joist
10	225	0.45	0.47	0.51	0.49
12	400	0.51	0.53	0.58	0.56
14	600	0.58	0.60	0.66	0.64
16	800	0.64	0.67	0.73	0.70
18	1000	0.70	0.73	0.80	0.77
20	1350	0.79	0.82	0.90	0.87
24	1600	0.87	0.90	1.00	0.96
28	2000	1.17	1.22	1.34	1.28
32	2500	1.30	1.35	1.49	1.43
40	3000	1.45	1.51	1.66	1.59
54	4000	1.83	1.90	2.09	2.01

Manhours include checking out of job storage, handling, job hauling, layout, measuring, and installing duct runs up to 300 linear feet.

For duct runs from 301 feet to 500 feet decrease manhours 30%. For duct runs from 501 feet to 1,000 feet decrease manhours 40%.

For 3-phase, 4-pole (1/2 neutral) bus duct increase manhours 20%.

For 3-phase, 4-pole (full neutral) bus duct increase manhours 25%.

For 4-pole, plug-in type bus duct increase manhours 20%.

Manhours exclude installation of duct fittings, hangers, enclosures and branch circuit or lighting panelboards, the cutting of holes or openings, welding, painting, and scaffolding. See respective tables for these time frames.

3-PHASE, 3-POLE BUS DUCT
COPPER PLUG-IN TYPE

For Installation Height to 20 Feet

MANHOURS PER LINEAR FOOT

Approx. Weight Per Foot (Pounds)	Amperes	Manhours			
		Vertical Surface	Horizontal Suspended	Concrete Ceilings	Open Truss or Bar Joist
10	225	0.47	0.49	0.54	0.52
12	400	0.53	0.55	0.61	0.58
14	600	0.60	0.62	0.69	0.66
16	800	0.67	0.70	0.77	0.74
18	1000	0.73	0.76	0.84	0.80
20	1350	0.82	0.85	0.94	0.90
24	1600	0.90	0.94	1.03	0.99
28	2000	1.22	1.27	1.40	1.34
32	2500	1.35	1.40	1.54	1.48
40	3000	1.51	1.57	1.73	1.66
54	4000	1.90	1.98	2.17	2.09

For Installation Height to 25 Feet

MANHOURS PER LINEAR FOOT

Approx. Weight Per Foot (Pounds)	Amperes	Manhours			
		Vertical Surface	Horizontal Suspended	Concrete Ceilings	Open Truss or Bar Joist
10	225	0.48	0.50	0.55	0.53
12	400	0.55	0.57	0.63	0.60
14	600	0.62	0.64	0.71	0.68
16	800	0.69	0.72	0.79	0.76
18	1000	0.75	0.78	0.86	0.82
20	1350	0.84	0.87	0.96	0.92
24	1600	0.93	0.97	1.06	1.02
28	2000	1.26	1.31	1.44	1.38
32	2500	1.39	1.45	1.59	1.53
40	3000	1.56	1.62	1.78	1.71
54	4000	1.98	2.06	2.27	2.17

Manhours include checking out of job storage, handling, job hauling, layout, measuring, and installing duct runs up to 300 linear feet.

For duct runs from 301 feet to 500 feet decrease manhours 30%. For duct runs from 501 feet to 1,000 feet decrease manhours 40%.

For 3-phase, 4-pole (1/2 neutral) bus duct increase manhours 20%.

For 3-phase, 4-pole (full neutral) bus duct increase manhours 25%.

For 4-pole, plug-in type bus duct increase manhours 20%.

Manhours exclude installation of duct fittings, hangers, enclosures and branch circuit or lighting panelboards, the cutting of holes or openings, welding, painting, and scaffolding. See respective tables for these time frames.

For heights above 25 feet increase manhours 3% for each additional 5 feet or fraction thereof.

3-PHASE, 3-POLE BUS DUCT FITTINGS COPPER PLUG-IN TYPE

For Installation Height to 10 Feet

MANHOURS EACH

Amperes	Manhours										
	A	B	C	D	E	F	G	H	I	J	K
225	4.18	5.28	6.49	4.73	0.46	4.18	5.28	4.18	2.40	3.30	4.18
400	4.73	5.83	7.15	5.28	0.51	4.73	5.83	4.73	2.64	3.96	4.73
600	5.28	6.49	7.92	5.83	0.57	5.28	6.27	5.28	2.90	4.73	5.28
800	5.83	7.26	8.69	6.49	0.62	5.83	7.26	5.83	3.15	5.50	5.83
1000	6.49	8.14	9.57	7.15	0.68	6.49	8.14	6.49	3.41	6.27	6.49
1350	7.26	9.46	10.56	7.92	0.75	7.26	9.46	7.26	3.65	7.15	7.26
1600	8.14	11.00	12.10	8.69	0.83	8.14	11.00	8.14	4.07	8.14	8.14
2000	9.46	12.54	14.08	9.57	0.92	9.46	12.54	9.46	4.77	9.46	9.46
2500	11.00	14.08	16.28	10.56	1.03	11.00	14.08	11.00	5.83	11.88	11.00
3000	13.42	16.50	18.92	11.88	1.14	13.42	16.50	13.42	7.15	14.52	13.42
4000	16.94	19.58	23.98	14.52	1.39	16.94	19.58	16.94	9.46	17.60	16.94

For Installation Height to 15 Feet

MANHOURS EACH

Amperes	Manhours										
	A	B	C	D	E	F	G	H	I	J	K
225	4.31	5.39	6.62	4.87	0.49	4.31	5.39	4.31	2.54	3.43	4.31
400	4.87	5.95	7.28	5.39	0.54	4.87	5.95	4.87	2.80	4.08	4.87
600	5.39	6.62	8.06	5.95	0.60	5.39	6.40	5.39	3.07	4.87	5.39
800	5.95	7.39	8.82	6.62	0.66	5.95	7.39	5.95	3.28	5.61	5.95
1000	6.62	8.28	9.72	7.28	0.72	6.62	8.28	6.62	3.55	6.40	6.62
1350	7.39	9.61	10.73	8.06	0.80	7.39	9.61	7.39	3.76	7.28	7.39
1600	8.28	11.18	12.27	8.82	0.88	8.28	11.18	8.28	4.19	8.28	8.28
2000	9.61	12.72	14.25	9.72	0.98	9.61	12.72	9.61	4.91	9.61	9.61
2500	11.18	14.75	16.48	10.73	1.09	11.18	14.25	11.18	5.95	12.04	11.18
3000	13.61	16.70	19.11	12.04	1.21	13.61	16.70	13.61	7.28	14.69	13.61
4000	17.14	19.78	24.22	14.69	1.47	17.14	19.78	17.14	9.61	17.77	17.14

Code:

A—90° Elbows
B—Tees
C—Crosses
D—Offsets
E—End Closures
F—Reducers (Unfused)

G—Offset Elbows
H—Expansion Sections
I—Panelboard Connections
J—Cable Tap Boxes
K—Transfer Tap Section

Manhours include checking out of job storage, handling, job hauling and installing of fittings as outlined. For duct runs up to 300 linear feet.

Manhours exclude installation of other items and other operations. See respective tables for these time frames.

For duct runs from 301 feet to 500 feet and from 501 feet to 1,000 feet, decrease manhours 30 and 40% respectively.

For 3-phase, 4-pole, (1/2 neutral) bus duct fittings increase manhours 20%.

For 3-phase, 4-pole (full neutral) bus duct fittings increase manhours 25%.

For 4-pole, plug-in type bus duct fittings increase manhours 20%.

3-PHASE, 3-POLE BUS DUCT FITTINGS
COPPER PLUG-IN TYPE
For Installation Height to 20 Feet

MANHOURS EACH

Amperes	Manhours										
	A	B	C	D	E	F	G	H	I	J	K
225	4.43	5.49	6.75	5.02	0.51	4.43	5.49	4.43	2.65	3.53	4.43
400	5.02	6.07	7.40	5.49	0.56	5.02	6.07	5.02	2.91	4.20	5.02
600	5.49	6.75	8.20	6.07	0.63	5.49	6.52	5.49	3.20	5.02	5.49
800	6.07	7.52	8.97	6.75	0.68	6.07	7.52	6.07	3.37	5.72	6.07
1000	6.75	8.43	9.88	7.40	0.75	6.75	8.43	6.75	3.65	6.52	6.75
1350	7.52	9.77	10.90	8.20	0.83	7.52	9.77	7.52	3.87	7.40	7.52
1600	8.43	11.35	12.44	8.97	0.91	8.43	11.35	8.43	4.32	8.43	8.43
2000	9.77	12.89	14.42	9.88	1.01	9.77	12.89	9.77	5.06	9.77	9.77
2500	11.35	14.42	16.67	10.90	1.14	11.35	14.42	11.35	6.07	12.21	11.35
3000	13.80	16.90	19.30	12.21	1.26	13.80	16.90	13.80	7.40	14.87	13.80
4000	17.38	19.97	24.46	14.87	1.53	17.38	19.97	17.38	9.77	17.95	17.38

For Installation Height to 25 Feet

MANHOURS EACH

Amperes	Manhours										
	A	B	C	D	E	F	G	H	I	J	K
225	4.52	5.55	6.82	5.12	0.52	4.52	5.55	4.52	2.70	3.61	4.52
400	5.12	6.13	7.46	5.55	0.57	5.12	6.13	5.12	2.97	4.33	5.12
600	5.55	6.83	8.26	6.13	0.64	5.55	6.59	5.55	3.26	5.12	5.55
800	6.13	7.57	9.02	6.82	0.70	6.13	7.57	6.13	3.44	5.78	6.13
1000	6.82	8.49	9.94	7.46	0.76	6.82	8.49	6.82	3.73	6.59	6.82
1350	7.57	9.82	10.97	8.26	0.84	7.57	9.82	7.57	3.95	7.46	7.57
1600	8.49	11.42	12.50	9.02	0.93	8.49	11.42	8.49	4.40	8.49	8.49
2000	9.82	12.96	14.48	9.94	1.03	9.82	12.96	9.82	5.16	9.82	9.82
2500	11.42	14.48	16.74	10.97	1.16	11.42	14.48	11.42	6.13	12.27	11.42
3000	13.87	16.96	19.36	12.27	1.28	13.87	16.96	13.87	7.46	14.93	13.87
4000	17.45	20.03	24.54	14.93	1.56	17.45	20.03	17.45	9.82	18.01	17.45

Code:

A—90° Elbows
B—Tees
C—Crosses
D—Offsets
E—End Closures
F—Reducers (Unfused)

G—Offset Elbows
H—Expansion Sections
I—Panelboard Connections
J—Cable Tap Boxes
K—Transfer Tap Section

Manhours include checking out of job storage, handling, job hauling and installing of fittings as outlined. For duct runs up to 300 linear feet.

Manhours exclude installation of other items and other operations. See respective tables for these time frames.

For duct runs from 301 feet to 500 feet and from 501 feet to 1,000 feet, decrease manhours 30 and 40% respectively.

For 3-phase, 4-pole, (1/2 neutral) bus duct fittings increase manhours 20%.

For 3-phase, 4-pole (full neutral) bus duct fittings increase manhours 25%.

For 4-pole, plug-in type bus duct fittings increase manhours 20%.

For heights above 25 feet increase manhours 3% for each additional 5 feet or fraction thereof.

3-PHASE, 3-POLE BUS DUCT
ALUMINUM PLUG-IN TYPE

For Installation Height to 10 Feet

MANHOURS PER LINEAR FOOT

Approx. Weight Per Foot (Pounds)	Amperes	Manhours			
		Vertical Surface	Horizontal Suspended	Concrete Ceilings	Open Truss or Bar Joist
8	225	0.32	0.33	0.37	0.35
9	400	0.36	0.37	0.41	0.40
10	600	0.42	0.44	0.48	0.46
11	800	0.47	0.49	0.54	0.52
12	1000	0.53	0.55	0.51	0.68
14	1350	0.58	0.60	0.66	0.64
16	1600	0.65	0.68	0.74	0.71
18	2000	0.73	0.76	0.84	0.80
22	2500	0.86	0.89	0.98	0.94
26	3000	1.12	1.16	1.28	1.23
32	4000	1.23	1.28	1.41	1.35

For Installation Height to 15 Feet

MANHOURS PER LINEAR FOOT

Approx. Weight Per Foot (Pounds)	Amperes	Manhours			
		Vertical Surface	Horizontal Suspended	Concrete Ceilings	Open Truss or Bar Joist
8	225	0.35	0.36	0.40	0.38
9	400	0.39	0.41	0.45	0.43
10	600	0.45	0.47	0.51	0.49
11	800	0.51	0.53	0.58	0.56
12	1000	0.57	0.59	0.65	0.63
14	1350	0.63	0.66	0.72	0.69
16	1600	0.70	0.73	0.80	0.77
18	2000	0.79	0.82	0.90	0.87
22	2500	0.93	0.97	1.06	1.02
26	3000	1.21	1.26	1.38	1.33
32	4000	1.33	1.38	1.52	1.46

Manhours include checking out of job storage, handling, job hauling, layout, measuring, and installing duct runs up to 300 linear feet.

For duct runs from 301 feet to 500 feet decrease manhours 30%. For duct runs from 501 feet to 1,000 feet decrease manhours 40%.

For 3-phase, 4-pole (1/2 neutral) bus duct increase manhours 20%.

For 3-phase, 4-pole (full neutral) bus duct increase manhours 25%.

For 4-pole, plug-in type bus duct increase manhours 20%.

Manhours exclude installation of duct fittings, hangers, enclosures and branch circuit or lighting panelboards, the cutting of holes or openings, welding, painting, and scaffolding. See respective tables for these time frames.

3-PHASE, 3-POLE BUS DUCT ALUMINUM PLUG-IN TYPE

For Installation Height to 20 Feet

MANHOURS PER LINEAR FOOT

Approx. Weight Per Foot (Pounds)	Amperes	Manhours			
		Vertical Surface	Horizontal Suspended	Concrete Ceilings	Open Truss or Bar Joist
8	225	0.36	0.37	0.41	0.40
9	400	0.41	0.43	0.47	0.45
10	600	0.47	0.49	0.54	0.52
11	800	0.53	0.55	0.61	0.58
12	1000	0.59	0.61	0.67	0.65
14	1350	0.66	0.69	0.76	0.72
16	1600	0.73	0.76	0.84	0.80
18	2000	0.82	0.85	0.94	0.90
22	2500	0.97	1.01	1.11	1.07
26	3000	1.26	1.31	1.44	1.38
32	4000	1.38	1.44	1.58	1.52

For Installation Height to 25 Feet

MANHOURS PER LINEAR FOOT

Approx. Weight Per Foot (Pounds)	Amperes	Manhours			
		Vertical Surface	Horizontal Suspended	Concrete Ceilings	Open Truss or Bar Joist
8	225	0.37	0.38	0.42	0.41
9	400	0.42	0.44	0.48	0.46
10	600	0.48	0.50	0.55	0.53
11	800	0.55	0.57	0.63	0.60
12	1000	0.61	0.63	0.70	0.67
14	1350	0.68	0.71	0.78	0.75
16	1600	0.75	0.78	0.86	0.82
18	2000	0.84	0.87	0.96	0.92
22	2500	1.00	1.04	1.14	1.10
26	3000	1.30	1.35	1.49	1.43
32	4000	1.42	1.48	1.62	1.56

Manhours include checking out of job storage, handling, job hauling, layout, measuring, and installing duct runs up to 300 linear feet.

For duct runs from 301 feet to 500 feet decrease manhours 30%. For duct runs from 501 feet to 1,000 feet decrease manhours 40%.

For 3-phase, 4-pole (1/2 neutral) bus duct increase manhours 20%.

For 3-phase, 4-pole (full neutral) bus duct increase manhours 25%.

For 4-pole, plug-in type bus duct increase manhours 20%.

Manhours exclude installation of duct fittings, hangers, enclosures and branch circuit or lighting panelboards, the cutting of holes or openings, welding, painting, and scaffolding. See respective tables for these time frames.

For heights above 25 feet increase manhours 3% for each additional 5 feet or fraction thereof.

3-PHASE, 3-POLE BUS DUCT FITTINGS ALUMINUM PLUG-IN TYPE

For Installation Height to 10 Feet

MANHOURS EACH

Amperes	Manhours										
	A	B	C	D	E	F	G	H	I	J	K
225	3.19	3.63	4.18	3.19	1.90	3.19	3.63	3.19	1.65	2.40	3.19
400	3.63	4.18	4.73	3.63	2.15	3.63	4.18	3.63	1.89	2.64	3.63
600	4.18	4.73	5.50	4.18	2.40	4.18	4.73	4.18	2.35	2.90	4.18
800	4.73	5.39	6.27	4.73	2.65	4.73	5.39	4.73	2.40	3.17	4.73
1000	5.28	6.16	7.15	5.28	2.90	5.28	6.16	5.28	2.75	3.48	5.28
1350	5.83	7.04	8.14	5.83	3.16	5.83	7.04	5.83	3.10	3.83	5.83
1600	6.49	8.03	9.46	6.49	3.51	6.49	8.03	6.49	3.45	4.25	6.49
2000	7.26	9.46	11.00	7.26	3.99	7.26	9.46	7.26	3.87	4.77	7.26
2500	8.58	11.00	13.20	8.58	4.82	8.58	11.00	8.58	4.80	5.94	8.58
3000	10.78	13.42	15.62	10.78	6.03	10.78	13.42	10.78	5.94	8.03	10.78
4000	14.52	16.28	19.58	14.52	7.99	14.52	16.28	14.52	8.25	11.88	14.52

For Installation Height to 15 Feet

MANHOURS EACH

Amperes	Manhours										
	A	B	C	D	E	F	G	H	I	J	K
225	3.32	3.74	4.31	3.32	2.01	3.32	3.74	3.32	1.75	2.54	3.32
400	3.74	4.31	4.87	3.74	2.28	3.74	4.31	3.74	2.00	2.80	3.74
600	4.31	4.87	5.61	4.31	2.54	4.31	4.87	4.31	2.49	3.07	4.31
800	4.87	5.50	6.40	4.87	2.81	4.87	5.50	4.87	2.54	3.30	4.87
1000	5.39	6.28	7.28	5.39	3.07	5.39	6.28	5.39	2.92	3.62	5.39
1350	5.95	7.16	8.28	5.95	3.29	5.95	7.16	5.95	3.22	3.94	5.95
1600	6.62	8.17	9.61	6.62	3.65	6.62	8.17	6.62	3.59	4.38	6.62
2000	7.39	9.61	11.18	7.39	4.11	7.39	9.61	7.39	3.99	4.91	7.39
2500	8.73	11.18	13.38	8.73	4.96	8.73	11.18	8.73	4.94	6.06	8.73
3000	10.95	13.61	15.81	10.95	6.15	10.95	13.61	10.95	6.06	8.17	10.95
4000	14.69	16.48	19.78	14.69	8.13	14.69	16.48	14.69	8.39	12.04	14.69

Code:

A—90° Elbows
B—Tees
C—Crosses
D—Offsets
E—End Closures
F—Reducers (Unfused)

G—Offset Elbows
H—Expansion Sections
I—Panelboard Connections
J—Cable Tap Boxes
K—Transfer Tap Section

Manhours include checking out of job storage, handling, job hauling and installing of fittings as outlined. For duct runs up to 300 linear feet.

Manhours exclude installation of other items and other operations. See respective tables for these time frames.

For duct runs from 301 feet to 500 feet and from 501 feet to 1,000 feet, decrease manhours 30 and 40% respectively.

For 3-phase, 4-pole, (1/2 neutral) bus duct fittings increase manhours 20%.

For 3-phase, 4-pole (full neutral) bus duct fittings increase manhours 25%.

For 4-pole, plug-in type bus duct fittings increase manhours 20%.

3-PHASE, 3-POLE BUS DUCT FITTINGS ALUNINUM PLUG-IN TYPE

For Installation Height to 20 Feet

MANHOURS EACH

Amperes	Manhours										
	A	B	C	D	E	F	G	H	I	J	K
225	3.42	3.85	4.43	3.42	2.09	3.42	3.85	3.42	1.82	2.65	3.42
400	3.85	4.43	5.02	3.85	2.37	3.85	4.43	3.85	2.08	2.91	3.85
600	4.43	5.02	5.72	4.43	2.65	4.43	5.02	4.43	2.59	3.20	4.43
800	5.02	5.61	6.52	5.02	2.92	5.02	5.61	5.02	2.65	3.39	5.02
1000	5.49	6.41	7.40	5.49	3.20	5.49	6.41	5.49	3.03	3.73	5.49
1350	6.07	7.29	8.43	6.07	3.38	6.07	7.29	6.07	3.32	4.06	6.07
1600	6.75	8.31	9.77	6.75	3.76	6.75	8.31	6.75	3.70	4.51	6.75
2000	7.52	9.77	11.35	7.52	4.23	7.52	9.77	7.52	4.11	5.06	7.52
2500	8.88	11.35	13.57	8.88	5.11	8.88	11.35	8.88	5.09	6.18	8.88
3000	11.13	13.80	16.00	11.13	6.27	11.13	13.80	11.13	6.18	8.31	11.13
4000	14.87	16.67	19.97	14.87	8.27	14.87	16.67	14.87	6.54	12.21	14.87

For Installation Height to 25 Feet

Amperes	Manhours										
	A	B	C	D	E	F	G	H	I	J	K
225	3.49	3.93	4.52	3.49	2.14	3.49	3.93	3.49	1.86	2.70	3.49
400	3.93	4.52	5.12	3.93	2.42	3.93	4.52	3.93	2.13	2.97	3.93
600	4.52	5.12	5.78	4.52	2.70	4.52	5.12	4.52	2.64	3.26	4.52
800	5.12	5.66	6.59	5.12	2.98	5.12	5.66	5.12	2.70	3.46	5.12
1000	5.55	6.47	7.46	5.55	3.26	5.55	6.47	5.55	3.09	3.80	5.55
1350	6.13	7.34	8.49	6.13	3.45	6.13	7.34	6.13	3.39	4.14	6.13
1600	6.82	8.38	9.82	6.82	3.83	6.82	8.38	6.82	3.77	4.60	6.82
2000	7.57	9.82	11.42	7.57	4.32	7.57	9.82	7.57	4.19	5.16	7.57
2500	8.95	11.42	13.64	8.95	5.22	8.95	11.42	8.95	5.25	6.24	8.95
3000	11.19	13.87	16.06	11.19	6.34	11.19	13.87	11.19	6.24	8.38	11.19
4000	14.93	16.74	20.03	14.93	8.33	14.93	16.74	14.93	8.61	12.27	14.93

Code:

A—90° Elbows
B—Tees
C—Crosses
D—Offsets
E—End Closures
F—Reducers (Unfused)

G—Offset Elbows
H—Expansion Sections
I—Panelboard Connections
J—Cable Tap Boxes
K—Transfer Tap Section

Manhours include checking out of job storage, handling, job hauling and installing of fittings as outlined. For duct runs up to 300 linear feet.

Manhours exclude installation of other items and other operations. See respective tables for these time frames.

For duct runs from 301 feet to 500 feet and from 501 feet to 1,000 feet, decrease manhours 30 and 40% respectively.

For 3-phase, 4-pole, (1/2 neutral) bus duct fittings increase manhours 20%.

For 3-phase, 4-pole (full neutral) bus duct fittings increase manhours 25%.

For 4-pole, plug-in type bus duct fittings increase manhours 20%.

For heights above 25 feet increase manhours 3% for each additional 5 feet or fraction thereof.

FUSIBLE AND NONFUSIBLE SWITCHES
250- or 600-Volt Plug-In and Bolt-On Types
For Heights to 10 Feet

MANHOURS EACH

| Amperes | Conductors | | Manhours | | | | | | | |
| | Quantity | Size | Fusible | | | | Non-Fusible | | | |
			A	B	C	D	E	F	G	H
30	1	#10	1.54	1.54	1.76	1.98	1.32	1.47	1.54	1.76
60	1	#6	1.87	2.15	2.35	2.42	1.60	1.87	1.87	2.15
100	1	#2	2.53	2.86	2.97	3.30	2.31	2.64	2.75	3.08
200	1	#3/0	4.51	4.95	5.17	5.61	4.07	4.51	4.73	5.17
400	1	250 MCM	11.00	11.55	12.10	12.65	10.45	11.07	11.55	12.17
600	1	500 MCM	16.06	17.16	18.26	19.36	15.40	16.50	17.60	18.70
800	2	250 MCM	22.00	23.10	24.20	25.30	20.90	22.00	23.10	24.20
1200	2	500 MCM	29.70	31.90	34.10	36.30	28.60	30.80	33.00	35.20

For Heights to 15 Feet

MANHOURS EACH

| Amperes | Conductors | | Manhours | | | | | | | |
| | Quantity | Size | Fusible | | | | Non-Fusible | | | |
			A	B	C	D	E	F	G	H
30	1	#10	1.69	1.69	1.94	2.18	1.45	1.62	1.69	1.94
60	1	#6	2.06	2.37	2.59	2.66	1.76	2.06	2.06	2.37
100	1	#2	2.78	3.15	3.27	3.63	2.54	2.90	3.03	3.39
200	1	#3/0	4.96	5.45	5.69	6.17	4.48	4.96	5.20	5.69
400	1	250 MCM	12.10	12.71	13.31	13.92	11.50	12.18	12.71	13.39
600	1	500 MCM	17.67	18.88	20.09	21.30	16.94	18.15	19.36	20.57
800	2	250 MCM	24.20	25.41	26.62	27.83	22.99	24.20	25.41	26.62
1200	2	500 MCM	32.17	35.09	37.51	39.93	31.46	33.88	36.30	38.72

Code:

A—2-Pole, 2-Wire   E—2-Pole, 2-Wire
B—2-Pole, 3-Wire   F—2-Pole, 3-Wire
C—3-Pole, 3-Wire   G—3-Pole, 3-Wire
D—3-Pole, 4-Wire   H—3-Pole, 4-Wire

Manhours include checking out of job storage, handling, job hauling, installing, and connecting copper conductors as outlined.

For aluminum conductors deduct 5% from the manhours.

Manhours exclude installation of other items, and scaffolding. See respective tables for these time frames.

FUSIBLE AND NONFUSIBLE SWITCHES
250- or 600-Volt Plug-In and Bolt-On Types
For Heights to 20 Feet

MANHOURS EACH

Amperes	Conductors		Manhours							
	Quantity	Size	Fusible				Non-Fusible			
			A	B	C	D	E	F	G	H
30	1	#10	1.83	1.83	2.09	2.35	1.57	1.75	1.83	2.09
60	1	#6	2.72	2.55	2.79	2.87	1.90	2.22	2.22	2.55
100	1	#2	3.01	3.40	3.53	3.92	2.74	3.14	3.27	3.66
200	1	#3/0	5.36	5.88	6.14	6.66	4.84	5.36	5.62	6.14
400	1	250 MCM	13.07	13.72	14.37	15.03	12.41	13.15	13.72	14.56
600	1	500 MCM	19.08	20.39	21.69	23.00	18.30	19.60	20.91	22.22
800	2	250 MCM	26.14	27.44	28.75	30.06	24.83	26.14	27.44	28.75
1200	2	500 MCM	35.28	37.90	40.51	43.12	33.98	36.59	39.20	41.82

Code:

A—2-Pole, 2-Wire	E—2-Pole, 2-Wire
B—2-Pole, 3-Wire	F—2-Pole, 3-Wire
C—3-Pole, 3-Wire	G—3-Pole, 3-Wire
D—3-Pole, 4-Wire	H—3-Pole, 4-Wire

Manhours include checking out of job storage, handling, job hauling, installing, and connecting copper conductors as outlined.

For aluminum conductors deduct 5% from the manhours.

Manhours exclude installation of other items, and scaffolding. See respective tables for these time frames.

PLUG-IN MOLDED CASE CIRCUIT BREAKERS
For Heights to 10 Feet

MANHOURS EACH

Amperage Size	Type of Frame	Pressure Wire Connectors	Manhours			
			A	B	C	D
15-50	E, EF, EH	#10	1.65	1.93	1.93	2.20
70-100	E, EF, EH	#3	1.92	2.20	2.48	2.75
70-125	FJ, FK	#2	2.48	2.75	3.03	3.58
150-225	FJ, FK	#2/0	4.95	5.50	5.78	6.33
125-225	JJ, JK	#3/0	6.60	7.15	7 43	8.25
250-400	JJ, JK	350 MCM	11.00	12.10	12.65	13.75
70-100	JL	#3	4.40	4.68	4.95	5.50
125-225	JL	#3/0	6.60	7.15	7.43	8.25
250-400	JL	350 MCM	11.00	12.10	12.65	13.75
125-300	KM	#4/0	8.80	9.35	9.63	10.36
350-600	KM	500 MCM	15.40	17.05	17.60	19.25
700-800	KM	3-350 MCM	23.10	26.40	28.24	31.35
800-1000	KP	4-300 MCM	–	–	30.80	34.10
1200	KP	4-500 MCM	–	–	40.70	47.30
70-100	CJ	#3	6.60	6.88	7.15	7.98
150-225	CJ	#2/0	8.80	9.35	9.63	10.73
250-400	CJ	350 MCM	13.20	14.30	14.85	16.50
600	CM	2-350 MCM	–	–	24.20	26.40
700-800	CM	3-350 MCM	–	–	30.25	33.19
800-1000	CP	5-300 MCM	–	–	35.20	40.70
1200-1400	CP	5-350 MCM	–	–	38.50	44.00
1600	CP	5-500 MCM	–	–	47.30	56.11
1800-2000	CR	5-600 MCM	–	–	58.67	68.94
800-1000	HP	5-300 MCM	–	–	35.20	40.70
1200-1400	HP	5-350 MCM	–	–	38.50	44.00
1600	HP	5-500 MCM	–	–	47.30	56.11
1800-2000	HR	5-600 MCM	–	–	58.67	68.94

Code:

A—2-Pole, 2-Wire

B—2-Pole, 3-Wire

C—3-Pole, 3-Wire

D—3-Pole, 4-Wire

Manhours include checking out of job storage, handling, job hauling, and installing circuit breakers as listed.

Manhours exclude installation of ducts, boxes or other devices, and scaffolding. See respective tables for these time frames.

PLUG-IN MOLDED CASE CIRCUIT BREAKERS

For Heights to 15 Feet

MANHOURS EACH

Amperage Size	Type of Frame	Pressure Wire Connectors	Manhours			
			A	B	C	D
15-50	E, EF, EH	#10	1.82	2.12	2.12	2.42
70-100	E, EF, EH	#3	2.11	2.42	2.73	3.03
70-125	FJ, FK	#2	2.73	3.03	3.33	3.94
150-225	FJ, FK	#2/0	5.45	6.05	6.36	6.96
125-225	JJ, JK	#3/0	7.26	7.87	8.17	9.08
250-400	JJ, JK	350 MCM	12.10	13.31	13.92	15.13
70-100	JL	#3	4.84	5.15	5.45	6.05
125-225	JL	#3/0	7.26	7.87	8.17	9.08
250-400	JL	350 MCM	12.10	13.31	13.92	15.13
125-300	KM	#4/0	9.68	10.29	10.59	11.40
350-600	KM	500 MCM	16.94	18.76	19.36	21.18
700-800	KM	3-350 MCM	25.41	29.04	31.06	34.49
800-1000	KP	4-300 MCM	–	–	33.88	37.51
1200	KP	4-500 MCM	–	–	44.77	52.03
70-100	CJ	#3	7.26	7.57	7.87	8.78
150-225	CJ	#2/0	9.68	10.29	10.59	11.80
250-400	CJ	350 MCM	14.52	15.73	16.34	18.15
600	CM	2-350 MCM	–	–	26.62	29.04
700-800	CM	3-350 MCM	–	–	33.28	36.51
800-1000	CP	5-300 MCM	–	–	38.72	44.77
1200-1400	CP	5-350 MCM	–	–	42.35	48.40
1600	CP	5-500 MCM	–	–	52.03	61.72
1800-2000	CR	5-600 MCM	–	–	64.54	75.83
800-1000	HP	5-300 MCM	–	–	38.72	44.77
1200-1400	HP	5-350 MCM	–	–	42.35	48.40
1600	HP	5-500 MCM	–	–	52.03	61.72
1800-2000	HR	5-600 MCM	–	–	64.54	75.83

Code:

    A—2-Pole, 2-Wire
    B—2-Pole, 3-Wire
    C—3-Pole, 3-Wire
    D—3-Pole, 4-Wire

Manhours include checking out of job storage, handling, job hauling, and installing circuit breakers as listed.

Manhours exclude installation of ducts, boxes or other devices, and scaffolding. See respective tables for these time frames.

PLUG-IN MOLDED CASE CIRCUIT BREAKERS
For Heights to 20 Feet

MANHOURS EACH

Amperage Size	Type of Frame	Pressure Wire Connectors	Manhours			
			A	B	C	D
15-50	E, EF, EH	#10	1.97	2.29	2.29	2.61
70-100	E, EF, EH	#3	2.28	2.61	2.95	3.27
70-125	FJ, FK	#2	2.95	3.27	3.60	4.26
150-225	FJ, FK	#2/0	5.89	6.53	6.87	7.52
125-225	JJ, JK	#3/0	7.84	8.50	8.82	9.81
250-400	JJ, JK	350 MCM	13.07	14.37	15.03	16.34
70-100	JL	#3	5.23	5.56	5.88	6.53
125-225	JL	#3/0	7.84	8.50	8.82	9.81
250-400	JL	350 MCM	13.07	14.37	15.03	16.34
125-300	KM	#4/0	10.45	11.11	11.43	12.31
350-600	KM	500 MCM	18.30	20.26	20.91	22.87
700-800	KM	3-350 MCM	27.44	31.36	33.54	37.25
800-1000	KP	4-300 MCM	–	–	36.59	40.51
1200	KP	4-500 MCM	–	–	47.70	56.19
70-100	CJ	#3	7.84	8.18	8.50	9.84
150-225	CJ	#2/0	10.45	11.11	11.44	12.74
250-400	CJ	350 MCM	15.68	16.99	17.65	19.60
600	CM	2-350 MCM	–	–	28.75	31.36
700-800	CM	3-350 MCM	–	–	35.95	39.43
800-1000	CP	5-300 MCM	–	–	41.82	47.70
1200-1400	CP	5-350 MCM	–	–	45.74	52.27
1600	CP	5-500 MCM	–	–	56.19	66.66
1800-2000	CR	5-600 MCM	–	–	69 70	81.90
800-1000	HP	5-300 MCM	–	–	41.82	48.35
1200-1400	HP	5-350 MCM	–	–	45.74	52.27
1600	HP	5-500 MCM	–	–	56.19	66.66
1800-2000	HR	5-600 MCM	–	–	69.70	81.90

Code:

A—2-Pole, 2-Wire
B—2-Pole, 3-Wire
C—3-Pole, 3-Wire
D—3-Pole, 4-Wire

Manhours include checking out of job storage, handling, job hauling, and installing circuit breakers as listed.

Manhours exclude installation of ducts, boxes or other devices, and scaffolding. See respective tables for these time frames.

3-PHASE, 3-POLE BUS DUCT
COPPER—FEEDER TYPE

For Installation Height to 10 Feet

MANHOURS PER LINEAR FOOT

Approx. Weight Per Foot (Pounds)	Amperes	Manhours			
		Vertical Surface	Horizontal Suspended	Concrete Ceilings	Open Truss or Bar Joist
12	800	0.53	0.55	0.61	0.58
14	1000	0.58	0.60	0.66	0.64
18	1350	0.65	0.68	0.74	0.71
22	1600	0.78	0.81	0.89	0.86
25	2000	0.92	0.96	1.05	1.01
32	2500	1.09	1.13	1.25	1.20
40	3000	1.34	1.39	1.53	1.47
54	4000	1.67	1.74	1.91	1.83
66	5000	2.02	2.10	2.31	2.22

For Installation Height to 15 Feet

MANHOURS PER LINEAR FOOT

Approx. Weight Per Foot (Pounds)	Amperes	Manhours			
		Vertical Surface	Horizontal Suspended	Concrete Ceilings	Open Truss or Bar Joist
12	800	0.57	0.59	0.65	0.63
14	1000	0.63	0.66	0.72	0.69
18	1350	0.70	0.73	0.80	0.77
22	1600	0.84	0.87	0.96	0.92
25	2000	0.99	1.03	1.13	1.09
32	2500	1.18	1.23	1.35	1.30
40	3000	1.45	1.51	1.66	1.59
54	4000	1.80	1.87	2.06	1.98
66	5000	2.18	2.27	2.49	2.39

Manhours include checking out of job storage, handling, job hauling, layout, measuring, and installing duct runs up to 300 linear feet.

For duct runs from 301 feet to 500 feet decrease manhours 40%. For duct runs from 501 feet to 1,000 feet decrease manhours 40%.

For 3-phase, 4-pole (1/2 neutral) bus duct increase manhours 20%.

For 3-phase, 4-pole (full neutral) bus duct increase manhours 25%.

For 4-pole, feeder type bus duct increase manhours 20%.

Manhours exclude installation of duct fittings, hangers, enclosures and branch circuit or lighting panelboards, the cutting of holes or openings, welding, painting, and scaffolding. See respective tables for these time frames.

3-PHASE, 3-POLE BUS DUCT
COPPER—FEEDER TYPE

For Installation Height to 20 Feet

MANHOURS PER LINEAR FOOT

Approx. Weight Per Foot (Pounds)	Amperes	Manhours			
		Vertical Surface	Horizontal Suspended	Concrete Ceilings	Open Truss or Bar Joist
12	800	0.59	0.61	0.67	0.65
14	1000	0.66	0.69	0.76	0.72
18	1350	0.73	0.76	0.84	0.80
22	1600	0.87	0.90	1.00	0.96
25	2000	1.03	1.07	1.18	1.13
32	2500	1.23	1.28	1.41	1.35
40	3000	1.51	1.57	1.73	1.56
54	4000	1.87	1.94	2.14	2.05
66	5000	2.27	2.36	2.60	2.49

For Installation Height to 25 Feet

MANHOURS PER LINEAR FOOT

Approx. Weight Per Foot (Pounds)	Amperes	Manhours			
		Vertical Surface	Horizontal Suspended	Concrete Ceilings	Open Truss or Bar Joist
12	800	0.61	0.63	0.70	0.67
14	1000	0.68	0.71	0.78	0.75
18	1350	0.75	0.78	0.86	0.82
22	1600	0.90	0.94	1.03	0.99
25	2000	1.06	1.10	1.21	1.16
32	2500	1.27	1.32	1.45	1.39
40	3000	1.56	1.62	1.78	1.71
54	4000	1.93	2.01	2.21	2.12
66	5000	2.34	2.43	2.68	2.57

Manhours include checking out of job storage, handling, job hauling, layout, measuring, and installing duct runs up to 300 linear feet.

For duct runs from 301 feet to 500 feet decrease manhours 40%. For duct runs from 501 feet to 1,000 feet decrease manhours 40%.

For 3-phase, 4-pole (1/2 neutral) bus duct increase manhours 20%.

For 3-phase, 4-pole (full neutral) bus duct increase manh urs 25%.

For 4-pole, feeder type bus duct increase manhours 20%.

Manhours exclude installation of duct fittings, hangers, enclosures and branch circuit or lighting panelboards, the cutting of holes or openings, welding, painting, and scaffolding. See respective tables for these time frames.

For heights above 25 feet increase manhours 3% for each additional 5 feet or fraction thereof.

3-PHASE, 3-POLE BUS DUCT FITTINGS COPPER—FEEDER TYPE

For Installation Height to 10 Feet

MANHOURS EACH

Amperes	Manhours														
	A	B	C	D	E	F	G	H	I	J	K	L	M	N	O
800	5.28	6.60	8.03	5.83	0.58	5.28	6.60	5.28	3.19	4.73	5.28	10.56	1.10	2.20	16.50
1000	5.83	7.37	8.80	6.49	0.65	5.83	7.37	5.83	3.63	5.39	5.83	11.88	1.38	2.75	22.00
1350	6.49	8.25	9.57	7.26	0.73	6.49	8.25	6.49	4.18	6.05	6.49	13.42	1.65	3.30	—
1600	7.81	9.46	10.78	8.14	0.81	7.81	9.46	7.81	4.73	7.15	7.81	15.40	1.93	3.85	—
2000	9.24	10.78	12.98	9.24	0.92	9.24	10.78	9.24	5.28	8.69	9.24	17.82	2.20	4.40	—
2500	10.89	12.98	15.84	10.67	1.07	10.89	12.98	10.89	5.83	10.78	10.78	21.56	2.75	5.50	—
3000	13.42	16.06	19.36	12.10	1.21	13.42	16.06	13.42	7.15	12.98	12.98	26.84	3.30	6.60	—
4000	16.72	19.80	23.98	14.52	1.45	16.72	19.80	16.72	9.46	16.50	16.50	33.22	4.40	8.80	—
5000	20.24	23.76	29.04	17.16	1.72	20.24	23.76	20.24	11.88	20.24	20.24	40.26	5.50	11.00	—

For Installation Height to 15 Feet

MANHOURS EACH

Amperes	Manhours														
	A	B	C	D	E	F	G	H	I	J	K	L	M	N	O
800	5.70	7.13	8.67	6.30	0.63	5.70	7.13	5.70	3.45	5.11	5.70	11.09	1.19	2.38	17.00
1000	6.30	7.96	9.50	7.01	0.70	6.30	7.96	6.30	3.92	5.82	6.30	12.47	1.49	2.97	22.22
1350	7.01	8.91	10.34	7.84	0.79	7.01	8.91	7.01	4.51	6.53	7.01	14.09	1.78	3.56	—
1600	8.43	10.22	11.32	8.79	0.87	8.43	10.22	8.43	5.11	7.72	8.43	15.86	2.08	4.16	—
2000	9.98	11.32	13.63	9.98	0.99	9.98	11.32	9.98	5.70	9.39	9.98	18.35	2.38	4.75	—
2500	11.43	13.63	16.32	11.20	1.16	11.43	13.63	11.43	6.30	11.32	11.32	21.78	2.97	5.94	—
3000	14.09	16.54	19.55	12.71	1.31	14.09	16.54	14.09	7.72	13.63	13.63	27.11	3.56	7.13	—
4000	17.22	20.00	24.22	15.24	1.57	17.22	20.00	17.22	10.22	17.00	17.00	35.55	4.75	9.50	—
5000	20.44	24.00	29.33	17.67	1.86	20.44	24.00	20.44	12.47	20.44	20.44	40.66	5.94	11.55	—

Code:

A—90° Elbows	I—Panelboard Connections
B—Tees	J—End Tap Boxes
C—Crosses	K—Transfer Tap Sections
D—Offsets	L—Weatherproof Service Heads
E—End Closures	M—Wall and Floor Flanges
F—Reducers (Unfused)	N—Fire and Weather Stops
G—Offset Elbows	O—Indoor Tap-offs
H—Expansion Sections	

Manhours include checking out job storage, handling, job hauling, and installing of fittings as outlined for duct runs up to 300 linear feet.

Manhours exclude installation of other items and other operations. See respective tables for these time frames.

For duct runs from 301 feet to 500 feet and from 501 feet to 1,000 feet, decrease manhours 30 and 40% respectively.

For 3-phase, 4-pole (1/2 neutral) bus duct fittings increase manhours 20%.

For 3-phase, 4-pole (full neutral) bus duct fittings increase manhours 25%.

For 4-pole, feeder type bus duct fittings increase manhours 20%.

3-PHASE, 3-POLE BUS DUCT FITTINGS
COPPER—FEEDER TYPE

For Installation Height to 20 Feet

MANHOURS EACH

Amperes	Manhours													
	A	B	C	D	E	F	G	H	I	J	K	L	M	N	O
800	5.93	7.41	9.02	6.55	0.65	5.93	7 41	5.93	3.58	5.31	5.93	11.42	1.24	1.47	17.16
1000	6.55	8.28	9.88	7.29	0.73	6.55	8.28	6.55	4.08	6.05	6.55	12.85	1.55	3.09	23.33
1350	7.29	9.27	10.75	8.15	0.82	7.29	9.27	7.29	4.69	6.80	7.29	14.51	1.85	3.71	—
1600	8.77	10.63	11.66	9.14	0.91	8.77	10.63	8.77	5.31	8.03	8.77	16.02	2.17	4.32	—
2000	10.38	11.66	14.04	10.38	1.03	10.58	11.66	10.38	5.93	9.76	10.38	18.54	2.47	4.94	—
2500	11.78	14.04	16.48	11.54	1.20	11.78	14.04	11.78	6.55	11.66	11.66	21.88	3.09	6.18	—
3000	14.51	16.71	19.65	13.09	1.36	14.51	16.71	14.51	8.03	14.04	14.04	27.24	3.71	7.41	—
4000	17.39	20.10	24.34	15.70	1.63	17.39	20.10	17.39	10.63	17.16	17.16	33.72	4.94	9.88	—
5000	20.54	24.12	29 48	17.85	1.93	20.54	24.12	20.54	12.85	20.54	20.54	40.87	6.18	11.90	—

For Installation Height to 25 Feet

MANHOURS EACH

Amperes	Manhours													
	A	B	C	D	E	F	G	H	I	J	K	L	M	N	O
800	6.11	7.64	9.29	6.74	0.67	6.11	7.64	6.11	3.69	5.47	6.11	11.54	1.27	2.55	17.25
1000	6.74	8.53	10.18	7.51	0.75	6.74	8.53	6.74	4.20	6.24	6.74	13.11	1.60	3.18	22.39
1350	7.51	9.54	11.07	8.40	0.84	7.51	9.54	7.51	4.84	7.00	7.51	14.80	1.91	3.82	—
1600	9.04	10.94	11.89	9.41	0.94	9.04	10.94	9.04	5.47	8.27	9.04	16.10	2.23	4.45	—
2000	10.69	11.89	14.32	10.69	1.06	10.69	11.85	10.69	6.11	10.05	10.69	18.65	2.55	5.09	—
2500	12.01	14.32	16.56	11.77	1.24	12.01	14.32	12.01	6.74	11.89	11.89	21.94	3.18	6.36	—
3000	14.80	16.79	19.70	13.35	1.40	14.80	16.79	14.80	8.27	14.32	14.32	27.31	3.82	7.64	—
4000	17 48	20.15	24.40	16.02	1.68	17.48	20.15	17.48	10.94	17.25	17.25	33.80	5.09	10.18	—
5000	20.60	24.18	29.55	17.94	1.99	20.60	24.18	20.60	13.11	20.60	20.60	40.97	6.36	12.13	—

Code:

A—90° Elbows
B—Tees
C—Crosses
D—Offsets
E—End Closures
F—Reducers (Unfused)
G—Offset Elbows
H—Expansion Sections

I—Panelboard Connections
J—End Tap Boxes
K—Transfer Tap Sections
L—Weatherproof Service Heads
M—Wall and Floor Flanges
N—Fire and Weather Stops
O—Indoor Tap-offs

Manhours include checking out of job storage, handling, job hauling, and installing of fittings as outlined for duct runs up to 300 linear feet.

Manhours exclude installation of other items and other operations. See respective tables for these time frames.

For duct runs from 301 feet to 500 feet and from 501 feet to 1,000 feet, decrease manhours 30 and 40% respectively.

For 3-phase, 4-pole (1/2 neutral) bus duct fittings increase manhours 20%.

For 3-phase, 4-pole (full neutral) bus duct fittings increase manhours 25%.

For 4-pole, feeder type bus duct fittings increase manhours 20%.

3-PHASE, 3-POLE BUS DUCT
ALUMINUM—FEEDER TYPE
For Installation Height to 10 Feet

MANHOURS PER LINEAR FOOT

Approx. Weight Per Foot (Pounds)	Amperes	Manhours			
		Vertical Surface	Horizontal Suspended	Concrete Ceilings	Open Truss or Bar Joist
9	800	0.42	0.44	0.48	0.46
10	1000	0.47	0.49	0.54	0.52
12	1350	0.54	0.56	0.62	0.59
15	1600	0.62	0.64	0.71	0.68
17	2000	0.73	0.76	0.84	0.80
22	2500	0.87	0.90	1.00	0.96
30	3000	1.09	1.13	1.25	1.20
42	4000	1.43	1.49	1.64	1.57
54	5000	1.80	1.87	2.06	1.98

For Installation Height to 15 Feet

MANHOURS PER LINEAR FOOT

Approx. Weight Per Foot (Pounds)	Amperes	Manhours			
		Vertical Surface	Horizontal Suspended	Concrete Ceilings	Open Truss or Bar Joist
9	800	0.45	0.47	0.51	0.49
10	1000	0.51	0.53	0.58	0.56
12	1350	0.58	0.60	0.66	0.64
15	1600	0.67	0.70	0.77	0.74
17	2000	0.79	0.82	0.90	0.87
22	2500	0.94	0.98	1.08	1.03
30	3000	1.18	1.23	1.35	1.30
42	4000	1.54	1.60	1.76	1.69
54	5000	1.94	2.02	2.22	2.13

Manhours include checking out of job storage, handling, job hauling, layout, measuring, and installing duct runs up to 300 linear feet.

For duct runs from 301 feet to 500 feet decrease manhours 30%. For duct runs from 501 feet to 1,000 feet decrease manhours 40%.

For 3-phase, 4-pole (1/2 neutral) bus duct increase manhours 20%.

For 3-phase, 4-pole (full neutral) bus duct increase manhours 25%.

For 4-pole, plug-in type bus duct increase manhours 20%.

Manhours exclude installation of duct fittings, hangers, enclosures and branch circuit or lighting panelboards, the cutting of holes or openings, welding, painting, and scaffolding. See respective tables for these time frames.

3-PHASE, 3-POLE BUS DUCT ALUMINUM—FEEDER TYPE

For Installation Height to 20 Feet

MANHOURS PER LINEAR FOOT

Approx. Weight Per Foot (Pounds)	Amperes	Manhours			
		Vertical Surface	Horizontal Suspended	Concrete Ceilings	Open Truss or Bar Joist
9	800	0.47	0.49	0.54	0.52
10	1000	0.53	0.55	0.61	0.68
12	1350	0.60	0.62	0.69	0.66
15	1600	0.70	0.73	0.80	0.77
17	2000	0.82	0.85	0.94	0.90
22	2500	0.98	1.02	1.12	1.08
30	3000	1.23	1.28	1.41	1.35
42	4000	1.60	1.66	1.73	1.86
54	5000	2.02	2.10	2.31	2.22

For Installation Height to 25 Feet

MANHOURS PER LINEAR FOOT

Approx. Weight Per Foot (Pounds)	Amperes	Manhours			
		Vertical Surface	Horizontal Suspended	Concrete Ceilings	Open Truss or Bar Joist
9	800	0.48	0.50	0.55	0.53
10	1000	0.55	0.57	0.63	0.60
12	1350	0.62	0.64	0.71	0.68
15	1600	0.72	0.75	0.82	0.79
17	2000	0.84	0.87	0.96	0.92
22	2500	1.01	1.05	1.16	1.11
30	3000	1.27	1.32	1.45	1.39
42	4000	1.65	1.72	1.89	1.81
54	5000	2.08	2.16	2.38	2.28

Manhours include checking out of job storage, handling, job hauling, layout, measuring, and installing duct runs up to 300 linear feet.

For duct runs from 301 feet to 500 feet decrease manhours 30%. For duct runs from 501 feet to 1,000 feet decrease manhours 40%.

For 3-phase, 4-pole (1/2 neutral) bus duct increase manhours 20%.

For 3-phase, 4-pole (full neutral) bus duct increase manhours 25%.

For 4-pole, plug-in type bus duct increase manhours 20%.

Manhours exclude installation of duct fittings, hangers, enclosures and branch circuit or lighting panelboards, the cutting of holes or openings, welding, painting, and scaffolding. See respective tables for these time frames.

For heights above 25 feet increase manhours 3% for each additional 5 feet or fraction thereof.

3-PHASE, 3-POLE BUS DUCT FITTINGS
ALUMIMUM—FEEDER TYPE
For Installation Height to 10 Feet

MANHOURS each

Amperes	Manhours														
	A	B	C	D	E	F	G	H	I	J	K	L	M	N	O
800	4.18	5.28	6.49	4.73	0.43	4.18	5.28	4.18	2.53	3.41	4.18	9.46	1.10	2.20	16.50
1000	4.73	5.83	7.26	5.28	0.53	4.73	5.83	4.73	2.86	3.96	4.73	10.78	1.38	2.75	22.00
1350	5.39	6.49	8.36	5.94	0.59	5.39	6.49	5.39	3.30	4.73	5.39	11.88	1.65	3.30	–
1600	6.16	7.26	9.57	6.49	0.65	6.61	7.26	6.16	3.96	5.50	6.16	14.52	1.93	3.85	–
2000	7.26	8.36	11.11	7.26	0.73	7.26	8.36	7.26	4.73	6.27	7.26	17.16	2.20	4.40	–
2500	8.69	10.34	13.20	8.69	0.87	8.69	10.34	8.69	5.50	7.59	8.69	20.24	2.75	5.50	–
3000	10.89	13.20	15.84	10.78	1.08	10.89	13.20	10.89	6.27	9.57	10.89	23.98	3.30	6.60	–
4000	14.30	16.72	19.58	14.30	1.43	14.30	16.72	14.30	7.92	12.98	14.30	29.04	4.40	8.80	–
5000	18.04	20.24	23.20	18.04	1.80	18.04	20.24	18.04	10.12	16.72	18.04	34.10	5.50	11.00	–

For Installation Height to 15 Feet

MANHOURS each

Amperes	Manhours														
	A	B	C	D	E	F	G	H	I	J	K	L	M	N	O
800	4.51	5.70	7.01	5.11	0.46	4.51	5.70	4.51	2.73	3.68	4.51	10.22	1.19	2.38	17.00
1000	5.11	6.30	7.84	5.70	0.57	5.11	6.30	5.11	3.09	4.28	5.11	11.32	1.49	2.97	22.22
1350	5.82	7.01	9.03	6.42	0.64	5.82	7.01	5.82	4.56	5.11	5.82	12.47	1.78	3.56	–
1600	6.65	7.84	10.34	7.01	0.70	6.65	7.84	6.65	4.28	5.94	6.65	15.24	2.08	4.16	–
2000	7.84	9.03	11.67	7.84	0.79	7.84	9.03	7.84	5.11	6.77	7.84	17.67	2.38	4.75	–
2500	9.39	10.86	13.86	9.39	0.94	9.39	10.86	9.39	5.94	8.20	9.39	20.44	2.97	5.94	–
3000	11.43	13.86	16.32	11.32	1.17	11.43	13.86	11.43	6.77	10.34	11.43	24.22	3.56	7.13	–
4000	15.02	17.22	19.78	15.02	1.54	15.02	17.22	15.02	8.55	13.63	15.02	29.33	4.75	9.50	–
5000	18.58	20.44	23.43	18.58	1.94	18.58	20.44	18.58	10.63	17.22	18.58	34.44	5.94	11.55	–

Code:

A—90° Elbows
B—Tees
C—Crosses
D—Offsets
E—End Closures
F—Reducers (Unfused)
G—Offset Elbows
H—Expansion Sections

I—Panelboard Connections
J—End Tap Boxes
K—Transfer Tap Sections
L—Weatherproof Service Heads
M—Wall and Floor Flanges
N—Fire and Weather Stops
O—Indoor Tap-offs

Manhours include checking out of job storage, handling, job hauling, and installing of fittings as outlined for duct runs up to 300 linear feet.

Manhours exclude installation of other items and other operations. See respective tables for these time frames.

For duct runs from 301 feet to 500 feet and from 501 feet to 1,000 feet, decrease manhours 30 and 40% respectively.

For 3-phase, 4-pole (1/2 neutral) bus duct fittings increase manhours 20%.

For 3-phase, 4-pole (full neutral) bus duct fittings increase manhours 25%.

For 4-pole, feeder type bus duct fittings increase manhours 20%.

For Heights above 25 feet increase manhours 3% for each additional 5 feet or fraction thereof.

3-PHASE, 3-POLE BUS DUCT FITTINGS ALUMINUM—FEEDER TYPE

For Installation Height to 20 Feet

MANHOURS each

Amperes	Manhours														
	A	B	C	D	E	F	G	H	I	J	K	L	M	N	O
800	4.69	5.93	7.29	5.31	0.48	4.69	5.93	4.69	2.84	3.83	4.69	10.63	1.24	2.47	17.16
1000	5.31	6.55	8.15	5.93	0.60	5.31	6.55	5.31	3.21	4.45	5.31	11.66	1.55	3.09	23.33
1350	6.05	7.29	9.39	6.67	0.66	6.05	7.29	6.05	3.71	5.31	6.05	12.85	1.85	3.71	–
1600	6.92	8.15	10.75	7.29	0.73	6.92	8.15	6.92	4.45	6.18	6.92	15.70	2.17	4.32	–
2000	8.15	9.39	12.02	8.15	0.82	8.15	9.39	8.15	5.31	7.04	8.15	17.85	2.47	4.94	–
2500	9.76	11.18	14.28	9.76	0.98	9.76	11.18	9.76	6.18	8.53	9.76	20.54	3.09	6.18	–
3000	11.78	14.28	16.48	11.66	1.21	11.78	14.28	11.78	7.04	10.75	11.78	24.34	3.71	7 41	–
4000	15.47	17.39	19.87	15.47	1.61	15.47	17.39	15.47	8.90	14.04	15.47	29.48	4.94	9.88	–
5000	18.58	20.54	23.55	18.77	2.02	18.58	20.54	18.58	10.94	17.39	18.77	34.61	6.18	11.90	–

For Installation Height to 25 Feet

MANHOURS each

Amperes	Manhours														
	A	B	C	D	E	F	G	H	I	J	K	L	M	N	O
800	4.84	6.11	7.51	5.47	0.50	4.84	6.11	4.84	2.93	3.96	4.84	10.94	1.27	2.55	17.25
1000	5.47	6.74	8.40	6.11	0.61	5.47	6.74	5.47	3.31	4.58	5.47	11.89	1.60	3.18	22.39
1350	6.24	7.51	9.67	6.87	0.68	6.24	7.51	6.24	3.82	5.47	6.24	13.11	1.91	3.82	–
1600	7.13	8.40	11.07	7.51	0.75	7.13	8.40	7.13	4.58	6.36	7.13	16.02	2.23	4.45	–
2000	8.40	9.67	12.26	8.40	0.84	8.40	9.67	8.40	5.47	7.25	8.40	17.94	2.55	5.09	–
2500	10.05	11.41	14.56	10.05	1.01	10.05	11.41	10.05	6.36	8.78	10.05	20.60	3.18	6.36	–
3000	12.01	14.56	16.56	11.89	1.25	12.01	14.56	12.01	7.25	11.07	12.01	24.40	3.82	7.64	–
4000	15.77	17.48	19.92	15.77	1.65	15.77	17.48	15.77	9.16	14.32	15.77	29.55	5.09	10.18	–
5000	18.86	20.60	23.61	18.86	2.08	18.86	20.60	18.86	11.16	17 48	18.86	34.70	6.36	12.13	–

Code:

A—90° Elbows	I—Panelboard Connections
B—Tees	J—End Tap Boxes
C—Crosses	K—Transfer Tap Sections
D—Offsets	L—Weatherproof Service Heads
E—End Closures	M—Wall and Floor Flanges
F—Reducers (Unfused)	N—Fire and Weather Stops
G—Offset Elbows	O—Indoor Tap-offs
H—Expansion Sections	

Manhours include checking out of job storage, handling, job hauling, and installing of fittings as outlined for duct runs up to 300 linear feet.

Manhours exclude installation of other items and other operations. See respective tables for these time frames.

For duct runs from 301 feet to 500 feet and from 501 feet to 1,000 feet, decrease manhours 30 and 40% respectively.

For 3-phase. 4-pole (1/2 neutral) bus duct fittings increase manhours 20%.

For 3-phase, 4-pole (full neutral) bus duct fittings increase manhours 25%.

For 4-pole, feeder type bus duct fittings increase manhours 20%.

For Heights above 25 feet increase manhours 3% for each additional 5 feet or fraction thereof.

Section 8

ELECTRIC HEATING AND VENTILATING

It is not the intent of this section to cover all items as may be required for a heating or ventilating system, but rather to allow the electrical estimator time frames in manhours for the installation of electrical equipment and their components that may form a part of these systems.

The manhour units represent average time requirements for the installation of the items at the locations as described and in accordance with the notes appearing with the tables and include time allowance to complete all necessary labor for the particular operation.

BASEBOARD CONVECTORS

MANHOURS PER SECTION

Section Length (Inches)	Prewired Baseboard				Custom Wired Baseboard			
	50 to 100 LF.	100 to 200 LF.	200 to 300 LF.	Over 300 LF.	50 to 100 LF.	100 to 200 LF.	200 to 300 LF.	Over 300 LF.
12	0.97	0.66	0.44	0.33	0.65	0.44	0.29	0.22
24	1.14	0.78	0.52	0.39	0.82	0.56	0.37	0.28
36	1.14	0.78	0.52	0.39	0.82	0.56	0.37	0.28
48	1.29	0.88	0.59	0.44	0.97	0.66	0.44	0.33
60	1.29	0.88	0.59	0.44	0.97	0.66	0.44	0.33
72	1.47	1.00	0.67	0.50	1.14	0.78	0.52	0.39
96	1.61	1.10	0.73	0.55	1.29	0.88	0.59	0.44
120	1.79	1.22	0.81	0.61	1.47	1.00	0.67	0.50
144	1.94	1.32	0.88	0.66	1.61	1.10	0.73	0.55

Manhours include checking out of job storage, handling, job hauling, distributing and installing items as outlined.

Prewired section manhours include the wiring connecting of sections together.

Custom wired section includes the installation of the sections only and do not include sections connecting wiring.

Manhours exclude the installation of circuit conduit, boxes, wiring, and circuit connections. See respective tables for these time frames.

RADIANT BASEBOARDS

MANHOURS PER SECTION

Section Length (Inches)	Prewired Radiant Baseboard				Blank Baseboard Section			
	50 to 100 LF.	100 to 200 LF.	200 to 300 LF.	Over 300 LF.	50 to 100 LF.	100 to 200 LF.	200 to 300 LF.	Over 300 LF.
12	1.11	0.78	0.55	0.39	0.51	0.34	0.23	0.17
24	1.25	0.87	0.62	0.44	0.66	0.44	0.29	0.22
36	1.25	0.87	0.62	0.44	0.66	0.44	0.29	0.22
48	1.42	0.99	0.71	0.50	0.84	0.56	0.37	0.28
60	1.42	0.99	0.71	0.50	0.84	0.56	0.37	0.28
72	1.56	1.09	0.78	0.55	0.99	0.66	0.44	0.39
96	1.73	1.21	0.87	0.61	1.17	0.78	0.52	0.39
120	1.88	1.31	0.94	0.66	–	–	–	–
144	2.05	1.43	1.02	0.72	–	–	–	–

BASEBOARD ACCESSORIES

MANHOURS FOR FOLLOWING LINEAR FEET OF BASEBOARD

Item Description	Manhours Each for Following Linear Feet			
	50 to 100 LF.	100 to 200 LF.	200 to 300 LF.	Over 300 LF.
Splice Plate or Joining Strip	0.17	0.12	0.10	0.06
End Section Cap or Plate	0.17	0.12	0.10	0.06
Inside Corner	0.91	0.66	0.53	0.33
Outside Corner	0.91	0.66	0.53	0.33
Receptable Sec. – Ground 15 Amp 120 V.	1.21	0.88	0.70	0.44
Receptacle Sec. – Ground 20 Amp 120 V.	1.21	0.88	0.70	0.44
Air Cond. Receptacle 240 V.	1.52	1.10	0.88	0.55
Baseboard Sec. – 1-Pole Thermostat	1.21	0.88	0.70	0.44
Baseboard Sec. – 2-Pole Thermostat	1.52	1.10	0.88	0.55

Manhours include checking out of job storage, handling, job hauling, distributing, and installing items as outlined.

Prewired section manhours include the wiring connecting of sections together.

Inside and outside corner manhours include an allowance for cutting to length one section of baseboard.

Manhours exclude the installation of circuit conduit, boxes, wiring, and circuit connections. See respective tables for these time frames.

INDOOR TYPE RADIANT HEAT CABLE

MANHOURS PER SPOOL

Rated Wattage	Approximate Spool Length (Feet)	Slab on Ground	Manhours for Precise Layout				Plaster or Suspended Ceiling
			Elevated Concrete Slab Height to				
			10'	15'	20'	25'	
200	80	0.50	0.51	0.53	0.55	0.59	0.61
300	110	0.56	0.57	0.59	0.62	0.66	0.68
400	145	0.61	0.62	0.64	0.67	0.72	0.74
500	180	0.66	0.67	0.69	0.73	0.78	0.80
600	220	0.70	0.71	0.74	0.77	0.83	0.75
700	250	0.75	0.77	0.79	0.83	0.89	0.91
800	290	0.80	0.82	0.84	0.88	0.94	0.97
900	330	0.85	0.87	0.89	0.94	1.00	1.03
1000	360	0.89	0.91	0.94	0.98	1.05	1.08
1100	400	0.95	0.97	1.00	1.05	1.12	1.15
1200	435	1.00	1.02	1.05	1.10	1.18	1.22
1300	470	1.09	1.11	1.15	1.20	1.29	1.33
1400	505	1.20	1.22	1.26	1.32	1.42	1.46
1500	545	1.30	1.33	1.37	1.43	1.53	1.58
1600	580	1.47	1.50	1.54	1.62	1.74	1.79
1700	615	1.50	1.53	1.58	1.65	1.77	1.82
1800	650	1.60	1.63	1.68	1.77	1.89	1.95
1900	680	1.69	1.72	1.78	1.86	1.99	2.05
2000	720	1.80	1.84	1.89	1.99	2.12	2.19
2200	800	2.00	2.04	2.10	2.21	2.36	2.43
2400	880	2.20	2.27	2.33	2.45	2.62	2.70
2600	960	2.40	2.45	2.52	2.65	2.83	2.92
2800	1030	2.60	2.65	2.73	2.87	3.07	3.16
3000	1100	2.80	2.86	2.94	3.09	3.30	3.40
3200	1170	3.00	3.06	3.15	3.31	3.54	3.65
3400	1240	3.20	3.26	3.36	3.53	3.78	3.89
3600	1310	3.40	3.47	3.57	3.75	4.01	4.13
3800	1380	3.60	3.67	3.78	3.97	4.25	4.38
4000	1450	3.80	3.88	3.99	4.19	4.49	4.62
4200	1525	4.00	4.08	4.20	4.41	4.72	4.86
4400	1600	4.20	4.28	4.41	4.63	4.96	5.11
4600	1670	4.40	4.49	4.62	4.85	5.19	5.35
4800	1745	4.60	4.69	4.83	5.07	5.43	5.59
5000	1820	4.80	4.90	5.04	5.30	5.67	5.84

Manhours include checking out of job storage, handling, job hauling, distributing, installing, and splicing connections.

Manhours exclude installation of circuit conduit, boxes, circuit wiring, or connections. See respective tables for these time frames.

600-VOLT OUTSIDE HEATING CABLE
Buried in Concrete

MANHOURS PER 100 LINEAR FEET

Type of Conductor	Number of Conductors	Wattage Range Per LF.	AWG No.	Repetitive	Non-Repetitive	Precise
Plastic Jacketed	1	5-10	26	0.70	0.90	1.10
Plastic Jacketed	1	5-10	19	0.80	1.05	1.32
Plastic Jacketed	1	5-10	14	1.00	1.30	1.64
Plastic Jacketed	2	5	26	0.80	1.05	1.32
Plastic Jacketed	2	5	19	0.92	1.20	1.52
Lead Jacketed	1	7	26	1.00	1.30	1.64
Lead Jacketed	1	7	19	1.32	1.76	2.20
Tinned Copper Braid Jacket	1	10	26	0.92	1.20	1.52
Tinned Copper Braid Jacket	1	10	19	1.05	1.41	1.76

Manhours include checking out of job storage, hob hauling, handling, distributing, layout, installing, and splicing connections.

Manhours exclude installation of circuit conduit, boxes, circuit wiring, connections, and placement of concrete. See respective tables for these time frames.

Manhours are based on installation of up to 500 linear feet of cable. For runs of 500 to 1,000 linear feet reduce manhours 8% and for runs greater than 1,000 linear feet reduce manhours 12%.

OUTDOOR MAT TYPE HEATING CABLE

MANHOURS PER MAT

Mat Width by Mat Length	Repetitive	Non-Repetitive	Precise
0'8" x 5'0"	0.34	0.46	0.57
0'8" x 10'0"	0.35	0.47	0.58
0'8" x 15'0"	0.38	0.50	0.63
0'8" x 20'0"	0.41	0.55	0.69
0'8" x 25'0"	0.45	0.60	0.75
0'8" x 30'0"	0.53	0.71	0.89
0'8" x 35'0"	0.57	0.76	0.95
0'8" x 40'0"	0.61	0.81	1.01
1'0" x 5'0"	0.35	0.47	0.58
1'0" x 10'0"	0.38	0.50	0.63
1'0" x 15'0"	0.41	0.55	0.69
1'0" x 20'0"	0.45	0.60	0.75
1'0" x 25'0"	0.53	0.71	0.89
1'0" x 30'0"	0.57	0.76	0.95
1'0" x 35'0"	0.61	0.81	1.01
1'0" x 40'0"	0.65	0.86	1.08
1'6" x 5'0"	0.41	0.55	0.69
1'6" x 10'0"	0.45	0.60	0.75
1'6" x 15'0"	0.53	0.71	0.89
1'6" x 20'0"	0.57	0.76	0.95
1'6" x 25'0"	0.61	0.81	1.01
1'6" x 30'0"	0.65	0.86	1.08
1'6" x 35'0"	0.68	0.91	1.14
1'6" x 40'0"	0.71	0.95	1.19
2'0" x 5'0"	0.53	0.71	0.89
2'0" x 10'0"	0.57	0.76	0.95
2'0" x 15'0"	0.61	0.81	1.01
2'0" x 20'0"	0.65	0.86	1.08
2'0" x 25'0"	0.68	0.91	1.14
2'0" x 30'0"	0.71	0.95	1.19
2'0" x 35'0"	0.76	1.02	1.27
2'0" x 40'0"	0.80	1.06	1.33

Manhours include checking out of job storage, handling, job hauling, distributing, layout, installing, and splicing connection.

Manhours exclude installation of circuit conduit, boxes, circuit wiring, circuit connection, and placement of concrete. See respective tables for these time frames.

OUTDOOR MAT TYPE HEATING CABLE

MANHOURS PER MAT

Mat Width by Mat Length	Repetitive	Non-Repetitive	Precise
2'6" x 5'0"	0.61	0.81	1.01
2'6" x 10'0"	0.65	0.86	1.08
2'6" x 15'0"	0.68	0.91	1.14
2'6" x 20'0"	0.71	0.94	1.18
2'6" x 25'0"	0.76	1.02	1.27
2'6" x 30'0"	0.80	1.06	1.33
2'6" x 35'0"	0.84	1.12	1.40
2'6" x 40'0"	0.87	1.16	1.45
3'0" x 5'0"	0.68	0.91	1.14
3'0" x 10'0"	0.71	0.94	1.18
3'0" x 15'0"	0.76	1.02	1.27
3'0" x 20'0"	0.80	1.06	1.33
3'0" x 25'0"	0.84	1.12	1.40
3'0" x 30'0"	0.87	1.16	1.45
3'0" x 35'0"	0.91	1.22	1.52
3'0" x 40'0"	0.95	1.26	1.58
4'0" x 5'0"	0.76	1.02	2.27
4'0" x 10'0"	0.80	1.06	1.33
4'0" x 15'0"	0.84	1.12	1.40
4'0" x 20'0"	0.87	1.16	1.45
4'0" x 25'0"	0.91	1.22	1.52
4'0" x 30'0"	0.95	1.26	1.58
4'0" x 35'0"	0.99	1.32	1.65
4'0" x 40'0"	1.03	1.38	1.72
5'0" x 5'0"	0.84	1.12	1.40
5'0" x 10'0"	0.87	1.16	1.45
5'0" x 15'0"	0.91	1.22	1.52
5'0" x 20'0"	0.95	1.26	1.58
5'0" x 25'0"	0.99	1.32	1.65
5'0" x 30'0"	1.03	1.38	1.72
5'0" x 35'0"	1.06	1.42	1.77
5'0" x 40'0"	1.10	1.47	1.84
6'0" x 5'0"	0.91	1.22	1.52
6'0" x 10'0"	0.95	1.26	1.58
6'0" x 15'0"	0.99	1.32	1.65
6'0" x 20'0"	1.03	1.38	1.72
6'0" x 25'0"	1.06	1.42	1.77
6'0" x 30'0"	1.10	1.47	1.84
6'0" x 35'0"	1.14	1.52	1.90
6'0" x 40'0"	1.18	1.57	1.96

Manhours include checking out of job storage, handling, job hauling, distributing, layout, installing, and splicing connection.

Manhours exclude installation of circuit conduit, boxes, circuit wiring, circuit connection, and placement of concrete. See respective tables for these time frames.

SURFACE MOUNTED AND SUSPENDED CEILING MOUNTED RADIANT HEATING PANELS

Panel Size (Feet)	MANHOURS EACH Surface Mounted	Suspended Ceiling Mounted
2 x 2	0.65	0.52
2 x 3	0.80	0.65
2 x 4	0.92	0.79
2 x 5	1.08	0.95
2 x 6	1.21	1.08
2 x 8	1.36	1.23
4 x 4	1.08	0.95
4 x 6	1.36	1.21
4 x 8	1.65	1.52
4 x 10	2.05	1.96
4 x 12	2.73	2.44

Manhours include checking out of job storage, handling, job hauling, distributing, installing, and connecting panels as outlined.

Labor for nailing, stapling or glueing surface mounted panels is included.

Manhours do not include placement of wood furring strips, T-bars, wall angles, support wires, circuit conduit, boxes, circuit wiring, circuit connections, and scaffolding. See respective tables for these time frames.

ELECTRIC FLOOR AND WALL HEATERS

MANHOURS EACH

Wattage	Floor Insert Type			Wall Convector Type		
	1 to 10 Each	10 to 20 Each	20 or More	1 to 10 Each	10 to 20 Each	20 or More
400	1.19	1.08	0.92	0.83	0.75	0.68
500	–	–	–	0.91	0.83	0.75
600	–	–	–	0.98	0.89	0.80
800	1.34	1.22	1.04	1.07	0.97	0.87
1000	1.50	1.36	1.16	1.13	1.03	0.93
1200	1.66	1.51	1.28	1.22	1.11	1.00
1400	1.82	1.65	1.40	1.30	1.18	1.06
1600	1.97	1.79	1.52	1.35	1.23	1.11
1800	2.13	1.94	1.65	1.45	1.32	1.19
2000	2.29	2.08	1.77	1.54	1.40	1.26
2500	2.67	2.43	2.07	1.67	1.52	1.37
3000	3.07	2.79	2.37	1.82	1.65	1.49
3500	3.47	3.15	2.68	1.97	1.79	1.61
4000	3.78	3.44	2.92	2.13	1.94	1.75
4500	–	–	–	2.29	2.08	1.87
5000	4.58	4.16	3.54	2.44	2.22	2.00
5500	–	–	–	2.61	2.37	2.13
6000	6.15	5.59	4.75	2.76	2.51	2.26
6500	–	–	–	2.92	2.65	2.39
7000	–	–	–	3.07	2.79	2.51
7500	–	–	–	3.22	2.93	2.64
8000	9.30	8.45	7.18	3.39	3.08	2.77

Manhours include checking out of job storage, handling, job hauling, distributing, installing, and connecting to preinstalled circuit.

Wall convector type heaters are mounted at floor level.

Manhours exclude installation of circuit conduit, boxes, circuit wiring, and circuit connection to main feeder. See respective tables for these time frames.

RECESSED AND RADIANT WALL HEATERS

MANHOURS EACH

Wattage	Recessed with Fans			Radiant Without Fans		
	1 to 10 Each	10 to 20 Each	20 or More	1 to 10 Each	10 to 20 Each	20 or More
400	0.91	0.83	0.71	0.63	0.57	0.51
500	1.07	0.97	0.82	0.68	0.62	0.56
600	1.22	1.11	0.94	0.75	0.68	0.61
800	1.38	1.25	1.06	0.83	0.75	0.68
1000	1.54	1.40	1.19	0.91	0.83	0.75
1200	1.67	1.52	1.29	0.98	0.89	0.80
1400	1.82	1.65	1.40	1.07	0.97	0.87
1600	1.97	1.79	1.52	1.18	1.07	0.96
1800	2.13	1.94	1.65	1.34	1.22	1.10
2000	2.53	2.30	1.96	1.50	1.36	1.22
2500	2.85	2.59	2.20	1.82	1.65	1.49
3000	2.92	2.65	2.25	2.13	1.94	1.75
3500	3.23	2.94	2.50	2.50	2.27	2.04
4000	3.54	3.22	2.74	2.76	2.51	2.26
4500	3.86	3.51	2.98	3.07	2.79	2.51
5000	4.16	3.78	3.21	3.39	3.08	2.77
*1000	2.73	2.48	2.11	1.82	1.65	1.49
*2000	2.63	3.30	2.81	2.73	2.48	2.23
*3000	4.54	4.13	3.51	3.63	3.30	2.97
*4000	5.45	4.95	4.21	4.54	4.13	3.72
*5000	6.36	5.78	4.91	5.45	4.95	4.46

*Explosion proof heaters for hazardous locations.

Manhours include checking out of job storage, handling, job hauling, distributing, installing, and connecting to preinstalled circuit.

Manhours exclude installation of circuit conduit, boxes, circuit wiring, and circuit connection to main feeder. See respective tables for these time frames.

RADIANT AND INFRARED CEILING HEATERS

MANHOURS EACH

Item Description	Ceiling Height to											
	10'			15'			20'			25'		
	1	2	3	1	2	3	1	2	3	1	2	3
Recessed Single Infrared Lamp	1.08	1.11	1.16	1.13	1.16	1.22	1.16	1.19	1.25	1.20	1.24	1.30
Recessed Double Infrared Lamp	1.35	1.39	1.42	1.42	1.46	1.54	1.46	1.50	1.58	1.51	1.56	1.63
Recessed Triple Infrared Lamp	1.63	1.68	1.76	1.71	1.76	1.85	1.76	1.81	1.90	1.82	1.87	1.97
Comb. Two Lamp Heater & Exhaust	1.90	1.96	2.05	2.00	2.06	2.16	2.06	2.12	2.28	2.13	2.19	2.30
Recessed Radiant Heater — 1000 W.	1.63	1.68	1.76	1.71	1.76	1.85	1.76	1.81	1.90	1.82	1.87	1.97
Radiant Heater with Fan — 1000 W.	2.05	2.11	2.22	2.15	2.21	2.33	2.21	2.28	2.39	2.29	2.36	2.48
Radiant Heater with Fan — 1500 W.	2.43	2.50	2.63	2.55	2.63	2.76	2.63	2.71	2.84	2.72	2.80	2.94
Comb. Heater, Fan & Light — 1000 W.	2.70	2.78	2.92	2.84	2.93	3.07	2.93	3.02	3.17	3.03	3.12	3.28
Comb. Heater, Exhaust & Light — 1000 W.	2.70	2.78	2.92	2.84	2.93	3.07	2.93	3.02	3.17	3.03	3.12	3.28
Comb. Heater, Exhaust, Light & Fan — 1000 W.	3.38	3.48	3.66	3.55	3.66	3.84	3.66	3.77	3.96	3.79	3.90	4.10
Surface Mounted Heater & Fan — 1000 W.	1.63	1.68	1.76	1.71	1.76	1.85	1.76	1.81	1.90	1.82	1.87	1.97

Codes:

    1—Exposed Concrete Ceilings
    2—Plaster or Suspended Ceilings
    3—Tile Ceilings

Manhours include checking out of job storage, handling, job hauling, distributing, installing, and connecting to preinstalled circuit.

Manhours exclude installation of circuit conduit, boxes, circuit wiring, and circuit connection to main feeder. See respective tables for these time frames.

RADIANT AND INFRARED COMMERCIAL CEILING HEATERS

MANHOURS EACH

Item Description	Wattage	1 to 10 Each	10 to 20 Each	20 or More
Radiant Indoor Heaters	500	1.50	1.36	1.22
	1000	1.82	1.65	1.49
	1500	2.40	2.18	1.96
	2000	3.00	2.73	2.46
	2500	3.36	3.05	2.75
	3000	4.19	3.81	3.43
	3500	4.82	4.38	3.94
	4000	5.45	4.95	4.46
Indoor-Outdoor Infrared Heaters	1000	3.00	2.73	2.46
	2000	4.50	4.09	3.68
	3000	6.00	5.46	4.91
	4000	7.50	6.82	6.14
	5000	9.00	8.18	7.36
	6000	12.00	10.91	9.82
	8000	15.00	13.64	12.28

Manhours include checking out of job storage, handling, job hauling, distributing, installing, and connecting to preinstalled circuit.

Manhours exclude installation of circuit conduit, boxes, circuit wiring, and circuit connection to main feeder See respective tables for these time frames.

DUCT INSERT HEATERS

MANHOURS EACH

KW Rating	1 to 10 Each	10 to 20 Each	20 or More
1.0	0.95	0.83	0.66
2.0	1.09	0.95	0.76
3.0	1.24	1.08	0.86
4.0	1.40	1.22	0.98
6.0	1.74	1.51	1.21
8.0	2.04	1.77	1.42
10.0	2.58	2.24	1.79
15.0	2.75	2.39	1.91
20.0	3.14	2.73	2.18
30.0	3.92	3.41	2.73
40.0	4.70	4.09	3.27
50.0	5.49	4.77	3.82
60.0	6.28	5.46	4.37
70.0	7.05	6.13	4.90
80.0	7.84	6.82	5.46
90.0	8.64	7.51	6.00
100.0	9.44	8.21	6.57
150.0	10.98	9.55	7.64
200.0	12.55	10.91	8.73
400.0	15.69	13.64	10.91
600.0	18.98	16.50	13.20
800.0	22.26	19.36	15.49
1000.0	25.55	22.22	17.78

Manhours include checking out of job storage, handling, job hauling, distributing, rigging, picking, setting, aligning, and connecting to preinstalled circuit.

Manhours exclude installation of circuit conduit, boxes, circuit wiring, circuit connection to main feeder and scaffolding. See respective tables for these time frames.

ELECTRIC FURNACES

MANHOURS EACH

Item	KW Rating	Slab on Grade	Heights to			
			10'	15'	20'	25'
Forced Air Units	10	6.82	7.50	8.03	8.43	8.68
	15	7.54	8.29	8.87	9.32	9.60
	20	8.25	9.08	9.71	10.20	10.50
	25	8.97	9.87	10.56	11.09	11.42
	30	9.68	10.65	11.39	11.96	12.32
	35	10.40	11.44	12.24	12.85	13.24
	40	10.89	11.98	12.82	13.46	13.86
Hydronic Units	10	6.82	7.50	8.03	8.43	8.68
	20	8.25	9.08	9.71	10.20	10.50
	30	9.57	10.53	11.26	11.83	12.18
	40	11.11	12.22	13.08	13.73	14.14
	50	13.64	15.00	16.05	16.86	17.36
	100	14.67	16.14	17.27	18.13	18.67
	200	17.97	19.77	21.15	22.21	22.87
	400	22.00	24.20	25.89	27.19	28.00
	600	27.13	29.84	31.93	33.53	34.53
	800	31.53	34.68	37.11	38.97	40.14
	1000	37.40	41.14	44.00	46.22	47.61
	1500	48.40	53.24	56.97	59.82	61.61
	2000	60.50	66.55	71.21	74.77	77.00
	2500	72.60	79.86	85.45	89.72	92.41
	3000	86.53	95.18	101.85	106.94	110.15

Manhours include checking out of job storage, handling, job hauling, distributing, rigging, picking, setting, aligning, and connecting items as outlined.

Manhours exclude installation of circuit conduit, boxes, circuit wiring or connecting to main feeder, and scaffolding. See respective tables for these time frames.

HYDRONIC HEATER ELEMENTS

MANHOURS EACH

Item Description	Ground Floor	Heights to		
		10'	20'	25'
3 Element Units				
1 0 – 5 KW – 15 Pounds	1.02	1.09	1.15	1.18
1 0 – 10 KW – 18 Pounds	1.18	1.26	1.33	1.37
1 0 – 15 KW – 20 Pounds	1.49	1.59	1.67	1.72
1 0 – 20 KW – 24 Pounds	1.93	2.07	2.17	2.23
3 0 – 5 KW – 15 Pounds	1.33	1.42	1.49	1.54
3 0 – 10 KW – 18 Pounds	1.49	1.59	1.67	1.72
3 0 – 15 KW – 20 Pounds	1.79	1.92	2.01	2.07
3 0 – 20 KW – 24 Pounds	2.23	2.39	2.51	2.58
6 Element Units				
1 0 – 10 KW – 28 Pounds	1.49	1.59	1.67	1.72
1 0 – 15 KW – 30 Pounds	1.79	1.92	2.01	2.07
1 0 – 20 KW – 34 Pounds	2.08	2.23	2.34	2.41
1 0 – 25 KW – 38 Pounds	2.38	2.55	2.67	2.75
1 0 – 30 KW – 40 Pounds	2.67	2.86	3.00	3.09
1 0 – 40 KW – 50 Pounds	2.97	3.18	3.34	3.44
3 0 – 10 KW – 28 Pounds	1.79	1.92	2.01	2.07
3 0 – 15 KW – 30 Pounds	2.08	2.23	2.34	2.41
3 0 – 20 KW – 34 Pounds	2.38	2.55	2.67	2.75
3 0 – 25 KW – 38 Pounds	2.67	2.86	3.00	3.09
3 0 – 30 KW – 40 Pounds	2.97	3.18	3.34	3.44
3 0 – 40 KW – 50 Pounds	3.27	3.50	3.67	3.78
9 Element Units				
1 0 – 20 KW – 50 Pounds	2.38	2.55	2.67	2.75
1 0 – 30 KW – 60 Pounds	2.97	3.18	3.34	3.44
1 0 – 40 KW – 75 Pounds	3.56	3.81	4.00	4.12
1 0 – 50 KW – 100 Pounds	4.15	4.44	4.66	4.80
3 0 – 20 KW – 50 Pounds	2.67	2.86	3.00	3.09
3 0 – 30 KW – 60 Pounds	3.27	3.50	3.67	3.78
3 0 – 40 KW – 75 Pounds	3.86	4.13	4.34	4.47
3 0 – 50 KW – 100 Pounds	4.46	4.77	5.01	5.16
9 Moisture Resistant Element Units				
1 0 – 20 KW – 55 Pounds	2.67	2.86	3.00	3.09
1 0 – 30 KW – 65 Pounds	3.27	3.50	3.67	3.78
1 0 – 40 KW – 80 Pounds	3.86	4.13	4.34	4.47
1 0 – 50 KW – 110 Pounds	4.46	4.77	5.01	5.16
3 0 – 20 KW – 55 Pounds	2.97	3.18	3.34	3.44
3 0 – 30 KW – 65 Pounds	3.56	3.81	4.00	4.12
3 0 – 40 KW – 80 Pounds	4.15	4.44	4.66	4.80
3 0 – 50 KW – 110 Pounds	4.76	5.09	5.35	5.51

Manhours include checking out of job storage, handling, job hauling, distributing, and installing elements as outlined.

Manhours exclude installation of heaters and wiring. See respective tables for these time frames.

HYDRONIC HEATER ELEMENTS

MANHOURS EACH

Item Description	Ground Floor	Heights to		
		10'	20'	25'
12 Element Units				
1　0 – 18 KW – 　65 Pounds	2.97	3.18	3.34	3.44
1　0 – 20 KW – 　75 Pounds	3.63	3.88	4.08	4.20
1　0 – 24 KW – 　85 Pounds	4.66	4.99	5.24	5.39
1　0 – 36 KW – 115 Pounds	4.95	5.30	5.56	5.73
3　0 – 18 KW – 　65 Pounds	3.30	3.53	3.71	3.82
3　0 – 20 KW – 　75 Pounds	3.45	3.69	3.88	3.99
3　0 – 24 KW – 　85 Pounds	4.62	4.94	5.19	5.35
3　0 – 36 KW – 115 Pounds	5.28	5.65	5.93	6.11
18 Element Units				
1　0 – 20 KW – 100 Pounds	3.47	3.71	3.90	4.02
1　0 – 30 KW – 110 Pounds	3.76	4.02	4.22	4.35
1　0 – 40 KW – 125 Pounds	4.79	5.13	5.38	5.54
1　0 – 50 KW – 130 Pounds	5.24	5.61	5.89	6.06
3　0 – 20 KW – 100 Pounds	3.96	4.24	4.45	4.58
3　0 – 30 KW – 110 Pounds	4.79	5.13	5.38	5.54
3　0 – 40 KW – 125 Pounds	5.57	5.96	6.26	6.46
3　0 – 50 KW – 130 Pounds	6.60	7.06	7.42	7.64
27 Element Units				
3　0 – 45 KW – 135 Pounds	6.60	7.06	7.42	7.64
3　0 – 60 KW – 165 Pounds	8.25	8.83	9.27	9.55
3　0 – 75 KW – 200 Pounds	9.90	10.59	11.12	11.46
36 Element Units				
3　0 – 40 KW – 135 Pounds	6.60	7.06	7.42	7.64
3　0 – 50 KW – 150 Pounds	7.43	7.95	8.35	8.60
3　0 – 60 KW – 170 Pounds	8.25	8.83	9.27	9.55
3　0 – 70 KW – 190 Pounds	9.08	9.72	10.20	10.51
3　0 – 80 KW – 210 Pounds	9.90	10.59	11.12	11.46
3　0 – 90 KW – 230 Pounds	11.55	12.36	12.98	13.37
3　0 – 100 KW – 250 Pounds	13.20	14.12	14.83	15.28
45 Element Units				
3　0 – 　75 KW – 200 Pounds	9.90	10.59	11.12	11.46
3　0 – 100 KW – 225 Pounds	13.20	14.12	14.83	15.28
3　0 – 125 KW – 270 Pounds	16.50	17.66	18.54	19.09

Manhours include checking out of job storage, handling, job hauling, distributing, and installing elements as outlined.

Manhours exclude installation of heaters and wiring. See respective tables for these time frames.

FORCED AIR ELECTRIC HEAT ACCESSORIES

Electrostatic Air Filters

MANHOURS EACH

Duct Size (Inches)	CFM Range	Approximate Weight (Pounds)	Ground Floor	Heights to		
				10′	20′	25′
20 x 24	500–1,500	50	4.37	4.68	4.91	5.06
24 x 30	1,500–2,500	70	5.83	6.24	6.55	6.75
30 x 40	2,500–3,500	120	7.29	7.80	8.19	8.44
36 x 48	3,500–5,000	160	8.75	9.36	9.83	10.13

Domestic Humidifuers

MANHOURS EACH

Type	GPD Rating	Ground Floor	Heights to		
			10′	20′	25′
Disc Type	10	2.92	3.12	3.28	3.38
	15	3.37	3.61	3.79	3.90
	20	4.37	4.68	4.91	5.06
	25	5.10	5.46	5.73	5.90
	30	5.83	6.24	6.55	6.75
	35	6.56	7.02	7.37	7.59
	40	7.00	7.49	7.86	8.10
Spray Type	10	2.92	3.12	3.28	3.38
	20	4.37	4.68	4.91	5.06
	30	5.83	6.24	6.55	6.75
	40	7.00	7.49	7.86	8.10
	50	8.02	8.58	9.01	9.28
	60	8.75	9.36	9.83	10.13
	70	9.47	10.13	10.64	10.96
	80	10.20	10.91	11.46	11.80
	100	11.66	12.48	13.10	13.49

Manhours include checking out of job storage, handling, job hauling, rigging, picking, setting, aligning, and connecting items as outlined.

Manhours exclude installation of circuit conduit, boxes, circuit wiring or connecting to main feeder, and scaffolding. See respective tables for these time frames.

AIR TO AIR HEAT PUMPS

MANHOURS EACH

Type	Heating Range BTU	Cooling Range Tons	Weight Range Pounds	Manhours for Heights to				
				Ground	10'	15'	20'	25'
Single Package	12,000–18,000	1 – 1-1/2	180–240	16.5	17.3	18.0	18.6	18.9
	20,000–24,000	1-3/4 – 2	270–320	19.8	20.8	21.6	22.3	22.7
	24,000–30,000	2-1/4 – 2-1/2	340–360	24.6	25.8	26.9	27.7	28.2
	34,000–40,000	3 – 3-1/2	370–430	33.0	34.7	36.0	37.1	37.9
	46,000–50,000	3-3/4 – 4	560–640	41.4	43.5	45.2	46.6	47.5
	60,000–66,000	4-1/2 – 5	650–700	49.5	52.0	54.1	55.7	56.8
	80,000–90,000	6 – 7	800–900	66.0	69.3	72.1	74.2	75.7
	96,000–110,000	8	1,000–1,200	82.5	86.6	90.1	92.8	94.6
Split System	12,000–18,000	1 – 1-1/2	60–80	13.2	13.9	14.4	14.8	15.1
	20,000–24,000	1-3/4 – 2	90–100	16.5	17.3	18.0	18.6	18.9
	26,000–30,000	2-1/4 – 2-1/2	120–140	19.8	20.8	21.6	22.3	22.7
	36,000–42,000	3 – 3-1/2	150–180	24.6	25.8	26.9	27.7	28.2
	45,000–48,000	3-3/4 – 4	200–230	33.0	34.7	36.0	37.1	37.9
	52,000–60,000	4-1/2 – 5	240–270	41.4	43.5	45.2	46.6	47.5
	70,000–80,000	6 – 6	300–350	49.5	52.0	54.1	55.7	56.8
	90,000–100,000	8	380–430	66.0	69.3	72.1	74.2	75.7
	100,000–120,000	10	450–500	74.3	78.0	81.1	83.6	85.1
	120,000–140,000	12	550–600	88.0	92.4	96.1	99.0	101.0
	140,000–160,000	14	650–700	101.8	106.9	111.2	114.5	116.8
	160,000–190,000	16	750–800	115.5	121.3	126.1	129.9	132.5
	190,000–210,000	18	850–900	130.6	137.1	142.6	146.9	149.8
	210,000–240,000	20	950–1,000	145.8	153.1	159.2	164.0	167.3

Manhours include checking out of job storage, handling, job hauling, rigging, picking, setting, aligning, and connecting items as outlined.

Manhours exclude preparation of structural openings and installation of sheet metal work, piping connections, circuit conduit, boxes, circuit wiring, connecting to main feeder, and scaffolding. See respective tables for these time frames.

STANDARD UNIT HEATERS WITH BLOWERS

MANHOURS EACH

KW Rating	Manhours Required for Heights to			
	10'	15'	20'	25'
1	2.3	2.5	2.6	2.7
2	2.7	2.9	3.0	3.1
3	3.4	3.6	3.8	3.9
4	4.1	4.4	4.6	4.7
5	4.8	5.1	5.4	5.6
6	5.5	5.9	6.2	6.4
7	6.1	6.5	6.9	7.1
8	6.8	7.3	7.6	7.9
9	7.5	8.0	8.4	8.7
10	8.2	8.8	9.2	9.5
12	9.5	10.2	10.7	11.0
14	10.9	11.7	12.2	12.6
16	12.3	13.2	13.8	14.2
18	13.6	14.6	15.3	15.7
20	16.5	17.7	18.5	19.1
25	19.1	20.4	21.5	22.1
30	21.6	23.1	24.3	25.0
35	24.4	26.1	27.4	28.2
40	27.3	29.2	30.7	31.6

Manhours include checking out of job storage, handling, job hauling, distributing, rigging, picking, setting, aligning, and connecting unit.

Manhours exclude installation of circuit conduit, boxes, circuit wiring, and scaffolding. See respective tables for these time frames.

For installation of 10 to 20 units at same elevation decrease manhours 10%.

For installation of 20 units or more at same elevation decrease manhours 15%.

EXPLOSION PROOF UNIT HEATERS WITH BLOWERS

MANHOURS EACH

KW Rating	Manhours Required for Heights to			
	10'	15'	20'	25'
1	2.7	2.9	3.0	3.1
2	3.4	3.6	3.8	3.9
3	4.1	4.4	4.6	4.7
4	4.8	5.1	5.4	5.6
5	5.5	5.9	6.2	6.4
6	6.1	6.5	6.9	7.1
7	6.8	7.3	7.6	7.9
8	7.5	8.0	8.4	8.7
9	8.2	8.8	9.2	9.5
10	9.5	10.2	10.7	11.0
12	10.9	11.7	12.2	12.6
14	12.3	13.2	13.8	14.2
16	13.6	14.6	15.3	15.7
18	16.5	17.7	18.5	19.1
20	19.1	20.4	21.5	22.1
25	21.6	23.1	24.3	25.0
30	24.4	26.1	27.4	28.2
35	27.3	29.2	30.7	31.6
40	30.1	32.2	33.8	34.8

Manhours include checking out of job storage, handling, job hauling, distributing, rigging, picking, setting, aligning, and connecting unit.

Manhours exclude installation of circuit conduit, boxes, circuit wiring and scaffolding. See respective tables for these time frames.

For installation of 10 to 20 units at same elevation decrease manhours 10%.

For installation of 20 units or more at same elevation decrease manhours 15%.

POWERED ROOF VENTILATORS

MANHOURS EACH

HP Rating	Weight Range Pounds	Fan Size Range Inches	CFM Range	For Heights to			
				10'	15'	20'	25'
1/8	50-100	12-18	1,000-2,000	13.6	14.6	15.3	15.7
1/8	80-150	20-24	2,000-4,000	16.5	17.7	18.5	19.1
1/4	100-180	12-18	1,500-3,000	19.1	20.4	21.5	22.1
1/4	150-200	20-24	3,000-5,000	21.8	23.3	24.5	25.2
1/2	200-300	16-24	3,000-5,000	24.6	26.3	27.6	28.5
1/2	250-450	30-36	7,000-8,000	27.3	29.2	30.7	31.6
3/4	300-500	24-36	7,000-10,000	32.8	35.1	36.9	38.0
1	350-600	36-48	9,000-14,000	38.1	40.8	42.8	44.1
1-1/2	400-800	36-48	10,000-18,000	43.8	46.9	49.2	50.7
2	600-1,000	40-60	15,000-25,000	49.1	52.5	55.2	56.8
3	700-1,200	40-60	20,000-30,000	54.6	58.4	61.3	63.2

Manhours include checking out of job storage, handling, job hauling, distributing, rigging, picking, setting. aligning, and connecting units.

Manhours exclude installation of circuit conduit, boxes, circuit wiring, sheet metal work, preparation of structural openings, and scaffolding. See respective tables for these time frames.

For installation of from 10 to 20 units at same elevation decrease manhours 10%.

For installation of 20 or more units at same elevation decrease manhours 15%.

NONHEATING BLOWERS

MANHOURS EACH

Item	HP Range	CFM Range	Ground Floor	For Heights to			
				10'	15'	20'	25'
Floor Mounted	To 3/4	To 8,000	6.8	7.4	7.9	8.3	8.6
	5/8-1	6,000-12,000	10.9	11.9	12.7	13.3	13.7
	1-1/4-1-3/4	10,000-18,000	16.1	17.5	18.8	19.7	20.3
	2-2-1/2	15,000-25,000	24.6	26.8	28.7	30.1	31.0
	3	20,000-30,000	32.8	35.8	38.3	40.2	41.4
	5	25,000-50,000	40.9	44.6	47.7	50.1	51.6
	7-1/2	30,000-60,000	48.8	53.2	56.9	59.8	61.6
	10	40,000-80,000	57 4	62.6	66.9	70.3	72.4
	15	50,000-100,000	68.2	74.3	79.5	83.5	86.0
Floor Mounted Twins	2-5	35,000-75,000	61.4	66.9	71.6	75.2	77 4
	2-7-1/2	45,000-90,000	73.5	80.1	85.7	90.0	92.7
In-Line Mounted	To 3/4	To 8,000	10.7	11.7	12.5	13.1	13.5
	5/8-1	6,000-12,000	16.5	18.0	19.2	20.2	20.8
	1-1/4-1-3/4	10,000-18,000	22.2	24.2	25.9	27.2	28.0
	2-2-1/2	15,000-25,000	30.1	32.8	35.1	36.9	38.0
	3	20,000-30,000	41.1	44.8	47.9	50.3	51.8
	5	25,000-50,000	51.7	56.4	60.3	63.3	65.2
	7-1/2	30,000-60,000	62.9	68.6	73.4	77.0	79.3
	10	40,000-80,000	73.3	79.9	85.5	89.8	92.5
	15	50,000-100,000	84.7	92.3	98.8	103.7	106.8
In-Line Twin Mounted	2-5	35,000-75,000	76.3	83.2	89.0	93.4	96.2
	2-7-1/2	45,000-90,000	93.5	101.9	109.0	114.5	117.9

Manhours include checking out of job storage, handling, job hauling, distributing, rigging, picking, setting, aligning, and connecting unit.

Manhours exclude installation of circuit conduit, boxes, circuit wiring, sheet metal work, and preparation of structural openings. See respective tables for these time frames.

EXHAUST FANS

MANHOURS EACH

Item	Horsepower Range	Manhours for Heights to			
		10'	15'	20'	25'
Flush Mounted Ceiling Exhaust Fans	To 1/8	1.36	1.46	1.53	1.57
	1/4–1/2	2.05	2.19	2.30	2.37
	5/8–3/4	2.73	2.92	3.07	3.16
	7/8–1·1/4	4.53	4.85	5.09	5.24
	1·1/2–2	5.46	5.84	6.13	6.32
Horizontal Discharge Wall Exhaust Fans	To 1/8	1.36	1.46	1.53	1.57
	1/4–1/2	1.69	1.81	1.90	1.96
	5/8–3/4	2.05	2.19	2.30	2.37
	7/8–1·1/4	2.73	2.93	3.07	3.16
	1·1/2–2	4.09	4.38	4.60	4.73

Manhours include checking out of job storage, handling, job hauling, distributing, setting, aligning, and connecting.

Manhours exclude installation of sheet metal work, preparation of structural openings, circuit conduit, boxes, circuit wiring, and scaffolding. See respective tables for these time frames.

ELECTRIC HEATING CONTROLS

MANHOURS EACH

Item	Manhours for Installation on			
	1	2	3	4
Thermostats				
Low Voltage Single Pole Wall Mounted	1.36	1.50	1.57	1.54
Low Voltage Double Pole Wall Mounted	1.65	1.82	1.91	1.87
Low Voltage Two Circuit Wall Mounted	2.05	2.26	2.37	2.32
Low Voltage Three Circuit Heat Cool Comb.	2.73	3.00	3.15	3.09
Line Voltage Single Pole Wall Mounted	1.65	1.82	1.91	1.87
Line Voltage Double Pole Wall Mounted	1.91	2.10	2.21	2.16
Line Voltage Two Circuit Wall Mounted	2.44	2.68	2.82	2.76
Line Voltage Three Circuit Heat Cool Comb.	3.28	3.61	3.79	3.71
Bulb Line Voltage Single Pole	2.05	2.26	2.37	2.32
Bulb Line Voltage Double Pole	2.31	2.54	2.67	2.61
Bulb Line Voltage Two Circuit	2.87	3.16	3.31	3.25
Explosion Proof Low Voltage Single Pole	2.73	3.00	3.15	3.09
Explosion Proof Low Voltage Double Pole	3.41	3.75	3.94	3.86
Interval Timer — Up to 4 On-Off Cycles				
24-Hour Dial, 15 Amp	7 43	8.17	8.58	8.41
24-Hour Dial, 40 Amp	11.88	13.07	13.72	13.45
7-Day Dial, 40 Amp	14.85	16.34	17.15	16.81
Humidistat — 120 Volt				
Up to 20 Amp	2.73	3.00	3.15	3.09
25 to 40 Amp	4.09	4.50	4.72	4.63
Low Voltage Thermal Relay				
Up to 25 Amp	1.36	1.50	1.57	1.54
25 to 40 Amp	2.05	2.26	2.37	2.32
Control				
Sequence — Up to 25 Amp	5.46	6.00	6.31	6.18
Sequence — 30 to 40 Amp	8.18	9.00	9.45	9.26
4 Step — 20 Amp Per Step	14.85	16.34	17.15	16.81
6 Step — 20 Amp Per Step	22.83	25.11	26.37	25.84
8 Step — 20 Amp Per Step	29.70	32.67	34.30	33.62
10 Step — 20 Amp Per Step	37.13	40.84	42.89	42.03

Code:

1—Steel Columns   3—Poured Concrete Walls
2—Hollow Plaster or Dry Wall   4—Tile or Masonry Walls

Manhours include checking out of job storage, handling, job hauling, distributing, mounting, wiring connection, and testing.

Manhours exclude installation of line or low voltage interconnecting wiring. See respective tables for these time frames.

Section 9

PANELBOARDS AND ACCESSORIES

This section includes labor operations, expressed in man-hour units, for the installation of branch circuit and distribution and power panelboards, and their accessories as may be required in a process or industrial plant.

Various manhour units are included for the installation of cabinets, branch units, switch and fuse type branches, motor controls, starters, circuit and cable connections, and lug connections.

Consideration has been given to all labor operations as may be required for the individual items in accordance with the notes appearing with the tables.

Other labor operations that may be required to support the total installed time requirements are covered in other sections of this manual.

BRANCH CIRCUIT PANELBOARDS
Standard Type

MANHOURS EACH

No. of 0-30 Amp Circuit	Flush Mounted				Surface Mounted					
	1	2	3	4	1	2	3	4	5	6
8	6.70	7.20	8.60	8.10	5.80	6.30	7.20	6.70	5.30	6.70
10	8.40	9.00	10.80	10.20	7.20	7.80	9.00	8.40	6.60	8.40
12	10.10	10.80	13.00	12.20	8.60	9.30	10.70	10.00	7.90	10.00
14	11.80	12.60	15.10	14.20	10.10	11.00	12.70	11.80	9.30	11.80
16	13.40	14.30	17.20	16.20	11.50	12.40	14.30	13.30	10.50	13.30
18	15.10	16.20	19.40	18.20	13.00	14.00	16.10	15.00	11.90	15.00
20	16.80	18.00	21.60	20.30	14.40	15.60	17.90	16.70	13.20	16.70
22	18.50	19.80	23.80	22.40	15.80	17.10	19.70	18.40	14.50	18.40
24	20.20	21.60	25.90	24.30	17.30	18.80	21.60	20.20	15.90	20.20
26	21.80	23.30	28.00	26.30	18.70	20.20	23.20	21.60	17.10	21.60
28	23.50	25.10	30.10	28.30	20.10	21.70	25.00	23.30	18.40	23.30
30	25.20	27.00	32.40	30.50	21.60	23.40	26.90	25.10	19.80	25.10
32	26.90	28.80	34.60	32.50	23.00	24.90	28.60	26.70	21.10	26.70
34	27.80	29.70	35.60	33.50	23.90	25.80	29.70	27.70	21.90	27.70
36	29.50	31.60	37.90	35.60	25.30	27.40	31.50	29.40	23.20	29.40
38	31.40	33.60	40.30	37.90	26.70	28.90	33.20	31.00	24.50	31.00
40	32.80	35.10	42.10	39.60	28.10	30.40	35.00	32.70	25.80	32.70
42	34.40	36.80	44.10	–	29.50	31.90	36.70	–	–	34.20

Code:

1 – Plug-in, Standard Width Breaker
2 – Bolt-in, Standard Width Breaker – Class A
3 – Bolt-in, Standard Width Breaker – Class C, D, or E
4 – Standard Width Switch and Fuse Block
5 – Plug-in, Column Mounted Breaker
6 – Bolt-in, Column Mounted Breaker – Class C, D, or E

Manhours include checking out of job storage, handling, hauling, distributing and installing standard panelboards and cabinets having fuses only or fuses and switches in the branches and having mains with lugs only, main fuses, main switches or main switches and fuses or circuit breakers for single phase, 3-wire or 3-phase, 4-wire systems.

Time has been allowed for the removal, temporary storage and reinstallation of the board insides, punch necessary knock-outs, locate and installation of box, installation of main pipe terminals, and installation of cover plate.

Manhours exclude installation of sub-feeder terminals, sub-feeder pipe entrances and fasteners or supports. See respective tables for these time frames.

BRANCH CIRCUIT PANELBOARDS
MISCELLANEOUS INSTALLATIONS AND
CONNECTIONS

Standard Branch Circuit Type Individual Manhours Units

MANHOURS EACH

Panelboard Type	0-30 Amp Branches — Main Lugs Only					
	Surface Mounted			Flush Mounted		
	1-Pole	2-Pole	3-Pole	1-Pole	2-Pole	3-Pole
Plug-In Standard Width Breaker	0.72	1.44	2.16	0.94	1.68	2.52
Plug-In, Column Mounted Breaker	0.66	1.32	1.98	–	–	–
Bolt-In, Standard Width Breaker — Class A	0.78	1.56	2.34	0.90	1.80	2.70
Bolt-In, Std. Width Breaker — Class C, D, or E	0.90	1.80	2.70	1.10	2.20	3.30
Bolt-In, Col. Mounted Breaker — Class C, D, or E	0.84	1.68	2.52	–	–	–
Standard Width Switch and Fuse Block	0.84	1.68	2.52	1.00	2.00	3.00

Branch manhours are applicable to calculating time required for complete panelboards only and include installation of cabinets, covers and branch circuit, and feeder conductor connections.

Circuit Wire Connections to Spare Branches in Place

Circuit Amperage	Wire Size	Separately			Same Time with Other Circuits		
		2-Wire	3-Wire	4-Wire	2-Wire	3-Wire	4-Wire
0-15	14	0.60	0.84	1.20	0.30	0.45	0.64
20	12	0.72	1.10	1.47	0.36	0.54	0.72
25	10	0.96	1.44	1.93	0.48	0.72	0.96
30	8	1.20	1.80	2.41	0.60	0.90	1.21

Manhours include time required for wire terminal connections, cableing conductors, circuit identification, testing and load balancing, and providing information for circuit directory.

Placing Branch Units in Blank Spaces in Panelboards
and Circuit Connections

Branch Size	Panelboard Type and Branch								
	Plug-In Breaker Standard Width			Bolt-In Class A Breaker Standard Width			Bolt-In Class C, D or E Standard Width Breaker		
	1-Pole	2-Pole	3-Pole	1-Pole	2-Pole	3-Pole	1-Pole	2-Pole	3-Pole
0-15 Amp	1.38	1.92	2.57	1.54	2.19	2.96	1.78	2.62	3.56
20 Amp	1.50	2.16	2.83	1.70	2.48	3.27	1.93	2.90	3.86
25 Amp	1.74	2.54	3.33	1.93	2.84	3.75	2.18	3.27	4.35
30 Amp	1.98	2.87	3.76	2.14	3.15	4.16	2.37	3.56	4.73

Manhours include time required to remove space cover, obtaining and installing added branch unit and circuit wire connections all in accordance to above note for circuit wire connections to spare branches in place.

BRANCH CIRCUIT PANELBOARDS
MAIN CIRCUIT BREAKER OR SWITCH ADDER
MANHOURS

1-Phase 3-Wire—Surface or Flush Mounted Panelboards

MANHOURS EACH

Number of 0-30 Amps Circuits	Main Breaker Size Amps	Types of Circuit Breaker Panelboard					Main Switch Size Amps	6
		1	2	3	4	5		
8	50	4.30	2.30	4.30	0.80	0.80	60	3.00
10	50	4.30	2.30	4.30	0.30	0.30	–	–
12	100	4.30	3.10	4.30	0.80	0.80	60	2.60
14	100	5.60	3.10	4.30	0.30	0.30	–	–
16	100	5.60	3.10	4.30	1.20	1.20	100	2.10
18	100	5.60	3.10	4.30	0.40	0.40	–	–
20	100	5.00	3.10	4.80	1.30	1.30	100	1.60
22	225	14.80	–	10.40	7.50	7.50	–	–
24	225	14.80	–	10.40	7.20	7.20	200	5.30
26	225	12.90	–	11.20	6.60	6.60	–	–
28	225	12.90	–	11.20	6.20	6.20	200	5.00
30	215	12.90	–	11.20	5.30	5.30	–	–
32	225	12.90	–	12.10	3.20	3.20	200	4.80
34	225	12.90	–	12.10	3.60	3.60	–	–
36	225	12.90	–	12.10	3.00	3.00	200	4.50
38	225	12.90	–	12.30	1.40	1.40	–	–
40	225	12.90	–	12.30	0.70	0.70	200	4.60
42	225	12.90	–	12.30	0.70	0.70	–	–

Code:

1—Plug-In, Standard Width Breaker
2—Plug-In, Column Mounted Breaker
3—Bolt-In, Standard Width Breaker – Class A
4—Bolt-In, Standard Width Breaker – Class C
5—Bolt-In, Standard Width Breaker – Class D or E
6—Standard Width Switch and Fuse Block

Manhours include necessary time as may be required for the addition of extra main circuit breakers or switches to the panelboard units. Time has been allowed to compensate for the increased weight of the cabinet, cover, and main unit.

Manhours exclude additional feeder connection time which should not be required.

These manhours can be added to the units as shown on the standard branch panelboards, table or listed separately.

BRANCH CIRCUIT PANELBOARDS
MAIN CIRCUIT BREAKER OR SWITCH ADDER
MANHOURS

3-Phase, 4-Wire Surface or Flush Mounted Panelboards

MANHOURS EACH

Number of 0-30 Amps Circuits	Main Breaker Size amps	Type of Circuit Breaker Panelboards						Switch and Fuse	
		1	2	3	4	5	6	Main Switch Size Amps	7
8	50	4.30	3.10	4.60	0.90	3.40	3.90	60	3.00
10	50	4.30	3.10	4.60	0.90	3.40	3.90	–	–
12	50 or 100	4.30	3.40	4.60	0.90	3.40	3.90	60	2.60
14	50 or 100	6.30	3.40	4.60	1.30	3.40	3.90	–	–
16	100	6.30	3.40	4.60	1.40	3.40	3.90	60	2.10
18	100	6.30	3.40	4.60	1.60	4.00	3.90	–	–
20	100	7.40	3.40	5.00	1.70	4.20	3.90	100	1.60
22	100	7.40	3.40	5.00	2.10	4.90	3.90	–	–
24	100	7.40	3.40	5.00	2.20	5.20	3.90	100	5.30
26	100	5.30	3.40	5.20	2.50	5.70	3.90	–	–
28	100	6.30	3.40	5.20	2.60	5.90	3.90	100	5.00
30	100	6.30	3.40	5.20	3.00	6.50	3.90	–	–
32	225	14.40	–	12.60	5.20	6.70	–	200	4.80
34	225	14.40	–	12.60	4.00	7.30	–	–	–
36	225	14.40	–	12.60	3.30	7.70	–	200	4.50
38	225	14.40	–	12.60	1.90	8.20	–	–	–
40	225	14.40	–	12.60	1.10	8.60	–	200	4.60
42	225	14.40	–	12.60	1.10	9.20	–	–	–

Code:

  1—Plug-In Standard Width Breaker
  2—Plug-In Column Mounted Breaker
  3—Bolt-In Standard Width Breaker — Class A
  4—Bolt-In Standard Width Breaker — Class C
  5—Bolt-In Standard Width Breaker — Class D or E
  6—Bolt-In Column Mounted Breaker — Class D or E
  7—Standard Width Switch and Fuse Block

Manhours include necessary time as may be required for the additional of extra main circuit breakers or switches to the panelboard units. Time has been allowed to compensate for the increased weight of the cabinet, cover, and main unit.

Manhours exclude additional feeder connection time which should not be required.

These units can be added to the units as shown on the standard branch panelboards, table or listed separately.

BRANCH CIRCUIT PANELBOARDS SUB-FEEDER CIRCUIT BREAKER OR SWITCH ADDER MANHOURS

1-Phase, 3-Wire Surface or Flush Mounted Panelboards

MANHOURS EACH

Sub-Feeder Breaker Size Amps	Type of Circuit Breaker panelboards						Switch and Fuse Sub-Feeder Switch Size amps	
	1	2	3	4	5	6		7
25−50-Amp #8-#6 Wire	6.00	4.20	6.00	2.70	2.70	5.10	40−60 Amp #8-#6	4.80
50-70 Amp #6-#2 Wire 90-100	7.50	5.40	7.20	3.90	3.90	5.40	65−100 Amp #4-#2	4.80
Amp #1-#1/0 Wire	9.30	7.20	9.00	5.40	5.40	7.20	10-−125 Amp #1-#1/0	9.00
150-175 Amp #1/0-#4/0 Wire	18.90	−	17.10	13.20	13.20	−	125−175 Amp #1/0-#4/0	10.80
200-225 Amp 250-300 MCM Wire	21.30	−	20.70	15.60	15.60	−	200−240 Amp 250-300 MCM	15.60

3-Phase, 4-Wire Surface or Flush Mounted Panelboards

25−50 Amp #8-#6 Wire	6.60	5.40	6.90	3.30	5.70	6.30	40−60 Amp #8-#6	5.40
50−70 Amp #6-#2 Wire	9.60	6.60	8.40	4.80	6.90	6.60	65−100 Amp #4-#2	5.40
90−100 Amp #1-#1/0 Wire	12.00	9.00	10.80	8.70	12.00	9.60	100−125 Amp #1-#1/0	10.50
150−175 Amp #2/0-#4/0 Wire	22.20	−	20.40	15.60	15.60	−	125−175 Amp #1/0-#4/0	12.60
200−225 Amp 250-300 MCM Wire	25.80	−	24.00	16.80	20.70	−	200−240 Amp 250-300 MCM	18.60

Code:

1 −Plug-In, Standard Width Breaker
2 −Plug-In, Column Mounted Breaker
3 −Bolt-In, Standard Width Breaker − Class A
4 −Bolt-In, Standard Width Breaker − Class C
5 −Bolt-In, Standard Width Breaker − Class D or E
6 −Bolt-In, Column Mounted Breaker − Class D or E
7 −Standard Width Switch and Fuse Block

Manhours include necessary time for the addition of sub-feeder circuit breaker or switch, increased feeder and sub-feeder connections and handling of additional weight of cabinet and cover.

These units can be added to the units as shown on the standard branch panelboards, table or listed separately.

DISTRIBUTION AND POWER PANELBOARDS
Cabinet Installation

MANHOURS EACH

Weight Range (Pounds)	Wall Construction In Place	Wall Construction Not In Place
25-50	4.40	5.00
51-74	6.60	7.40
76-100	8.80	9.90
101-125	11.00	12.40
126-150	13.20	15.10
151-175	15.40	17.90
176-200	17.60	20.30
201-250	21.60	25.40
251-300	23.80	28.00
301-350	29.70	35.50
351-400	34.00	40.80
401-450	39.00	46.10

Manhours include checking out of job storage, handling, job hauling, distributing, locating and setting of the back box, panel interior, and cover.

Manhours exclude installation of back box fastenings, special supports, and conductor connections. See respective tables for these time frames.

DISTRIBUTION AND POWER PANELBOARDS
CIRCUIT BREAKER TYPE
Panelboard Branches

MANHOURS EACH

Trip Amperage Range	Maximum Lug Capacity	Number of Pole Branch and Conductor Connections				
		1-Pole 2-Wire	2-Pole 2-Wire	2-Pole 3-Wire	3-Pole 3-Wire	3-Pole 4-Wire
15-50	#4	0.80	0.93	1.11	1.39	1.66
70-100	#1	1.64	1.85	2.21	2.49	3.05
70-125	#1/0	–	3.00	3.39	3.58	4.22
125-225	#1/0	–	3.00	3.30	3.45	3.85
150-225	#300 MCM	–	3.30	3.90	4.17	4.81
150-225	#4/0	–	3.30	3.90	3.90	4.60
250	#4/0	–	3.30	4.20	4.20	4.68
400	#4/0	–	–	–	4.60	5.50
125-225	#250 MCM	–	3.90	4.80	5.10	5.75
250-400	2-#250 MCM	–	4.50	5.40	5.70	6.33
250-400	1-#600 MCM	–	4.50	5.40	5.70	6.33
400	2-#500 MCM	–	6.60	8.70	9.00	10.64
600	2-#500 MCM	–	7.20	9.00	9.30	10.93
600	1-#600 MCM	–	7.20	9.00	9.30	10.93
800	3-#500 MCM	–	9.00	9.60	10.20	12.00
800	2-#600 MCM	–	9.00	9.60	10.20	12.00
1,000	3-#600 MCM	–	9.60	13.20	13.80	16.68

Manhours include checking out of job storage, handling, job hauling, distributing and installing the individual component items comprising the interior of a distribution panelboard and giving consideration to a prorated amount of bus and mounting paw assembly weight, and the conductor connections.

Manhours exclude installation of the cabinet or back box and cover. See respective table for this time frame.

DISTRIBUTION AND POWER PANELBOARDS
CIRCUIT BREAKER TYPE
Other Panelboard Sections
Terminating Conductors in Main Lugs of Panelboards

MANHOURS EACH

Main Amperage Rating	Main Lugs Only			Neutral Block Section
	Poles and Connections			
	2-Pole, 3-Wire	3-Pole, 3-Wire	3-Pole, 4-Wire	
200	2.75	3.00	3.55	1.35
400	3.85	4.10	5.20	1.90
600	6.60	6.85	8.80	1.45
800	7.40	7.70	9.90	3.00
1200	11.00	11.55	14.85	4.40

Manhour units for use of terminating conductors in main lugs of panelboard.

Main Circuit Breaker
Addition to Panelboard Branches and Sub-Branches

MANHOURS EACH

Amperage Rating	Poles and Connections		
	2-Pole, 3-Wire	3-Pole, 3-Wire	3-Pole, 4-Wire
225	4.45	4.70	5.50
400	5.90	6.15	7.35
600	9.10	9.35	11.25
800	9.60	10.15	12.35

Manhour units for use when a main circuit breaker is to be included in a panelboard.

DISTRIBUTION AND POWER PANELBOARDS
CIRCUIT BREAKER TYPE

Circuit Wire Connections to Spare Branches in Place

MANHOURS EACH

Trip Amperage Range	Maximum Lug Capacity	Number of Pole Branches and Conductor Connections				
		1-Pole 2-Wire	2-Pole 2-Wire	2-Pole 3-Wire	3-Pole 3-Wire	3-Pole 4-Wire
15-50	#4	0.60	0.60	0.90	0.90	1.20
70-100	#1	1.30	1.30	2.00	2.00	2.60
70-125	#1/0	–	1.40	2.20	2.20	2.90
125-225	#1/0	–	1.40	2.20	2.20	2.90
150-225	#300 MCM	–	2.20	3.20	3.20	4.30
150-225	#4/0	–	1.70	2.50	2.50	3.40
250	#4/0	–	1.70	2.50	2.50	3.40
400	#4/0	–	–	–	2.50	3.40
125-225	#250 MCM	–	1.90	2.90	2.90	3.80
250-400	2-#250 MCM	–	2.40	3.60	3.60	4.80
250-400	1-#600 MCM	–	2.40	3.60	3.60	4.80
125	#110	–	1.40	3.20	3.20	4.30
400	2-#500 MCM	–	3.60	5.40	5.40	7.20
400	1-#600 MCM	–	3.60	5.40	5.40	7.20
600	2-#500 MCM	–	3.60	5.40	5.40	7.20
600	1-#600 MCM	–	3.60	5.40	5.40	7.20
800	3-#500 MCM	–	4.80	7.20	7.20	9.60
800	2-#600 MCM	–	4.80	7.20	7.20	9.60
1,000	3-#600 MCM	–	7.20	10.80	10.80	14.40

Manhours include cabling conductors, make up of wire terminal connections, circuit identification, testing and load balancing, and furnishing information for circuit directory.

Manhours exclude installation of cabinet or back box and covers and panelboard internals. See respective tables for these time frames.

DISTRIBUTION AND POWER PANELBOARDS CIRCUIT BREAKER TYPE

Replace Blank Spaces with
Branch Units in Panelboard

MANHOURS EACH

Trip Amperage Range	Maximum Lug Capacity	Number of Pole Branches and Conductor Connections				
		1-Pole 2-Wire	2-Pole 2-Wire	2-Pole 3-Wire	3-Pole 3-Wire	3-Pole 4-Wire
15-50	#4	1.20	1.20	1.80	1.80	2.40
70-100	#1	2.60	2.60	4.00	4.00	5.30
125-225	#1/0	–	3.10	4.30	4.30	5.70
70-225	#1/0	–	3.30	4.50	4.50	5.90
150-225	#300 MCM	–	4.90	7.10	7.10	9.20
150-225	#4/0	–	3.60	5.00	5.00	6.70
125-225	#250 MCM	–	4.40	6.40	6.40	8.30
250-400	2-#250 MCM	–	5.40	7.80	7.80	10.20
250-400	1-#600 MCM	–	5.40	7.80	7.80	10.20
125	#1/0	–	4.50	5.70	5.70	7.20
250	2-#250 MCM	–	6.00	8.40	8.40	10.80
250	1-#600 MCM	–	6.00	8.40	8.40	10.80
400	2-#500 MCM	–	8.40	12.00	12.00	15.60
400	1-#600 MCM	–	8.40	12.00	12.00	15.60
600	2-#500 MCM	–	8.40	12.00	12.00	15.60
600	1-#600 MCM	–	8.40	12.00	12.00	15.60
800	3-#500 MCM	–	10.80	15.60	15.60	20.40
800	2-#600 MCM	–	10.80	15.60	15.60	20.40
1,000	3-#600 MCM	–	15.60	22.80	22.80	30.00

Manhours include checking out of job storage, handling, job hauling, distributing, installing, cabling conductors, make up of wire terminal connections, circuit identification, testing, load balancing, and furnishing information for circuit directory for replacing blank spaces with branch units in panelboards.

Manhours exclude installation of cabinet or back box and covers and other panelboard internals. See respective tables for these time frames.

DISTRIBUTION AND POWER PANELBOARDS

Switch and Fuse Type Panelboard Branches

MANHOURS EACH

Amperage	Voltage	Type of Unit Mounting	Switch and Fuse Branches				Space for Future Branch	
			2-Pole 2-Wire	2-Pole 3-Wire	3-Pole 3-Wire	3-Pole 4-Wire	2-Pole	3-Pole
30-Comp	250	Double	0.85	1.20	1.50	1.80	0.30	0.30
30.	250	Double	0.95	1.50	1.80	2.10	0.40	0.40
30-Std.	250	Single	1.20	1.80	2.10	2.70	0.40	0.40
60	250	Double	1.70	2.00	2.30	2.85	0.40	0.40
60	250	Single	2.00	2.30	2.60	3.15	0.40	0.40
100	250	Double	2.00	2.70	2.85	3.45	0.40	0.70
100	600	Double	2.20	2.75	3.00	3.60	0.70	0.70
100	250	Single	2.20	2.75	3.00	3.85	0.40	0.70
100	600	Single	2.45	3.00	3.30	4.10	0.70	0.70
200	250-600	Single	3.30	3.85	4.10	4.95	0.90	0.90
400	250-600	Single	6.30	8.15	8.65	10.50	1.30	1.30
600	250-600	Single	7.85	8.65	10.25	12.05	1.55	1.55

Switch and Fuse Type Other Panelboard Sections

MANHOURS EACH

Amperage	Main Switch Section			Neutral Block Section	Main Lug Section		
	2-Pole 3-Wire	3-Pole 3-Wire	3-Pole 4-Wire		2-Pole 3-Wire	3-Pole 3-Wire	3-Pole 4-Wire
100	3.90	4.20	5.10	–	–	–	–
200	4.80	5.10	5.70	1.50	3.00	3.30	3.90
400	9.50	10.05	12.05	2.00	4.00	4.30	5.45
600	10.05	11.75	13.80	2.60	6.90	7.20	9.20
800	10.45	12.10	13.90	3.00	7.45	7.70	9.90
1,200	13.75	14.30	17.60	4.40	11.00	11.55	14.85

Panelboard branch manhours include checking out of job storage, handling, job hauling, distributing and installing the individual component items comprising the interior of a switch and fuse type distribution panelboard and giving consideration to a prorated amount of bus and mounting pan assembly weight and the conductor connections.

Other panelboard section manhours include time requirements for making connections for items as outlined.

Manhours exclude installation of the cabinet or back box and cover. See respective table for this time frame.

DISTRIBUTION AND POWER PANELBOARDS
SWITCH AND FUSE TYPE
Circuit Wire Connections to Spare Branches in Place

MANHOURS EACH

Amperage	Voltage	Type of Unit Mounting	Maximum Lug Capacity	Switch and Fuse Branch			
				2-Pole 2-Wire	2-Pole 3-Wire	3-Pole 4-Wire	3-Pole 4-Wire
30-Compact	250	Double	#8	0.50	0.70	0.75	1.00
30-Standard	250	Double	#8	0.50	0.70	0.75	1.00
30	250	Single	#8	0.50	0.70	0.75	1.00
60	250	Double	#4	1.15	1.70	1.75	2.30
60	250	Single	#4	1.15	1.70	1.75	2.30
100	250	Double	#1	1.50	2.25	2.35	3.00
100	600	Double	#1	1.50	2.25	2.35	3.00
100	250	Single	#1	1.50	2.25	2.35	3.00
100	600	Single	#1	1.50	2.25	2.35	3.00
200	250-600	Single	#250 MCM	1.55	2.30	2.45	3.10
400	250-600	Single	2-#500 MCM	3.80	5.70	5.85	7.60
600	250-600	Single	2-#500 MCM	3.80	5.70	5.85	7.60

Circuit wire connection manhours include cabling conductors, wire terminal connections, testing load balancing, pro rata allowance for circuit identification, and providing information for circuit directory.

Manhours exclude installation of the components and cabinet or back box and covers. See respective tables for these time frames.

DISTRIBUTION AND POWER PANELBOARDS
SWITCH AND FUSE TYPE

Replacing Blank Spaces with Branch Units
and Connecting Circuit Conductors

MANHOURS EACH

Amperage	Voltage	Type of Unit Mounting	Maximum Lug Capacity	Switch and Fuse Branches			
				2-Pole 2-Wire	2-Pole 3-Wire	3-Pole 3-Wire	3-Pole 4-Wire
30-Compact	250	Double	#8	1.00	1.55	1.60	1.90
30-Standard	250	Double	#8	1.00	1.55	1.60	1.90
30	250	Single	#8	1.00	1.55	1.60	1.90
60	250	Double	#4	3.00	4.50	4.60	6.00
60	250	Single	#4	3.00	4.50	4.60	6.00
100	250	Double	#1	3.90	5.85	5.95	7.80
100	600	Double	#1	4.65	6.60	6.75	8.55
100	250	Single	#1	3.90	5.85	5.95	7.80
100	600	Single	#1	4.65	6.60	6.75	8.55
200	250-600	Single	#250 MCM	5.70	7.80	8.00	9.90
400	250-600	Single	2-#500 MCM	9.00	13.00	13.20	16.95
600	250-600	Single	2-#500 MCM	9.60	13.55	13.75	17.50

Manhours include checking out of job storage, handling, job hauling, distributing, installing, cabling conductors, make up of wire terminal connections, circuit identification, testing, load balancing and furnishing information for circuit directory for replacing blank spaces with switch and fuse type units in panelboards.

Manhours exclude installation of cabinet or back box and covers. See respective table for this time frame.

DISTRIBUTION AND POWER PANELBOARDS
Motor Control Units for
Circuit Breaker, Switch, and Fuse Panelboards
Magnetic Starter Sections

MANHOURS EACH

Amperage Rating	N.E.M.A. Rating	Wire Size	Circuit Breaker	Switch and Fuse
20	0	#12	3.00	3.00
30	1	#10	3.30	3.30
50	2	# 6	4.20	4.20
100	3	# 2	5.75	5.75

Manhours are for the installation of the starter unit, push button station and pilot light for a complete magnetic starter unit to a distribution panel board.

Circuit Wire Connections to Spare Starters in Place

MANHOURS EACH

Amperage Rating	N.E.M.A. Rating	Wire Size	Circuit Breaker	Switch and Fuse
20	0	#12	0.90	0.90
30	1	#10	1.10	1.10
50	2	# 6	1.45	1.45
100	3	# 2	2.30	2.30

Manhours include circuit conductor connections, cabling conductors, wire terminal connections, circuit identification and testing for spare starters which are in place.

Placing Complete Starter Units in Blank Spaces

MANHOURS EACH

Amperage Rating	N.E.M.A. Rating	Wire Size	Circuit Breaker	Switch and Fuse
20	0	#12	4.80	4.80
30	1	#10	5.40	5.40
50	2	# 6	7.20	7.20
100	3	# 2	10.35	10.35

Manhours are for complete installation of starter units in blank spaces and include time requirements for installation of push button station, pilot light, circuit conductor connections, cabling conductors, wire internal connections, testing, and circuit identification.

DISTRIBUTION AND POWER PANELBOARDS

Cable and Lug Connectors

MANHOURS EACH

Wire Size	Cable Connector	Lug Connector
#12	0.35	0.25
#10	0.50	0.25
#8	0.60	0.25
#6	0.70	0.35
#4	0.85	0.35
#2	0.95	0.35
#1	1.40	0.70
#1/0	1.50	0.90
#2/0	1.70	1.10
#3/0	1.80	1.30
#4/0	2.10	1.50
#250 MCM	2.40	1.85
#350 MCM	2.75	2.20
#500 MCM	3.35	2.65
#600 MCM	3.95	3.10
#750 MCM	4.95	3.85
#1000 MCM	5.75	4.40

Manhours units for cable connectors should be applied to the panelboard time frames when additional cable taps are anticipated in the panel gutters. The lug manhour units are for application when larger lugs are to be installed in existing panelboards.

Manhours exclude installation of cabinet or back box and covers and panelboard components. See respective tables for these time frames.

Section 10

SWITCHBOARDS AND ACCESSORIES

This section includes manhour requirements for the installation of components as may be required for the assembly of low voltage distribution switchboards.

Labor units, in manhours, are included for installation operations of switchboard sections, meters and various wire, cable, busway stub, main lug, and neutral block connections.

Certain requirements for the installation of support items that may be necessary for the complete labor time requirements may be found in other sections of this manual.

HANDLING LOW VOLTAGE SWITCHBOARD SECTIONS

MANHOURS EACH

Switchboard Sections Weight Range (Pounds)	Installation Conditions	
	Normal	Difficult
0-200	5.40	8.10
201-300	6.30	9.50
301-400	7.20	10.80
401-500	8.10	12.20
501-600	9.80	17.20
601-700	10.80	18.90
701-800	11.80	20.70
801-900	12.70	22.20
901-1,000	13.70	24.00
1,001-1,100	15.70	29.00
1,101-1,200	16.80	31.00
1,201-1,300	17.90	33.10
1,301-1,400	18.90	35.00
1,401-1,500	20.50	38.90
1,501-1,600	21.60	41.00
1,601-1,700	22.70	43.10
1,701-1,800	23.80	45.20
1,801-1,900	25.30	49.40
1,901-2,000	26.40	50.10

Normal installation conditions are considered to be those without obstructions or adverse situations.

Difficult installation conditions are those where obstructions may exist and involve extra rigging, handling, and manual movement.

Manhour units are for the installation of a totally deenergized switchboard. If work is to be performed on an energized panelboard these manhours should be substantially increased.

Manhours include checking out of job storage, handling, job hauling, layout, rigging, picking, setting, leveling, aligning, bolting enclosures together, and bolting busbars together.

Manhours exclude installation of supports or fasteners, pull boxes, special knockouts, raceway, or conductor terminations, welding, and painting. See respective tables for these time frames.

MOLDED CASE CIRCUIT BREAKERS
Switchboard Branch Conductor Connectors

MANHOURS EACH

Trip Amperage Range	Maximum Lug Capacity	Number of Pole Branches and Conductor Connections				
		1-Pole 2-Wire	2-Pole 2-Wire	2-Pole 3-Wire	3-Pole 3-Wire	3-Pole 4-Wire
15-50	#4	0.60	0.70	0.80	1.00	1.20
70-100	#1	1.20	1.30	1.70	1.90	2.40
70-125	#1/0	–	1.30	1.70	1.90	2.40
125-225	#1/0	–	2.45	2.70	2.90	3.60
125-225	#250 MCM	–	2.45	2.70	2.90	3.60
150-225	#4/0	–	2.45	2.70	2.90	3.60
150-225	#300 MCM	–	2.45	2.70	2.90	3.60
250-400	2-#250 MCM	–	2.70	3.30	3.60	4.50
250-400	1-#600 MCM	–	2.70	3.30	3.60	4.50
125	#1/0	–	2.45	2.70	2.90	3.60
250	#4/0	–	2.45	3.20	3.40	3.90
250	2-#500 MCM	–	2.45	3.20	3.40	3.90
250	1-#600 MCM	–	2.45	3.20	3.40	3.90
400	#4/0	–	–	–	3.60	4.00
400	2-#500 MCM	–	2.70	3.30	3.60	4.00
400	1-#600 MCM	–	2.70	3.30	3.60	4.00
600	2-#500 MCM	–	3.70	4.30	4.60	5.60
600	1-#600 MCM	–	3.70	4.30	4.60	5.60
800	3-#500 MCM	–	5.20	5.75	6.10	7.50
800	2-#600 MCM	–	5.20	5.75	6.10	7.50
1,000	3-#600 MCM	–	6.85	7.40	7.70	8.70

Manhours include make up of wire terminal connections, circuit identification, lacing, terminations, load balancing, testing, and furnishing information for circuit directory.

Manhours exclude installation of switchboard sections. See respective tables for these time frames.

MOLDED CASE CIRCUIT BREAKERS
Circuit Wire Connections to Spare Branches in Place

MANHOURS EACH

Trip Amperage Range	Maximum Lug Capacity	Number of Pole Branches and Conductor Connections				
		1-Pole 2-Wire	2-Pole 2-Wire	2-Pole 3-Wire	3-Pole 3-Wire	3-Pole 4-Wire
15-50	#4	0.55	0.60	0.80	0.90	1.10
70-100	#1	1.10	1.20	1.50	1.75	2.15
70-125	#1/0	—	1.20	1.50	1.75	2.15
125-225	#1/0	—	2.20	2.40	2.65	3.25
125-225	#250 MCM	—	2.20	2.40	2.65	3.25
150-225	#4/0	—	2.20	2.40	2.65	3.25
150-225	#300 MCM	—	2.20	2.40	2.65	3.25
250-400	2-#250 MCM	—	2.45	3.05	3.25	4.00
250-400	1-#600 MCM	—	2.45	3.05	3.25	4.00
125	#1/0	—	2.20	2.40	2.65	3.25
250	#4/0	—	2.20	2.90	3.10	3.50
250	2-#500 MCM	—	2.20	2.90	3.10	3.50
250	1-#600 MCM	—	2.20	2.90	3.10	3.50
400	#4/0	—	—	—	3.25	4.00
400	2-#500 MCM	—	2.45	3.00	3.25	4.00
400	1-#600 MCM	—	2.45	3.00	3.25	4.00
600	2-#500 MCM	—	3.50	4.00	4.25	5.30
600	1-#600 MCM	—	3.50	4.00	4.25	5.30
800	3-#500 MCM	—	5.00	5.50	5.70	6.70
800	2-#600 MCM	—	5.00	5.50	5.70	6.70
1,000	3-#600 MCM	—	6.65	7.15	7.35	8.35

Manhours include cabling conductors, make up of wire terminal connections, circuit identification, testing, load balancing, and furnishing information for circuit directory.

Manhours exclude installation of switchboard sections. See respective tables for these time frames.

MOLDED CASE CIRCUIT BREAKERS
Replacing Blank Spaces with Branch Units in Place
and Connecting Conductors

MANHOURS EACH

Trip Amperage Range	Maximum Lug Capacity	Number of Pole Branches and Conductor Connections				
		1-Pole 2-Wire	2-Pole 2-Wire	2-Pole 3-Wire	3-Pole 3-Wire	3-Pole 4-Wire
15-50	#4	0.80	0.85	1.10	1.40	1.70
70-100	#1	1.60	1.75	2.30	2.60	3.20
70-125	#1/0	–	1.75	2.30	2.60	3.20
125-225	#1/0	–	3.00	3.30	3.60	4.20
125-225	#250 MCM	–	3.00	3.30	3.60	4.20
150-225	#4/0	–	3.00	3.30	3.60	4.20
150-225	#300 MCM	–	3.00	3.30	3.60	4.20
250-400	2-#250 MCM	–	3.50	4.30	4.80	6.00
250-400	1-#600 MCM	–	3.50	4.30	4.80	6.00
125	#1/0	–	1.70	2.30	2.60	3.20
250	#4/0	–	3.30	4.20	4.50	5.10
250	2-#500 MCM	–	3.30	4.20	4.50	5.10
250	1-#600 MCM	–	3.30	4.20	4.50	5.10
400	#4/0	–	–	–	4.80	6.00
400	2-#500 MCM	–	3.50	4.30	4.80	6.00
400	1-#600 MCM	–	3.50	4.30	4.80	6.00
600	2-#500 MCM	–	4.55	5.30	5.75	6.90
600	1-#600 MCM	–	4.55	5.30	5.75	6.90
800	3-#500 MCM	–	6.00	6.70	7.15	8.25
800	2-#600 MCM	–	6.00	6.70	7.15	8.25
1,000	3-#600 MCM	–	7.65	8.35	8.80	9.90

Manhours include checking out of job storage, handling, job hauling, distributing, installing branch units in blank spaces, cabling conductors, make up wire terminal connections, circuit identification, testing, load balancing and furnishing information for directory for replacing blank spaces with branch units in switchboards.

MAIN CIRCUIT BREAKER CONDUCTOR AND SWITCHBOARD BUSWAY STUB CONNECTIONS FOR AIR CIRCUIT BREAKERS, POWER PROTECTORS, AND PRESSURE SWITCHES

MANHOURS EACH

Item Description	Main Amperage Rating	Poles and Connections					
		2-Pole 3-Wire		3-Pole 3-Wire		3-Pole 4-Wire	
		A	B	A	B	A	B
Air Circuit Breakers or Power Protectors	600	7.15	6.30	7.70	6.30	8.25	6.60
	800	8.00	6.60	8.50	6.85	9.35	7.40
	1,000	10.45	8.00	10.75	8.25	12.10	8.80
	15-225	5.50	4.40	6.00	4.40	7.15	4.95
	35-600	6.85	6.85	8.80	7.15	10.70	7 70
	200-1,600	–	–	20.90	13.20	25.30	15.40
	2,000-3,000	–	–	35.20	22.00	44.00	24.20
	4,000	–	–	44.00	28.60	57.20	33.00
	800-1,600	–	–	20.90	11.00	25.30	13.20
	2,000-2,500	–	–	35.20	22.00	44.00	24.20
	3,000-4,000	–	–	44.00	28.60	57.20	33.00
Pressure Switches	2,000	–	–	26.40	17.60	30.80	19.80
	3,000	–	–	35.20	22.00	44.00	24.20
	4,000	–	–	44.00	28.60	57.00	33.00

Code:

A—Cable Conductor Connections

B—Busway Stub Connection

Cable conductor manhours include time required for connecting cable conductors to the line side of the main protective device. Busway stub connection manhours include time required to connect the switchboard stub to the line side of a main protective device, instead of cable conductors.

All manhours include time allowance for circuit identification, cabling, terminal connections, testing, load balancing, and furnishing information for directory.

Manhours exclude installation and positioning of switchboard. See respective table for this time frame.

SWITCH AND FUSE TYPE BRANCHES

MANHOURS EACH

Amperage	Voltage	Type of Unit Mounting	Switch and Fuse Branches				Space for Future Branch	
			2-Pole 2-Wire	2-Pole 3-Wire	3-Pole 3-Wire	3-Pole 4-Wire	2-Pole	3-Pole
30-Comp.	250	Double	0.85	1.20	1.50	1.80	0.30	0.30
30-Std.	250	Double	0.95	1.50	1.80	2.10	0.40	0.40
30	250	Single	1.20	1.80	2.10	2.70	0.40	0.40
60	250	Double	1.70	2.00	2.30	2.85	0.40	0.40
60	250	Single	2.00	2.30	2.60	3.15	0.40	0.40
100	250	Double	2.00	2.70	2.85	3.45	0.40	0.70
100	600	Double	2.20	2.75	3.00	3.60	0.70	0.70
100	750	Single	2.20	2.75	3.00	3.85	0.40	0.70
100	600	Single	2.45	3.00	3.30	4.10	0.70	0.70
200	250-600	Single	3.30	3.85	4.10	4.95	0.90	0.90
400	250-600	Single	6.30	8.15	8.65	10.50	1.30	1.30
600	250-600	Single	7.85	8.65	10.25	12.05	1.55	1.55

Manhours include checking out of job storage, handling, job hauling, distributing, and installing the individual switch and fuse section, and making connections.

Manhours exclude installation of other switchboard sections. See respective tables for these time frames.

SWITCH AND FUSE TYPE BRANCHES
Circuit Wire Connections to Spare Branches in Place

MANHOURS EACH

Amperage	Voltage	Type of Unit Mounting	Maximum Lug Capacity	Switch and Fuse Branch			
				2-Pole 2-Wire	2-Pole 3-Wire	3-Pole 3-Wire	3-Pole 4-Wire
30-Compact	250	Double	#8	0.50	0.70	0.75	1.00
30-Standard	250	Double	#8	0.50	0.70	0.75	1.00
30	250	Single	#8	0.50	0.70	0.75	1.00
60	250	Double	#4	1.15	1.70	1.75	2.30
60	250	Single	#4	1.15	1.70	1.75	2.30
100	250	Double	#1	1.50	2.25	2.35	3.00
100	600	Double	#1	1.50	2.25	2.35	3.00
100	250	Single	#1	1.50	2.25	2.35	3.00
100	600	Single	#1	1.50	2.25	2.35	3.00
200	250-600	Single	#250 MCM	1.55	2.30	2.45	3.10
400	250-600	Single	2-#500 MCM	3.80	5.70	5.85	7.60
600	250-600	Single	2-#500 MCM	3.80	5.70	5.85	7.60

Circuit wire connection manhours include cabling conductors, wire terminal connections, testing, load balancing, prorata allowance for circuit identification, and providing information for circuit directory.

Manhours exclude installation of switchboard components. See respective table for this time frame.

MISCELLANEOUS SWITCHBOARD SECTIONS CONDUCTOR TERMINATIONS AND TERMINAL CONNECTIONS

MANHOURS EACH

Item Description	Amperage Rating	Poles and Connections						Neutral Block Section
		2-Pole 3-Wire		3-Pole 3-Wire		3-Pole 4-Wire		
		A	B	A	B	A	B	
Miscellaneous Sections	800	8.00	7.40	8.50	7.70	9.35	9.90	3.00
	1,000	–	9.90	–	10.45	–	12.10	3.60
	15-225	5.50	–	6.00	–	7.15	–	–
	35-600	8.25	–	8.80	–	10.70	–	–
	200-1,600	–	–	20.90	16.50	25.30	22.00	5.50
	2,000-3,000	–	–	35.20	30.80	44.00	40.70	9.90
	4,000	–	–	44.00	41.80	57.20	53.90	12.10
	800-1,600	–	–	20.90	16.50	25.30	22.00	5.50
	2,000-2,500	–	–	35.20	30.80	44.00	40.70	9.90
	3,000-4,000	–	–	44.00	41.80	57.20	53.90	12.10
Pressure Switches	2,000	–	–	26.40	20.90	30.80	27.50	6.60
	3,000	–	–	35.20	30.80	44.00	40.70	9.90
	4,000	–	–	44.00	41.80	57.00	53.90	12.10

Code:

A—Main Switch Section

B—Main Lug Sections

Manhours when properly applied to all the component sections of a switchboard will equal the time required for the conductor terminations, cabling, terminal connections, circuit identification, load balancing, testing, and supplying information for directory.

Manhours exclude installation and position of switchboard components. See respective table for these time frames.

METERS FOR SWITCHBOARDS
3-PHASE, 4-WIRE

MANHOURS EACH

Meter Type	Manhours
Ammeter	2.20
Voltmeter	2.20
Watt Hour Meter	2.75
Synchroscope	2.75
Power Factor Meter	3.30
Transfer Switch	1.65

Manhours include installation of meters as outlined when installed and connected in the switchboard at the time of the switchboard installation.

Manhours exclude installation and positioning of other switchboard components. See respective table for these time frames.

REPLACE BLANK SPACES WITH BRANCH UNITS IN SWITCHBOARDS AND CONNECT CONDUCTORS

MANHOURS EACH

Amperage Rating	Voltage	Type of Unit Mounting	Maximum Lug Capacity	Switch and Fuse Branch			
				2-Pole 2-Wire	2-Pole 3-Wire	3-Pole 3-Wire	3-Pole 4-Wire
30-Compact	250	Double	#8	0.85	1.15	1.45	1.75
30-Standard	250	Double	#8	0.90	1.45	1.75	2.00
30	250	Single	#8	1.15	1.75	2.00	2.60
60	250	Double	#4	1.65	1.90	2.20	2.75
60	250	Single	#4	1.90	2.20	2.45	2.95
100	250	Double	#1	1.90	2.50	2.75	3.20
100	600	Double	#1	2.20	2.75	2.95	3.50
100	250	Single	#1	2.20	2.75	2.95	3.55
100	600	Single	#1	2.45	2.95	3.20	4.00
200	250-600	Single	#250 MCM	2.95	3.40	3.65	4.45
400	250-600	Single	2-#500 MCM	5.90	7.60	8.15	9.85
600	250-600	Single	2-#500 MCM	7.35	8.15	9.55	11.30

Manhours include checking out of job storage, handling, job hauling, distrbuting, installing, cabling conductors, make up of wire terminal connections, circuit identification, testing, load balancing, and furnishing information for circuit directory.

Manhours exclude installation of panelboard. See respective table for this time frame.

SWITCHBOARD PULL BOXES

MANHOURS EACH

Pull Box Size Length x Width x Depth	Steel Gauge	Approximate Weight (Pounds)	Manhours
56" x 36" x 12"	#10	258	12.60
56" x 36" x 18"	#10	288	13.50
70" x 42" x 12"	#10	348	15.30
70" x 42" x 18"	#10	385	16.20
84" x 36" x 18"	#10	406	18.00
84" x 36" x 24"	#10	484	19.80
84" x 42" x 12"	#10	412	18.00
84" x 42" x 18"	#10	472	19.80
84" x 42" x 24"	#10	533	21.60
112" x 36" x 30"	#10	786	34.20
105" x 42" x 18"	#10	660	27.00
105" x 42" x 24"	#10	720	30.60
140" x 42" x 18"	#10	840	36.00
140" x 42" x 24"	#10	914	37.80

Manhours include checking out of job storage, handling, job hauling, distributing, drilling and bolting box to switchboard, and removing and reinstalling access cover.

Manhours exclude fabrication, welding or modifying boxes, field cut knockouts, and installation of panelboard sections. See respective tables for these time frames.

Section 11

SWITCHES, STARTERS, CONTROLS, AND GUTTERS

Manhour units for the installation of such items as externally operated safety switches, circuit breakers, motor control devices, auxiliary gutters, junction boxes, terminal cabinets, service entrance switches, meter troughts, transformer cabinets, and meter test cabinets are included in this section.

Consideration has been given to all labor operations as required for the installation of the individual item in accordance with the notes appearing with the tables.

Various labor operations that may be required to support the complete installation of all items for the particular block of work may be found under other sections of this manual.

EXTERNALLY OPERATED SAFETY SWITCHES
GENERAL PURPOSE TYPE ENCLOSURE

250-Volt Fusible and Nonfusible

Separate Installation

MANHOURS EACH

Amperage Size	Maximum Lug Capacity	Fusible				Nonfusible			
		2-Pole 2-Wire	2-Pole 3-Wire	3-Pole 3-Wire	3-Pole 4-Wire	2-Pole 2-Wire	2-Pole 3-Wire	3-Pole 3-Wire	3-Pole 4-Wire
30	#14–#6	2.20	2.40	2.40	2.60	1.70	1.90	1.90	2.00
60	#14–#2	2.70	3.00	3.00	3.20	2.30	2.50	2.50	2.70
100	#14–#1/0	3.00	3.30	3.30	3.50	2.50	2.80	2.80	3.00
200	#6–#250 MCM	4.90	5.30	5.30	5.80	4.40	4.90	4.90	5.20
400	2-#4–#250 MCM	7.70	8.50	8.50	9.20	7.30	8.00	8.00	8.70
600	2-#2–#600 MCM	10.40	11.50	11.50	12.40	10.00	11.00	11.00	11.90
800	3-#1/0–#600 MCM	14.40	15.80	15.80	17.10	14.00	15.30	15.30	16.60
1,200	4-#1/0–#500 MCM	19.80	21.80	21.80	23.50	18.90	20.80	20.80	22.50

Combination Installation

MANHOURS EACH

Amperage Size	Maximum Lug Capacity	Fusible				Nonfusible			
		2-Pole 2-Wire	2-Pole 3-Wire	3-Pole 3-Wire	3-Pole 4-Wire	2-Pole 2-Wire	2-Pole 3-Wire	3-Pole 3-Wire	3-Pole 4-Wire
30	#14–#6	2.60	2.90	2.90	3.10	2.20	2.40	2.40	2.60
60	#14–#2	3.60	4.00	4.00	4.30	3.20	3.50	3.50	3.70
100	#14–#1/0	3.90	4.30	4.30	4.60	3.40	3.80	3.80	4.10
200	#6–#250 MCM	5.80	6.30	6.30	6.80	5.30	5.80	5.80	6.30
400	2-#4–#250 MCM	8.60	9.50	9.50	10.30	8.20	9.00	9.00	9.70
600	2-#2–#600 MCM	11.30	12.50	12.50	13.50	10.90	12.00	12.00	12.90
800	3-#1/0–#600 MCM	15.30	16.80	16.80	18.20	14.90	16.30	16.30	17.60
1,200	4-#1/0–#500 MCM	20.70	22.80	22.80	24.60	19.80	21.80	21.80	23.50

Manhours include checking out of job storage, handling, job hauling, distributing, positioning, and conductor terminations of switches as outlined.

Manhours exclude fastening or field cut knockouts. See respective tables for these time frames.

Separate installation involves the installation of the individual switch and does not require the detailed layout of the combination installation. The combination installation manhours reflect the additional layout time required.

EXTERNALLY OPERATED SAFETY SWITCHES
GENERAL PURPOSE TYPE ENCLOSURE

600-Volt Fusible and Nonfusible

Separate Installation

MANHOURS EACH

Amperage Size	Maximum Lug Capacity	Fusible				Nonfusible			
		2-Pole 2-Wire	2-Pole 3-Wire	3-Pole 3-Wire	3-Pole 4-Wire	2-Pole 2-Wire	2-Pole 3-Wire	3-Pole 3-Wire	3-Pole 4-Wire
30	#14−#6	2.60	2.90	2.90	3.10	2.20	2.40	2.40	2.60
60	#14−#2	3.20	3.50	3.50	3.70	2.70	3.00	3.00	3.20
100	#14−#1/0	3.40	3.80	3.80	4.10	3.00	3.30	3.30	3.50
200	#6−#250 MCM	5.80	6.30	6.30	6.80	5.30	5.80	5.80	6.30
400	2-#4−#250 MCM	8.60	9.50	9.50	10.30	8.20	9.00	9.00	9.70
600	2-#2−#600 MCM	11.30	12.50	12.50	13.50	10.90	12.00	12.00	13.00
800	3-#1/0−#600 MCM	15.30	16.80	16.80	18.20	14.90	16.30	16.30	17.60
1,200	4-#1/0−#500 MCM	20.70	22.80	22.80	24.60	19.80	21.80	21.80	23.50

Combination Installation

MANHOURS EACH

Amperage Size	Maximum Lug Capacity	Fusible				Nonfusible			
		2-Pole 2-Wire	2-Pole 3-Wire	3-Pole 3-Wire	3-Pole 4-Wire	2-Pole 2-Wire	2-Pole 3-Wire	3-Pole 3-Wire	3-Pole 4-Wire
30	#14−#6	3.10	3.40	3.40	3.60	2.60	2.90	2.90	3.10
60	#14−#2	4.10	4.50	4.50	4.80	3.60	4.00	4.00	4.30
100	#14−#1/0	4.30	4.80	4.80	5.10	3.90	4.30	4.30	4.60
200	#6−#250 MCM	6.20	6.80	6.80	7.40	5.80	6.30	6.30	6.80
400	2-#4−#250 MCM	9.10	10.00	10.00	10.80	8.60	9.50	9.50	10.30
600	2-#2−#600 MCM	11.80	13.00	13.00	14.00	11.30	12.50	12.50	13.50
800	3-#1/0−#600 MCM	15.80	17.30	17.30	18.70	15.30	16.80	16.80	18.20
1,200	4-#1/0−#500 MCM	21.20	23.30	23.30	25.10	20.70	22.80	22.80	24.60

Manhours include checking out of job storage, handling, job hauling, distributing, positioning, and conductor terminations of switches as outlined.

Manhours exclude fastening or field cut knockouts. See respective tables for these time frames.

Separate installation involves the installation of the individual switch and does not require the detailed layout of the combination installation. The combination installation manhours reflect the additional layout time required.

CIRCUIT BREAKERS

Separate and Combination Installation

MANHOURS EACH

Trip Amperage Range	Maximum Lug Capacity	Separate				Combination			
		2-Pole 2-Wire	2-Pole 3-Wire	3-Pole 3-Wire	3-Pole 4-Wire	2-Pole 2-Wire	2-Pole 3-Wire	3-Pole 3-Wire	3-Pole 4-Wire
15-50	#4	3.00	3.30	3.30	3.60	4.20	4.50	4.50	4.80
70-100	#1	3.60	4.40	4.40	5.00	4.80	5.60	5.60	6.20
70-125	#1/0	5.20	6.00	6.00	6.30	6.40	7.20	7.20	7.80
125-225	#1/0	5.00	5.80	5.80	6.50	6.20	7.00	7.00	7.70
125-225	#250 MCM	8.40	9.60	9.60	10.80	9.60	10.80	10.80	12.00
150-225	#300 MCM	7.00	8.00	8.00	9.20	8.20	9.20	9.20	10.30
150-225	#4/0	6.50	7.60	7.60	8.60	7.70	8.80	8.80	9.80
250-400	2-#250 MCM	10.80	12.60	12.60	14.40	13.20	15.00	15.00	16.80
250-400	1-#600 MCM	10.80	12.60	12.60	14.40	13.20	15.00	15.00	16.80
125	#1/0	6.60	7.50	7.50	8.10	7.80	8.70	8.70	9.30
250	#4/0	6.90	8.10	8.10	9.30	8.10	9.30	9.30	10.50
250	2-#250 MCM	8.40	9.60	9.60	10.80	10.80	12.00	12.00	13.20
250	1-#600 MCM	8.40	9.60	9.60	10.80	10.80	12.00	12.00	13.20
400	#4/0	−	−	11.40	12.90	−	−	13.80	15.30
400	2-#500 MCM	10.80	12.60	12.60	14.40	13.20	15.00	15.00	16.80
400	1-#600 MCM	10.80	12.60	12.60	14.40	13.20	15.00	15.00	16.80
600	2-#500 MCM	12.65	14.40	14.40	16.80	15.00	16.70	16.70	18.40
600	1-#600 MCM	12.65	14.40	14.40	16.80	15.00	16.70	16.70	18.40
800	3-#500 MCM	15.40	17.10	17.10	18.70	17.60	19.30	19.30	20.90
800	2-#600 MCM	15.40	17.10	17.10	18.70	17.60	19.30	19.30	20.90
1,000	3-#600 MCM	18.70	20.40	20.40	22.00	20.90	22.60	22.60	24.20

Manhours include checking out of job storage, handling, job hauling, distributing, positioning, and conductor terminations of switches as outlined.

Manhours exclude fastening or field cut knockouts. See respective tables for these time frames.

Separate installation involves the installation of the individual switch and does not require the detailed layout of the combination installation. The combination installation manhours reflect the additional layout time required.

MAGNETIC STARTERS
Reversing Magnetic Starters

MANHOURS EACH

NEMA Size	HP	Voltage	1-Phase				3-Phase			
			Separate		Combination		Separate		Combination	
			2-Pole	3-Pole	2-Pole	3-Pole	3-Pole	4-Pole	3-Pole	4-Pole
0	1-2	115-230	3.00	4.10	3.60	4.70	–	–	–	–
0	3-5	208-550	–	–	–	–	4.70	5.50	5.20	6.00
1	2-3	115-230	4.10	5.20	4.70	5.80	–	–	–	–
1	7-1/2 · 10	208-550	–	–	–	–	6.30	7.70	6.90	8.29
2	15-25	208-550	–	–	–	–	9.30	9.70	9.60	9.90
3	30-50	208-550	–	–	–	–	11.30	13.50	11.80	14.00
4	50-100	208-550	–	–	–	–	18.90	22.60	19.40	23.10
5	100-200	208-550	–	–	–	–	28.30	32.50	29.40	33.60

Nonreversing Magnetic Starters

MANHOURS EACH

NEMA Size	HP	Voltage	1-Phase 2-Pole		3-Phase			
			Separate	Combination	Separate		Combination	
					3-Pole	4-Pole	3-Pole	4-Pole
0	1	115	2.20	2.75	–	–	–	–
0	2	230	2.20	2.75	–	–	–	–
0	2-5	110-550	–	–	2.75	3.55	3.30	4.10
1	2	115	2.45	3.00	–	–	–	–
1	3	230	2.45	3.00	–	–	–	–
1	3-10	110-550	–	–	4.10	4.95	4.65	5.50
2	3	115	2.75	3.30	–	–	–	–
2	7-1/2	230	3.30	3.85	–	–	–	–
2	10	440-550	3.85	4.40	–	–	–	–
2	7-1/2 · 25	110-550	–	–	6.00	6.85	6.60	7 40
3	15-50	110-550	–	–	7.70	8.80	8.25	9.35
4	50-100	208-550	–	–	13.10	14.14	13.65	15.75
5	100-200	208-550	–	–	20.00	24.15	21.00	25.20

Manhours include checking out of job storage, handling, job hauling, distributing, positioning, and conductor terminations of starters as outlined.

Manhours exclude fastening or field cut knockouts. See respective tables for these time frames.

Separate installation involves the installation of the individual starter and does not require the detailed layout of the combination installation. The combination installation manhours reflect the additional layout time required.

MAGNETIC STARTERS
Circuit Breaker and Starter or
Switch and Starter Combination

MANHOURS EACH

NEMA Size	HP	Voltage	3-Phase, 3-Pole			
			Reversing		Nonreversing	
			Separate	Combination	Separate	Combination
0	1-5	208-550	8.25	8.50	5.50	6.00
1	5-10	208-550	11.00	11.55	8.25	8.80
2	10-25	208-550	14.30	14.85	10.45	11.00
3	20-50	208-550	18.15	18.70	14.30	14.85

Pushbutton Stations

MANHOURS EACH

Number of Positions	Stations		Pilot Light
	Separate	Combination	
2	0.80	1.10	1.65
3	1.10	1.35	1.90
4	1.35	1.65	2.20
5	1.65	1.90	2.45
6	1.90	2.20	2.75

Manual Motor Starting Switches

MANHOURS EACH

NEMA Size	Separate Installation			Combination Installation		
	1-Phase		3-Phase	1-Phase		3-Phase
	Toggle	Pushbutton	Pushbutton	Toggle	Pushbutton	Pushbutton
0	1.65	1.90	2.75	2.20	2.45	3.00
1	1.90	2.20	3.55	2.45	2.75	3.85
2	2.45	2.75	4.10	3.00	3.30	4.40

Manhours include checking out of job storage, handling, job hauling, distributing, positioning, and conductor terminations of starters, stations, and switches as outlined.

Manhours exclude fastening or field cut knockouts. See respective tables for these time frames.

Separate installation involves the installation of the individual item and does not require the detailed layout of the combination installation. The combination installation manhours reflect the additional layout time required.

DRUM SWITCHES

1-Speed Drum Switches

MANHOURS EACH

NEMA Size	Horsepower					Reversing Drum Switches		Nonreversing Drum Switches	
	1-Phase		2- or 3-Phase			Separate	Combination	Separate	Combination
	110V.	220V.	110V.	220V.	550V.				
1	1-1/2	3	3	5	7-1/2	1.65	2.20	–	–
1				15	15	–	–	8.80	11.00
2	–	–	–	10	10	2.20	2.75	–	–
2				100	100	–	–	19.80	22.00
3	–	–	–	20	40	4.40	4.93	–	–

2-Speed Drum Wwitches

MANHOURS EACH

NEMA Size	Horsepower			Reversing Drum Switches		Nonreversing Drum Switches	
	110 V.	220 V.	550 V.	Separate	Combination	Separate	Combination
00	3/4	3/4	3/4	2.55	4.65	1.90	3.00
0	1-1/2	2	2	4.10	5.20	2.45	3.55
1	3	5	7-1/2	4.95	6.00	3.30	4.40
2	–	15	15	5.50	6.60	3.85	4.95
3	–	20	40	8.25	9.90	6.00	7.70
4	–	40	75	14.30	16.50	11.00	13.20
5	–	75	150	23.10	25.30	19.80	22.00

Manhours include checking out of job storage, handling, job hauling, distributing, positioning, and conductor terminations for switches as outlined.

Manhours exclude fastenings or field cut knockouts. See respective tables for these time frames.

Separate installation involves the installation of the individual switch and does not require the detailed layout of the combination installation. The combination installation manhours reflect the additional layout time required.

CONTROL DEVICES

MANHOURS EACH

Time Switches				
General Description Operating Positions	Amperage	Voltage	Installation	
			Separate	Combination
S.P.S.T. – Off-On – 24 Hrs.	1.50	120	2.20	2.75
S.P.S.T. – Off-On – 24 Hrs.	6.5	240	2.20	2.75
S.P.D T – Intermittent – 24 Hrs.	10.0	120	3.00	3.55
S.P.D.T. – Intermittent – 24 Hrs.	5.0	240	3.00	3.55
D.P.S.T. – Industrial Type	55.0	120	3.55	4.10
D.P.S.T. – Industrial Type	55.0	240	3.55	4.10
S.P.S.T – Program Intervals	5.0	120	3.85	4.40
S.P.S.T. – T-Rated Astrodial	40.0	120/208	4.40	4.95
D.P.D.T. – 4-Pole – T-Rated	40.0	120/208	4.95	5.50
S.P.S.T. – 4-Pole – T-Rated	20.0	120/208	4.65	5.20

Enclosed Magnetic Relays			Pressure–Float–Limit Switches		
Number of Poles	Manhours		Type of Switch	Manhours	
	Separate	Combination		Separate	Combination
1	1.10	1.65	Pressure–S.P.	0.80	1.35
2	1.35	1.90	Pressure–D.P	1.35	1.90
3	1.65	2.20	Pressure–Four P	1.90	2.45
4	1.90	2.45	Float–S.P.	1.10	1.65
5	2.20	2.75	Float–D.P.	1.65	2.20
6	3.00	3.55	Float–Four P.	2.20	2.75
7	3.85	4.40	Limit–S.P	1.35	1.90
8	4.65	5.20	Limit–D.P.	1.90	2.45
–	–	–	Limit–Four P	2.45	3.00

Manhours include checking out of job storage, handling, job hauling, distributing, positioning, and conductor terminations of control devices as outlined.

Manhours exclude fastening or field cut knockouts. See respective tables for these time frames.

Separate installation involves the installation of the individual item and does not require the detailed layout of the combination installation. The combination installation manhours reflect the additional layout time required.

REDUCED VOLTAGE STARTERS

MANHOURS EACH FOR SEPARATE INSTALLATION

NEMA Size	Horsepower Range	Voltage Range	Nonreversing 3-Phase, 3-Pole		Reversing 3-Phase, 3-Pole	
			Resistor	Auto Transformer	Resistor	Auto Transformer
1	5-10	208-550	18.90	21.00	24.15	28.35
2	10-25	208-550	21.00	25.20	31.50	37.80
3	20-50	208-550	25.20	29.40	40.95	44.10
4	40-100	208-550	29.40	39.10	43.30	58.70
5	60-200	208-550	35.00	49.40	52.50	73.40

NEMA Size	Constant Horsepower Rating		Starter Only 2-Speed	Starter and Circuit Breaker 2-Speed	Starter and Disconnect Switch 2-Speed
	Horsepower	Voltage			
0	1-2	208-550	5.00	7.85	7.85
1	3-5	208-550	6.60	11.00	11.00
2	10-20	208-550	9.20	14.20	14.20
3	20-40	208-550	11.30	17.85	18.35
4	30-60	208-550	18.50	24.70	25.00
5	75-150	208-550	28.80	37.00	37.50

Multi-Speed Starters — Nonreversing, 3-Phase

Manhours include checking out of job storage, handling, job hauling, distributing, positioning of enclosure, separate installation, and conductor terminations.

Manhours exclude fastenings and field cut knockouts. See respective tables for these time frames.

AUXILIARY GUTTERS AND FITTINGS

MANHOURS EACH

Item and Size		Manhours
Gutters		
4″ x 4″ − 12″ Long		0.60
4″ x 4″ − 24″ Long		0.90
4″ x 4″ − 36″ Long		1.50
4″ x 4″ − 48″ Long		2.10
4″ x 4″ − 60″ Long		2.70
6″ x 6″ − 12″ Long		0.90
6″ x 6″ − 24″ Long		1.20
6″ x 6″ − 36″ Long		1.80
6″ x 6″ − 48″ Long		2.70
6″ x 6″ − 60″ Long		3.30
8″ x 8″ − 12″ Long		1.20
8″ x 8″ − 24″ Long		1.50
8″ x 8″ − 36″ Long		2.10
8″ x 8″ − 48″ Long		3.00
8″ x 8″ − 60″ Long		3.00
Fittings		
Connectors	−4″	0.30
Connectors	−6″	0.48
Connectors	−8″	0.60
Adapter	−4″	0.12
Adapter	−6″	0.18
Adapter	−8″	0.24
End Wall	−4″	0.12
End Wall	−6″	0.18
End Wall	−8″	0.24
Tee	−4″	0.90
Tee	−6″	1.02
Tee	−8″	1.08
90° Elbow	−4″	0.60
90° Elbow	−6″	0.72
90° Elbow	−8″	0.90
Hanger Assembly	−4″	0.60
Hanger Assembly	−6″	0.72
Hanger Assembly	−8″	0.85

Manhours include checking out of job storage, handling, job hauling, distributing, layout, assembling, and installing auxiliary gutters and fittings.

Manhours exclude fastenings and field cut knockouts. See respective tables for these time frames.

JUNCTION BOXES

MANHOURS EACH

Junction Box Size (Inches)	Approximate Weight (Pounds)	Manhours		Junction Box Size (Inches)	Approximate Weight (Pounds)	Manhours	
		Wall	Ceiling			Wall	Ceiling
6 x 6 x 4	5.4	1.10	1.35	48 x 60 x 24	449.8	26.00	29.90
8 x 8 x 4	6.8	1.20	1.50	60 x 60 x 18	467.6	27.30	31.20
10 x 10 x 4	9.3	1.50	1.80	48 x 60 x 30	502.6	27.50	31.25
12 x 12 x 4	12.0	1.80	2.10	60 x 60 x 24	524.4	28.75	32.50
12 x 12 x 6	14.5	2.25	3.00	48 x 60 x 36	555.4	30.00	33.75
12 x 18 x 4	16.6	2.25	3.00	60 x 60 x 30	580.0	32.50	36.25
12 x 12 x 6	19.4	2.55	3.30	60 x 72 x 24	606.3	35.00	38.75
12 x 24 x 6	23.5	2.85	3.60	60 x 72 x 30	647.0	38.75	42.50
18 x 24 x 6	32.8	3.75	4.85	72 x 72 x 24	679.3	41.25	45.00
18 x 24 x 8	36.8	4.10	5.25	60 x 72 x 36	730.0	43.20	46.80
18 x 30 x 6	39.3	4.10	5.25	72 x 72 x 30	741.0	44.40	48.00
18 x 30 x 8	43.7	4.10	5.25	72 x 84 x 24	753.9	45.60	49.20
24 x 30 x 8	54.2	4.85	6.00	72 x 72 x 36	804.1	46.80	50.40
24 x 36 x 8	98.9	5.25	6.30	72 x 84 x 30	816.3	48.00	51.60
24 x 42 x 8	103.3	5.95	7.00	84 x 84 x 24	857.2	51.60	55.20
30 x 30 x 12	106.0	7.00	8.40	72 x 84 x 36	878.7	52.80	56.40
30 x 36 x 12	123.0	8.40	9.80	84 x 84 x 30	925.0	53.10	56.70
36 x 36 x 12	140.8	9.80	11.55	84 x 96 x 24	960.7	53.80	57.30
30 x 42 x 12	144.5	10.15	11.90	84 x 84 x 36	992.8	55.00	58.50
36 x 36 x 18	159.2	10.50	12.25	72 x 84 x 48	1,003.5	55.20	58.65
36 x 42 x 24	263.1	15.60	18.20	84 x 96 x 30	1,034.5	57.50	60.95
42 x 42 x 18	291.2	19.50	23.40	84 x 96 x 36	1,098.3	59.80	63.25
42 x 42 x 24	305.6	20.80	24.70	84 x 84 x 48	1,128.4	62.10	65.55
36 x 42 x 30	315.1	22.10	26.00	96 x 96 x 30	1,129.8	62.10	65.55
42 x 48 x 24	341.0	23.00	27.30	96 x 96 x 36	1,203.6	66.00	69.30
48 x 48 x 24	379.0	24.35	28.25	96 x 96 x 42	1,277.4	70.40	73.70
48 x 48 x 30	384.7	24.70	28.60	96 x 96 x 48	1,351.2	77.00	80.30

Manhours include checking out of job storage, handling, job hauling, distributing, layout, and positioning in place to heights of 12 feet.

Manhours exclude fastenings and field cut knockouts. See respective tables for these time frames.

1- and 2-DOOR TERMINAL CABINETS

MANHOURS EACH

Cabinet Size (Inches)	Approximate Weight (Pounds)	Manhours Wall Mounted	Cabinet Size (Inches)	Approximate Weight (Pounds)	Manhours Wall Mounted
1-Door					
12 x 24 x 6	37	3.35	42 x 30 x 8	142	7.70
16 x 12 x 6	26	3.00	42 x 36 x 8	170	9.80
20 x 16 x 6	37	3.75	48 x 24 x 8	128	6.65
20 x 20 x 6	44	4.15	48 x 30 x 8	150	8.75
24 x 20 x 6	52	4.35	48 x 36 x 8	181	9.80
24 x 24 x 6	66	4.70	60 x 36 x 8	224	12.60
30 x 20 x 6	66	4.85	16 x 12 x 10	31	3.00
30 x 24 x 6	78	5.25	20 x 16 x 10	44	4.15
36 x 24 x 6	92	5.45	20 x 20 x 10	51	4.35
16 x 12 x 8	29	3.00	24 x 12 x 10	41	3.75
20 x 16 x 8	41	3.75	24 x 20 x 10	61	4.35
20 x 20 x 8	47	4.15	24 x 24 x 10	74	4.85
24 x 12 x 8	39	3.75	30 x 24 x 10	75	4.85
24 x 20 x 8	57	4.35	30 x 24 x 10	92	5.40
24 x 24 x 8	70	4.55	36 x 24 x 10	105	5.60
30 x 20 x 8	70	4.55	36 x 30 x 10	129	6.30
30 x 24 x 8	90	5.10	42 x 30 x 10	149	8.75
36 x 24 x 8	99	5.25	48 x 30 x 10	159	9.10
36 x 30 x 8	123	5.30	48 x 36 x 10	190	9 75
42 x 24 x 8	109	5.60	60 x 36 x 10	233	11.70
2-Door					
54 x 42 x 8	273	14.30	72 x 72 x 10	701	40.80
60 x 48 x 8	428	23.40	60 x 48 x 12	450	26.00
60 x 48 x 10	431	23.40	60 x 60 x 12	535	28.60
60 x 60 x 10	540	28.60	72 x 60 x 12	637	36.00
72 x 60 x 10	597	33.80	72 x 72 x 12	743	43.20

Manhours include checking out of job storage, handling, job hauling, distribution, layout, and positioning cabinet.

Manhours exclude fastenings, field cut knockouts, and terminal strip installation. See respective tables for these time frames.

CURRENT TRANSFORMER, METER TEST CABINETS, AND METER TROUGHS

Current Transformer Cabinets

MANHOURS EACH

Cabinet Size (Inches)	Amperage Capacity	Manhours
36 x 27 x 11	200-600	9.75
44 x 27 x 11	600-1,200	15.60
40 x 30-1/4 x 10	2-500 MCM Per Leg	11.70
52 x 32-1/4 x 12	4-500 MCM Per Leg	23.40
77 x 32-7/8 x 14-1/2	4-500 MCM Per Leg W/Cur. Limit	31.20

Meter Test Cabinets

MANHOURS EACH

Cabinet Size (Inches)	Type of Equipment Design	Manhours
14 x 9 x 3-3/8	10-Pole-Switch and Test Block	4.55
14 x 29 x 3-3/8	Same with Meter Adapter	7.80

Meter Troughs

MANHOURS EACH

Number of Meters	Manhours for	
	1-Phase	3-Phase
1	3.25	4.85
2	5.20	7.15
3	7.15	9.40
4	9.10	11.70
5	11.00	13.30
6	12.00	14.40

Manhours include checking out of job storage, handling, job hauling, distributing, and positioning of items as outlined.

Manhours exclude fastening and field cut knockouts. See respective tables for these time frames.

METER SERVICE ENTRANCE SWITCHES AND METER SOCKET-CIRCUIT BREAKER COMBINATION

Meter Service Entrance Switches

MANHOURS EACH

Amperage Size	Manhours Each			
	Fusable—No Current Transformers		Fusable—With Current Transformers	
	3-Pole, 3-Wire	3-Pole, 4-Wire	3-Pole, 3-Wire	3-Pole, 4-Wire
100	4.85	6.00	–	–
200	7.90	8.85	10.90	11.75
400	13.50	16.40	18.70	22.00
600	16.00	18.20	22.10	25.00
800	21.60	24.00	28.60	31.80
1,000	27.50	28.60	39 75	41.30

METER SOCKET-CIRCUIT BREAKER COMBINATION

MANHOURS EACH

Number of Meters	Manhours for	
	70-Amp Circuit Breaker	100-Amp Circuit Breaker
1-4	13.30	–
5-8	19.80	–
9-12	25.40	–
13-16	31.80	–
1-5	–	16.80
6-10	–	25.20
11-15	–	33.60
16-20	–	42.00
1-6	–	21.00
7-12	–	29.40
13-18	–	37.80
19-24	–	46.20

Manhours include checking out of job storage, handling, job hauling, distributing, installing, and conductor terminations for items as outlined.

Manhours exclude cabinet installation, fastening, and field cut knockouts. See respective tables for these time frames.

INTERCONNECTING CONDUCTORS, SPLIT BOLT CONNECTORS AND LUGS

MANHOURS EACH

Conductor or Wire Size	Interconnecting Conductors								Connectors	
	3 Conductors				4 Conductors				Cable	Lug
	4'	6'	8'	12'	4'	6'	8'	12'		
#14	0.28	0.44	0.60	0.88	0.39	0.60	0.83	1.21	–	–
#12	0.28	0.44	0.60	0.88	0.39	0.60	0.83	1.21	0.33	0.22
#10	0.28	0.44	0.60	0.88	0.39	0.60	0.83	1.21	0.44	0.22
#8	0.28	0.44	0.60	0.88	0.39	0.60	0.83	1.21	0.55	0.22
#6	0.28	0.44	0.60	0.88	0.39	0.60	0.83	1.21	0.66	0.33
#4	0.28	0.44	0.60	0.88	0.39	0.60	0.83	1.21	0.77	0.33
#2	0.28	0.44	0.60	0.88	0.39	0.60	0.83	1.21	0.88	0.33
#1	0.33	0.50	0.66	1.05	0.44	0.66	0.94	1.38	1.27	0.66
#1/0	0.44	0.66	0.83	1.27	0.55	0.83	1.10	1.65	1.38	0.83
#2/0	0.50	0.77	0.99	1.49	0.66	0.99	1.32	1.98	1.54	0.99
#3/0	0.66	0.94	1.27	1.87	0.83	1.27	1.71	2.53	1.65	1.21
#4/0	0.77	1.21	1.60	2.42	1.05	1.60	2.10	3.19	1.93	1.38
#250 MCM	0.88	1.32	1.76	2.59	1.16	1.76	2.31	3.41	2.20	1.70
#350 MCM	0.99	1.49	1.98	2.86	1.32	1.98	2.64	3.96	2.64	2.10
#500 MCM	1.10	1.65	2.20	3.30	1.43	2.20	2.92	4.40	3.10	2.53
#600 MCM	1.21	1.82	2.42	3.63	1.60	2.42	3.19	4.79	3.96	3.10
#750 MCM	1.43	2.20	2.92	4.40	1.93	2.92	3.85	5.83	4.95	3.85
#1000 MCM	1.93	2.86	3.80	5.72	2.53	3.80	5.06	7.59	5.78	4.40

Interconnecting conductor manhours for auxiliary gutters, switches, circuit breakers, and control devices give consideration to conductors interconnected between the auxiliary gutter and operating devices and allow for time required to measure, cut, and prepare conductors for termination and cabling.

Split bolt cable connector manhours are based on 2 conductors in a split bolt connection.

Lug connector manhours are based on 1 conductor in a lug.

All manhours include checking materials out of job storage, handling, job hauling, distributing, and layout.

Manhours exclude splicing conductors in gutter or connecting conductors to equipment. See respective tables for these time frames.

CONDUIT NIPPLES FOR SPLIT BOLT AND LUG CONNECTORS

MANHOURS EACH

Conduit Size (Inches)	Field Cut and Thread		Prefabricated 6″ Nipples
	First Nipple 6″ Long	Additional Nipples 6″ Long	
1/2	0.42	0.28	0.28
3/4	0.55	0.28	0.28
1	0.66	0.28	0.28
1-1/4	0.77	0.28	0.42
1-1/2	0.88	0.42	0.42
2	0.99	0.42	0.55
2-1/2	1.10	0.42	0.55
3	1.27	0.55	0.83
3-1/2	1.38	0.55	0.83
4	1.54	0.55	0.83

Manhours include checking out of job storage, handling, job hauling, distributing, cutting and threading when required and assembling in position with locknuts and bushings conduit nipples for split bolt and lug connectors.

Manhours exclude installation of split bolt or lug connector. See respective tables for these time requirements.

Section 12

MOTOR CONTROL CENTERS

This section covers all labor operations, in manhour units, as may be required for the complete installation of motor control centers.

The following manhour tables for cable connections of motor control centers are based on the following design characteristics which are outlined by the National Electrical manufacturers Association (NEMA) standards.

These standards consist of two basic classes and three basic types as further defined:

Class I control centers are essentially a grouping of motor starters, and/or control assemblies which can be handled without system analysis or systems engineering.

Class II control centers are basically designed as a complete control system which includes electrical interlocking between units in the control center or with outside devices as may be required by the particular installation. They require system analysis and engineering.

Class I–Type A units include factory installed power wiring from the busbars to the line terminals of each starter unit and all wiring within each starter unit. No interwiring is made by the manufacturer between units. Interwiring and load and control connections are to be made by the contractor at time of installation.

Class I–Type B unit construction is similar to Class I–type A, except the units include load and control wiring to terminal boards for size 3 starters or smaller, and control wiring to terminal boards for size 4 and 5 starter

319

units. These terminal boards are mounted next to each unit in a stationary part of the structure. no interwiring is included between units.

Class I—Type C unit construction is similar to Class I—Type B, except factory installed load and control wiring is provided from each starter unit to master terminal blocks grouped at either the top or bottom of each vertical section. All outgoing wires from any starter unit are carried to a master terminal board, except load side wiring for size 4 starters or larger. No interwiring is made between starter, units, sections, or master terminals.

Class II—Type B unit construction is similar to Class I—Type B, except interwiring is provided between starter and control assemblies in a single section or between sections.

Class II—Type C unit construction is similar to Class II—Type B, except the load and control terminals are located on the top or bottom of the section instead of adjacent to the individual units.

Certain labor operations required to support a totally installed project may be found in other sections of this manual.

HANDLING MOTOR CONTROL MODULAR UNITS

MANHOURS EACH

Modular Sections Weight Range (Pounds)	Installation Conditions	
	Normal	Difficult
0-400	4.50	8.00
401-450	5.40	8.50
451-500	6.40	13.50
501-550	6.90	14.25
551-600	7.90	14.65
601-650	8.40	17.40
651-650	9.70	18.30
701-750	10.30	20.25
751-800	11.00	20.70
801-850	11.80	22.80

Normal installation conditions are considered to be those without obstructions or adverse situations. Difficult installation conditions are those where obstructions may exist and involve extra rigging, handling, and manual movement.

Manhours include checking out of job storage, handling, job hauling, layout, rigging, picking, setting, leveling, aligning, bolting, steel enclosures, and bolting busbars together.

Manhours exclude installation of supports or fasteners, pull boxes, special knockouts, raceway or conductor terminations, welding, and painting. See respective tables for these time frames.

MODULAR CONTROL UNITS
Class I—Type A—No Terminal Blocks
Full Voltage—Nonreversing
Conductor Connections

MANHOURS EACH

Motor Horsepower	Amperage		NEMA Starter Size	Conductor Size	Manhours 3-Phase	
	440-550 V.	208-220 V.			440-550 V.	208-220 V
1-5	1.5	15	1	#14	2.10	2.40
7-1/2	9	22	1	#14−#10	2.10	2.40
10	11	27	1	#14−#8	2.10	2.40
15	16	40	2	#12−#6	2.40	2.70
20	21	52	2-3	#10−#4	2.40	2.70
25	26	64	2-3	#8−#2	2.40	3.60
30	31	78	2-3	#8−#1	2.40	3.60
50	50	125	3-4	#4−#3/0	2.70	4.50
75	74	185	3-5	#2−#350 MCM	3.60	4.80
100	98	246	4-5	#1/0−#500 MCM	3.90	5.10
480-Volt Motors						
125	124	310	4	#3/0	4.50	−
150	144	360	5	#4/0	4.50	−
200	192	480	5	#350 MCM	4.80	−

Modular unit consist of circuit breaker or switch and fuse disconnect switch, magnetic motor starter, start and stop station, two pilot lights, control transformer, and fuse.

Manhours include layout, conductor connections, identification, lacing, terminations, testing, and checking motor rotation.

Manhours exclude handling and positioning control modular unit and remote control station. See respective tables for these time frames.

MODULAR CONTROL UNITS
Class I—Type B—Terminal Blocks Provided
Full Voltage—Nonreversing
Conductor Connections

MANHOURS EACH

Motor Horsepower	Amperage		NEMA Starter Size	Conductor Size	Manhours 3-Phase	
	440-550 V.	208-220 V.			440-550 V.	208-220 V.
1·5	1.5	15	1	#14	1.50	1.80
7·1/2	9	22	1	#14—#10	1.50	1.80
10	11	27	1	#14—#8	1.50	1.80
15	16	40	2	#12—#6	1.80	2.10
20	21	52	2-3	#10—#4	1.80	2.10
25	26	64	2-3	#8—#2	1.80	3.00
30	31	78	2-3	#8—#1	1.80	3.90
50	50	125	3-4	#4—#3/0	2.10	4.20
75	74	185	3-5	#2—#350 MCM	3.00	4.50
100	98	246	4-5	#1/0—#500 MCM	3.30	
480-Volt Motors						
125	124	310	4	#3/0	3.90	—
150	144	360	5	#4/0	3.90	—
200	192	480	5	#350 MCM	4.20	—

Modular unit consist of circuit breaker or switch and fuse disconnect switch, magnetic motor starter, start and stop station, two pilot lights, control transformer, and fuse.

Manhours include layout, conductor connections, identification, lacing, terminations, testing, and checking motor rotation.

Manhours exclude handling and positioning control modular unit and remote control station. See respective tables for these time frames.

MODULAR CONTROL UNITS

Class I—Type C—Master Terminal Block
Full Voltage—Nonreversing
Conductor Connections

MANHOURS EACH

Motor Horsepower	Amperage		NEMA Starter Size	Conductor Size	Manhours 3-Phase	
	440-550 V.	208-220 V.			440-550 V.	208-220 V
1-5	1.5	15	1	#14	1.20	1.50
7-1/2	9	22	1	#14−#10	1.20	1.50
10	11	27	1	#14−#8	1.20	1.50
15	16	40	2	#12−#6	1.50	1.80
20	21	52	2-3	#10−#4	1.50	1.80
25	26	64	2-3	#8−#2	1.50	2.70
30	31	78	2-3	#8−#1	1.50	2.70
50	50	125	3-4	#4−#3/0	1.80	3.60
75	74	185	3-5	#2−#350 MCM	2.70	3.90
100	98	246	4-5	#1/0−#500 MCM	3.00	4.20
480-Volt Motors						
125	124	310	4	#3/0	3.60	−
150	144	360	5	#4/0	3.60	−
200	192	480	5	#350 MCM	3.90	−

Modular unit consist of circuit breaker or switch and fuse disconnect switch, magnetic motor starter, start and stop station, two pilot lights, control transformer, and fuse.

Manhours include layout, conductor connections, identification, lacing, terminations, testing, and checking motor rotation.

Manhours exclude handling and positioning control modular unit and remote control station. See respective tables for these time frames.

MODULAR CONTROL UNITS

Class I—Types A, B, & C—Terminal Blocks Provided
Circuit Wire Connections to Spare Units in Place
Type A—Full Voltage—Nonreversing

MANHOURS EACH

Motor Horsepower	Amperage		NEMA Starter Size	Conductor Size	Manhours 3-Phase	
	440-550 V.	208-220 V.			440-550 V.	208-220 V.
1-10	1.5-11	15-27	1	#14—#8	2.95	3.20
15-30	16-31	40	2	#12—#6	3.20	3.60
20-75	50-74	52-78	3	#10—#1	3.30	4.45
50-125	98-124	125	4	#1/0—#3/0	4.45	4.80
75-200	144-192	185-246	5	#4/0—#350 MCM	5.15	—

Type B—Full Voltage—Nonreversing

Motor Horsepower	Amperage		NEMA Starter Size	Conductor Size	Manhours 3-Phase	
	440-550 V	208-220 V.			440-550 V.	208-220 V.
1-10	1.5-11	15-27	1	#14—#8	2.10	2.40
15-30	16-31	40	2	#12—#6	2.40	3.00
20-75	50-74	52-78	3	#10—#1	2.70	3.90
50-125	98-124	125	4	#1/0—#3/0	3.90	4.20
75-200	144-192	185-246	5	#4/0—#350 MCM	4.50	—

Type C—Full Voltage—Nonreversing

Motor Horsepower	Amperage		NEMA Starter Size	Conductor Size	Manhours 3-Phase	
	440-550 V.	208-220 V.			440-550 V.	208-220 V
1-10	1.5-11	15-27	1	#14—#8	1.70	2.00
15-30	16-31	40	2	#12—#6	2.00	2.70
20-75	50-74	52-78	3	#10—#1	2.45	3.60
50-125	98-124	125	4	#1/0—#3/0	3.60	3.90
75-200	144-192	185-246	5	#4/0—#350 MCM	4.15	—

Manhours include removing spare space cover, circuit identification, cabling, making motor load and control connections, testing, and checking motor rotation for spare units in place.

Manhours exclude handling and positioning control modular unit, and remote control station. See respective tables for these time frames.

MODULAR CONTROL UNITS

Class I—Type B—Terminal Blocks Provided
Full Voltage—Nonreversing
Conductor Connections

MANHOURS EACH

Motor Horsepower	Amperage		NEMA Starter Size	Conductor Size	Manhours 3-Phase	
	440-550 V.	208-220 V.			440-550 V.	208-220 V
1-5	1.5	15	1	#14	–	–
7-1/2	9	22	1	#14—#10	2.10	2.40
10	11	27	1	#14—#8	2.10	2.40
15	16	40	2	#12—#6	2.40	2.70
20	21	52	2-3	#10—#4	2.40	2.70
25	26	64	2-3	#8—#2	2.40	3.60
30	31	78	2-3	#8—#1	2.40	3.60
50	50	125	3-4	#4—#3/0	2.70	4.50
75	74	185	3-5	#2—#350 MCM	3.60	4.80
100	98	246	4-5	#1/0—#500 MCM	3.90	5.10
480-Volt Motors						
125	124	310	4	#3/0	4.50	–
150	144	360	5	#4/0	4.50	–
200	192	480	5	#350 MCM	4.80	–

Modular unit consist of circuit breaker or switch and fuse disconnect switch, magnetic motor starter, start and stop station, two pilot lights, control transformer, and fuse.

Manhours include layout, conductor connections, identification, lacing, terminations, testing, and checking motor rotation.

Manhours exclude handling and positioning control modular unit and remote control station. See respective tables for these time frames.

MODULAR CONTROL UNITS

Class I—Type B—Terminal Blocks Provided
Reduced Voltage—Nonreversing
Conductor Connectors Autotransformer

MANHOURS EACH

Motor Horsepower	Amperage		NEMA Starter Size	Conductor Size	Manhours 3-Phase	
	440-550 V.	208-220 V.			440-550 V.	208-220 V
1-5	1.5	15	1	#14	–	–
7-1/2	9	22	1	#14–#10	–	–
10	11	27	1	#14–#8	–	–
15	16	40	2	#12–#6	2.40	2.70
20	21	52	2-3	#10–#4	2.40	2.70
25	26	64	2-3	#8–#2	2.40	3.60
30	31	78	2-3	#8–#1	2.40	3.60
50	50	125	3-4	#4–#3/0	2.70	4.50
75	74	185	3-5	#2–#350 MCM	3.60	4.80
100	98	246	4-5	#1/0–#500 MCM	3.90	5.10
480-Volt Motors						
125	124	310	4	#3/0	4.50	–
150	144	360	5	#4/0	4.50	–
200	192	480	5	#350 MCM	4.80	–

Modular units consist of circuit breaker or switch and fuse disconnect switch, magnetic motor starter, start and stop station, two pilot lights, control transformer, and fuse and automatic transformer starter.

Manhours include layout, conductor connections, identification, lacing, terminations, testing, and checking motor rotation.

Manhours exclude handling and positioning control modular unit and remote control station. See respective tables for these time frames.

MODULAR CONTROL UNITS

Class I—Type B—Terminal Blocks Provided
2-Speed—Full Voltage—Nonreversing
Conductor Connections

MANHOURS EACH

Motor Horsepower	Amperage		NEMA Starter Size	Conductor Size	Manhours 3-Phase	
	440-550 V.	208-220 V.			440-550 V.	208-220 V.
1-5	1.5	15	1	#14	–	–
7-1/2	9	22	1	#14–#10	2.40	2.70
10	11	27	1	#14–#8	2.40	2.70
15	16	40	2	#12–#6	2.70	3.00
20	21	52	2-3	#10–#4	2.70	3.00
25	26	64	2-3	#8–#2	2.70	3.90
30	31	78	2-3	#8–#1	2.70	3.90
50	50	125	3-4	#4–#3/0	3.00	4.80
75	74	185	3-5	#2–#350 MCM	3.90	5.10
100	98	246	4-5	#1/0–#500 MCM	4.20	5.40
480-Volt Motors						
125	124	310	4	#3/0	4.80	–
150	144	360	5	#4/0	4.80	–
200	192	480	5	#350 MCM	5.10	–

Modular unit consist of circuit breaker or switch and fuse disconnect switch, magnetic motor starter, start and stop station, two pilot lights, control transformer and fuse. The exception to this unit is one or two winding starter units.

Manhours include layout, conductor connections, identification, lacing, terminations, testing, and checking motor rotation.

Manhours exclude handling and positioning control modular unit and remote control station. See respective tables for these time frames.

ACCESSORIES FOR MODULAR CONTROL STARTER

MANHOURS EACH

Remote Push Button Stations and Selector Switches		Pilot Lights		Auxiliary Contacts for Interlocking		Control Devices	
Number of Positions	Manhours	Number of Indicating Positions	Manhours	Number of Contacts	Manhours	Device	Manhours
2	1.50	1	1.50	1	1.50	Transformer	2.40
3	1.80	2	1.80	2	2.40	Limit Switch	2.10
4	2.10	3	2.10	3	3.60	–	–

Main Lugs and Neutral Block Connections

Main Amperage Rating	Main Lugs Only Poles and Connections		Neutral Block Section
	3-Pole, 3-Wire	3-Pole, 4-Wire	
600	7.50	9.60	2.70
800	8.60	10.80	3.30
1,200	12.60	16.20	4.80
2,000	22.80	30.00	7.20
3,000	30.80	40.70	10.80
4,000	39.90	51.40	13.20

Manhours include conductor terminations and related labor operations as may be required for items as outlined.

Manhours exclude handling and positioning control modular units, remote control stations and other conductor connections. See respective tables for these time frames.

MODULAR CONTROL UNITS
Class II—Type B—Terminal Blocks Provided
Interlocking Auxiliary Contacts
Full Voltage—Nonreversing
Conductor Connections

MANHOURS EACH

Motor Horsepower	Amperage		NEMA Starter Size	Conductor Size	Manhours 3-Phase	
	440-550 V	208-220 V			440-550 V.	208-220 V.
1-5	1.5	15	1	#14	3.00	3.30
7-1/2	9	22	1	#14—#10	3.00	3.30
10	11	27	1	#14—#8	3.00	3.30
15	16	40	2	#12—#6	3.30	3.60
20	21	52	2-3	#10—#4	3.30	3.60
25	26	64	2-3	#8—#2	3.30	4.50
30	31	78	2-3	#8—#1	3.30	4.50
50	50	125	3-4	#4—#3/0	3.60	5.40
75	74	185	3-5	#2—#350 MCM	4.50	5.70
100	98	246	4-5	#1/0—#500 MCM	4.80	6.00
480-Volt Motors						
125	124	310	4	#3/0	5.40	–
150	144	360	5	#4/0	5.40	–
200	192	480	5	#350 MCM	5.70	–

Modular units consist of circuit breaker or switch and fuse disconnect switch, magnetic motor starter, start and stop station, two pilot lights, control transformer and fuse. These units include the addition of one set of interlocking contacts for each starter.

Manhours include layout, conductor connections, identification, lacing, terminations, testing, and checking motor rotation.

Manhours exclude handling and positioning control modular unit and remote control station. See respective tables for these time frames.

MODULAR CONTROL UNITS
Class II—Type C—Terminal Blocks Provided
Interlocking Auxiliary Contacts
Full Voltage—Nonreversing
Conductor Connections

MANHOURS EACH

Motor Horsepower	Amperage		NEMA Starter Size	Conductor Size	Manhours 3-Phase	
	440-550 V.	208-220 V.			440-550 V.	208-220 V.
1-5	1.5	15	1	#14	2.70	3.00
7-1/2	9	22	1	#14—#10	2.70	3.00
10	11	27	1	#14—#8	2.70	3.00
15	16	40	2	#12—#6	3.00	3.30
20	21	52	2-3	#10—#4	3.00	3.30
25	26	64	2-3	#8—#2	3.00	4.20
30	31	78	2-3	#8—#1	3.00	4.20
50	50	125	3-4	#4—#3/0	3.30	5.10
75	74	185	3-5	#2—#350 MCM	4.20	5.40
100	98	246	4-5	#1/0—#500 MCM	4.50	5.70
480-Volt Motors						
125	124	310	4	#3/0	5.10	—
150	144	360	5	#4/0	5.10	—
200	192	480	5	#350 MCM	5.40	—

Modular units consists of circuit breaker or switch and fuse disconnect switch, magnetic motor starter, start and stop station, two pilot lights, control transformer, and fuse. These units include the addition of one set of interlocking contacts for each starter.

Manhours include layout, conductor connections, identification, lacing, terminations, testing, and checking motor rotation.

Manhours exclude handling and positioning control modular unit and remote control station. See respective tables for these time frames.

MOTOR CONTROL CENTER PULL BOXES

MANHOURS EACH

Pull Box Size Length X Width X Depth	Steel Gauge	Approximate —Weight (Pounds)	Manhours
56" x 36" x 12"	#10	258	12.60
56" x 36" x 18"	#10	288	13.50
70" x 42" x 12"	#10	348	15.30
70" x 42" x 18"	#10	385	16.20
84" x 36" x 18"	#10	406	18.00
84" x 36" x 24"	#10	484	19.80
84" x 42" x 12"	#10	412	18.00
84" x 42" x 18"	#10	472	19.80
84" x 42" x 24"	#10	533	21.60
112" x 36" x 30"	#10	786	34.20
105" x 42" x 18"	#10	660	27.00
105" x 42" x 24"	#10	720	30.60
140" x 42" x 18"	#10	840	36.00
140" x 42" x 24"	#10	914	37.80

Manhours include checking out of job storage, handling, job hauling, distributing, drilling and bolting box to switchboard, and removing and reinstalling access cover.

Manhours exclude fabrication, welding, or modifying boxes, field cut knockouts and installation of panelboard sections. See respective tables for these time frames.

Section 13

MOTOR CONTROLS AND MOTORS

This section covers as nearly as possible all operations which may be encountered in the installation of motors and starters of various sizes.

The manhours listed are for labor only and have no bearing on materials which must be added in all cases if a complete labor and material estimate is to be obtained.

All labor for unloading, handling, hauling, and installing has been given due consideration.

We again stress the fact that the estimator should be thoroughly familiar with the Introduction of this manual before attempting to apply the manhours of the following tables.

MOTOR STARTING SWITCHES
For 30-Amp AC Motors

MANHOURS EACH

Type Switch & Motor	Wood		Brick		Concrete	
	Soldered	Solderless	Soldered	Solderless	Soldered	Solderless
2-Pole, Single Phase	4. 13	3. 75	5. 63	5. 10	7. 13	6. 45
3-Pole, 3-Phase or 3-Wire, 2-Phase	4. 88	4. 43	6. 38	5. 78	7. 88	7. 13
4-Pole, 4-Wire, 2-Phase	5. 25	4. 73	6. 75	6. 08	8. 25	7. 50

MOUNTING & CONNECTING
DIAL TYPE SPEED REGULATING RHEOSTATS
For 220-Volt 3-Phase Slip-Ring AC Induction Motors

MANHOURS EACH

Motor HP	Mounted on:		
	Wood	Brick	Concrete
1	5. 60	6. 60	8. 10
2 — 3	6. 30	7. 50	9. 00
5	7. 20	8. 40	10. 05
7 1/2 — 10	9. 45	10. 80	12. 45
15	11. 10	12. 60	14. 10

Manhours include checking out of job storage, handling, job hauling, distributing, mounting switch or rheostat, making connection and testing motor for rotation.

Manhours exclude installation of conduit, pulling of wire, and mounting of motor. See respective tables for these time frames.

STARTING COMPENSATORS
For 3-Phase Squirrel Cage Induction AC Motors

MANHOURS EACH

Voltage	Motor Horsepower	Mounted on:		
		Wood	Brick	Concrete
220	7-1/2 – 10	7.50	8.25	9.00
220	15	9.60	10.50	11.25
220	20 – 25	10.80	11.70	12.45
220	30	12.00	12.75	13.50
220	40	13.50	14.25	15.00
220	50	14.25	15.00	15.75
220	75	15.00	16.50	17.25
220	100	19.50	21.00	21.75
440	7-1/2 – 10 – 15	7.50	8.25	9.00
440	20 – 25	9.00	10.50	11.25
440	30	10.50	11.25	12.00
440	40 – 50	11.25	12.00	12.75
440	75	13.50	14.25	15.00
440	100	15.00	15.75	16.50

Manhours include checking out of job storage, handling, job hauling, distributing, mounting compensator, making connections at compensator and motor, and testing motor for rotation direction.

For 2-Phase, 4-Wire add 20% to manhours.

Manhours exclude installation of conduit, pulling of wire, and mounting of motor. See respective tables for these time frames.

30-AMP AC MAGNETIC SWITCHES

MANHOURS EACH

Number	Mounted on:		
of Poles	Wood	Brick	Concrete
3	7.05	7.50	8.70
4	8.10	8.70	9.90

3-POLE 220-VOLT AC MAGNETIC SWITCHES

MANHOURS EACH

Capacity of	Mounted on:		
Switch Amperes	Wood	Brick	Concrete
15	5.40	6.00	7.20
75	8.25	9.45	10.50
150	10.50	12.00	12.90
300	13.95	15.00	16.20

Manhours include checking out of job storage, handling, job hauling, distributing and installing.

30 AMP AC MAGNETIC SWITCHES: Manhours include mounting and connecting with thermal cutouts or relays used as starters for small squirrel-cage motors, for mounting and connecting push button control station and making connections at the motor and testing.

3-POLE 220-VOLT AC MAGNETIC SWITCHES: Manhours include mounting and connecting push button control station.

Manhours do not include connection at any other apparatus controlled by the switch.

DC MOTOR SWITCHES
Push Button Controlled Magnetic Switches and Line Switches

MANHOURS EACH

Voltage	Motor Horsepower	Mounted on:		
		Wood	Brick	Concrete
115	1 - 2 - 3	10.50	12.00	13.50
115	5	12.60	14.10	16.05
115	7-1/2 - 10	15.00	16.50	18.75
230	1 - 2 - 3 - 5	10.50	12.00	13.50
230	7-1/2 - 10	12.60	14.10	16.05
230	15	14.55	16.05	18.00

DC MOTOR RHEOSTATS & SWITCHES
Speed Regulating Rheostats (Controlled by Resistance in Armature Circuit) and Externally Operated Switches

MANHOURS EACH

Voltage	Motor Horsepower	Mounted on:		
		Wood	Brick	Concrete
115	1/2 - 3/4 - 1	7.05	7.95	9.60
115	1-1/2 - 2 - 3	7.80	9.00	10.50
115	5	10.50	12.00	13.50
115	7-1/2 - 10	13.50	15.00	16.50
230	1/2 - 3/4 - 1	7.05	7.95	9.60
230	1-1/2 - 2 - 3	7.80	9.00	10.50
230	5	8.25	9.60	11.10
	7-1/2 - 10	10.80	12.00	13.50

Manhours include checking out of job storage, handling, job hauling, distributing and mounting and connecting push button control station and connections at motor or mounting and making connections at switch, rheostat and motor.

Manhours exclude installation of conduit, pulling of wire and mounting motors. See respective tables for these time frames.

DC MOTOR SWITCHES

Starting Rheostats and Externally Operated Switches

NET MANHOURS EACH

Voltage	Motor Horsepower	Mounted on: Wood	Brick	Concrete
115	1/2 – 3/4 – 1	6.00	6.90	8.40
115	1-1/2 – 2 – 3	6.75	7.50	9.00
115	5	9.00	10.05	11.55
115	7-1/2	12.00	13.05	14.55
115	10	12.75	13.80	15.45
115	15	15.00	16.05	17.70
	20	16.20	17.05	19.05
	25	19.95	21.45	23.10
230	1/2 – 3/4 – 1	6.00	6.90	8.40
230	1-1/2 – 2 – 3 – 4 – 5	6.75	7.50	9.00
230	7-1/2	9.00	10.05	11.55
230	10	9.90	11.10	12.75
230	15	11.86	13.05	14.70
230	20	12.90	14.10	15.75
230	25	14.86	16.05	17.70
230	30-35	16.05	17.40	18.90
230	40	16.95	18.30	19.80
230	50	19.20	20.70	22.50

Manhours include checking out of job storage, handling, job hauling, distributing and mounting and connecting at switches, rheostats, and motor.

For speed regulator with shunt field weakening instead of plain starting rheostats add one (1) manhour in each case.

Manhours exclude installation of conduit, pulling of wires, and mounting motor. See respective tables for these time frames.

MOUNTING MOTORS

AC—60-Cycle—3- and 3-Phase—220, 440, or 550 Volts
1750, 1160, 875, 700, 575, 490, and 420 RPM's

MANHOURS EACH

Motor		Manhours for Mounting on												
	Weight	Floors					Walls or Machine				Ceilings			
Horse	Range	Height to					Height to				Height to			
Power	(Pounds)	Gr.	10'	15'	20'	25'	10'	15'	20'	25'	10'	15'	20'	25'
1	93-335	4.13	4.21	4.34	4.56	4.87	5.58	5.69	5.86	6.15	5.78	5.96	6.25	6.69
1-1/2	119-405	4.19	4.27	4.40	4.62	4.95	5.66	5.77	5.94	6.24	5.87	6.04	6.34	6.79
2	131-550	5.41	5.52	5.68	5.97	6.39	7.30	7.45	7.67	8.06	7.57	7.80	8.19	8.76
3	131-690	6.43	6.56	6.76	7.09	7.59	8.68	8.85	9.12	9.58	9.00	9.27	9.74	10.42
5	174-750	7.86	8.00	8.26	8.67	9.28	10.61	10.82	11.15	11.71	11.00	11.33	11.90	12.73
7-1/2	251-970	9.55	9.74	10.00	10.53	11.27	12.89	13.15	13.54	14.22	13.37	13.77	14.46	15.47
10	335-1070	11.51	11.74	12.09	12.70	13.54	15.54	15.85	16.32	17.14	16.11	16.60	17.43	18.65
15	405-1155	13.39	13.68	14.07	14.77	15.80	18.08	18.44	18.99	19.94	18.75	19.31	20.27	21.69
20	525-1235	14.83	15.13	15.58	16.36	17.50	20.02	20.42	21.03	22.09	20.76	21.38	22.45	24.03
25	550-2200	18.21	18.57	19.13	20.09	21.49	24.58	25.08	25.33	27.12	25.49	26.26	27.57	29.50
30	690-2200	19.48	19.87	20.47	21.49	23.00	26.30	26.82	27.63	29.01	27.27	28.09	29.49	31.56
35	750-2650	23.93	24.41	25.14	26.40	28.25	32.31	32.95	33.94	35.64	33.50	34.51	36.23	38.77
40	970-2650	25.03	25.53	26.30	27.61	29.54	33.79	34.47	35.50	37.28	–	–	–	–
50	1175-2650	26.66	27.19	28.00	29.41	31.47	35.99	36.71	37.31	39.70	–	–	–	–
60	1460-3100	32.09	32.73	33.71	35.40	37.88	43.72	44.60	45.94	48.24	–	–	–	–
75	1650-4100	40.84	41.66	42.91	45.05	48.21	–	–	–	–	–	–	–	–
100	2100-4750	50.00	51.00	52.53	55.16	59.02	–	–	–	–	–	–	–	–
125	2100-5400	61.08	62.30	64.17	67.38	72.10	–	–	–	–	–	–	–	–
440- or 500-Volt														
150	2200-5655	64.11	65.39	67.35	70.72	75.67	–	–	–	–	–	–	–	–
175	2500-6500	74.83	76.33	78.62	82.55	88.33	–	–	–	–	–	–	–	–
200	2500-6500	74.83	76.33	78.62	82.55	88.33	–	–	–	–	–	–	–	–

Manhours include checking out of job storage, handling, job hauling, rolling and placing or rigging and picking, setting, and aligning.

Manhours exclude installation of fasteners or supports, conduit, switches, wiring, and circuit connections. See respective tables for these time frames.

MOUNTING MOTORS

AC—60-Cycle—Variable Speed—2- and 3-Phase—220, 440, or 550 Volts

1750, 1160, 875, 700, 575, and 490 RPM's

MANHOURS EACH

Motor		Manhours for Mounting on												
	Weight	Floors					Walls or Machine				Ceilings			
Horse	Range	Height to					Height to				Height to			
Power	(Pounds)	Gr.	10′	15′	20′	25′	10′	15′	20′	25′	10′	15′	20′	25′
1	189-350	4.38	4.47	4.60	4.83	5.17	5.91	6.03	6.21	6.52	6.13	6.32	6.63	7.10
1-1/2	189-350	4.38	4.47	4.60	4.83	5.17	5.91	6.03	6.21	6.52	6.13	6.32	6.63	7.10
2	189-600	6.73	6.86	7.07	7.42	7.95	9.09	9.27	9.55	10.02	9.42	9.70	10.19	10.90
3	189-780	7.88	8.04	8.82	8.69	9.30	10.64	10.85	11.18	11.74	11.03	11.36	11.93	12.77
5	189-780	8.65	8.82	9.09	9.54	10.21	11.67	11.91	12.27	12.83	12.11	12.47	13.10	14.00
7-1/2	265-1100	10.31	10.52	10.83	11.87	12.17	13.92	14.20	14.62	15.35	14.43	14.87	15.61	16.70
10	355-1200	12.29	12.54	12.91	13.56	14.51	16.59	16.92	17 43	18.30	17.21	17.72	18.61	19.91
15	425-1400	14.69	14.98	15.43	16.20	17.34	19.83	20.23	20.83	21.38	20.57	21.18	22.24	23.80
20	575-1600	16.66	16.99	17.50	18.38	19.66	22.49	22.94	23.63	24.81	23.32	24.02	25.22	26.99
25	600-2100	18.96	19.34	19.92	20.92	22.38	25.60	26.11	26.39	28.24	26.54	27.34	28.71	30.72
30	830-2300	22.50	22.95	23.64	24.82	26.56	30.38	30.98	31.91	33.51	31.50	32.45	34.07	36.45
35	1070-2800	24.80	25.30	26.05	27.36	29.27	33.48	34.15	35.17	36.93	34.72	35.76	37.55	40.18
40	1365-2800	27.09	27.63	28.46	29.88	31.98	36.57	37.30	38.43	40.34	37.93	39.06	41.00	43.84
50	1365-3300	29.50	30.02	30.99	32.54	34.82	39.83	40.62	41.84	43.93	–	–	–	–
60	1410-4400	34.34	34.00	35.03	36.78	39.35	45.00	45.91	47.29	49.66	–	–	–	–
75	1920-4400	44.50	45.39	46.75	49.09	52.53	60.00	61.28	53.11	66.27	–	–	–	–
100	1350-2300	46.00	46.92	48.33	50.74	54.30	–	–	–	–	–	–	–	–
440- or 500-Volt														
125	2190-5080	56.75	57.89	59.62	62.60	66.93	–	–	–	–	–	–	–	–
150	2750-5780	63.00	64.26	66.19	69.50	74.36	–	–	–	–	–	–	–	–
175	2750-6588	73.96	75.44	77.70	81.59	87.30	–	–	–	–	–	–	–	–
200	2750-6588-	77.09	78.63	80.99	85.04	91.00	–	–	–	–	–	–	–	–

Manhours include checking out of job storage, handling, job hauling, rolling and placing or rigging and picking, setting, and aligning.

Manhours exclude installation of fasteners or supports, conduit, switches, wiring, and circuit connections. See respective tables for these time frames.

MOUNTING MOTORS

AC—25-Cycle—2- and 3-Phase—220, 440, or 550 Volts
1440, 720, and 475 RPM's

MANHOURS EACH

Motor		Manhours for Mounting on												
	Weight	Floors				Walls or Machine				Ceilings			
Horse	Range	Height to				Height to				Height to			
Power	(Pounds)	Gr.	10'	15'	20'	25'	10'	15'	20'	25'	10'	15'	20'	25'
1	119-174	3.23	3.29	3.39	3.56	3.81	4.36	4.45	4.58	4.81	4.52	4.66	4.89	5.23
1-1/2	131-251	3.75	3.83	3.94	4.14	4.43	5.06	5.16	5.32	5.58	5.25	5.41	5.68	6.08
2	131-251	3.75	3.83	3.94	4.14	4.43	5.06	5.16	5.32	5.58	5.25	5.41	5.68	6.08
3	174-405	5.00	5.10	5.25	5.52	5.80	6.75	6.89	7.09	7.45	7.00	7.21	7.57	8.10
5	251-525	6.25	6.38	6.57	6.89	7.38	8.44	8.61	8.86	9.31	8.75	9.01	9.46	10.13
7-1/2	405-690	9.69	9.88	10.18	10.70	11.44	13.08	13.34	13.74	14.43	13.57	13.97	14.67	15.70
10	525-750	10.21	10.41	10.73	11.26	12.05	13.78	14.06	14.48	15.20	14.29	14.72	15.46	16.54
15	550-1070	13.13	13.39	13.79	14.48	15.50	17.73	18.08	18.62	19.55	18.38	18.93	19.88	21.27
20	690-1155	15.84	16.16	16.64	17.47	18.70	21.38	21.81	22.47	23.59	22.18	22.84	23.98	25.66
25	690-1235	16.46	16.79	17.29	18.16	19.43	22.22	22.67	23.35	24.51	23.04	23.74	24.92	26.67
30	1070-2100	22.09	22.53	23.21	24.37	26.07	29.82	30.42	31.33	32.90	30.93	31.85	33.45	35.79
35	1155-2300	23.54	24.00	24.73	25.97	27.79	31.78	32.41	33.39	35.06	32.96	33.94	35.64	38.14
40	1460-2350	24.79	25.29	26.04	27.35	29.26	33.47	34.14	35.16	36.92	–	–	–	–
50	1650-2800	32.09	33.75	34.76	36.50	39.06	43.32	44.62	45.96	48.26	–	–	–	–
60	1650-2800	33.34	34.00	35.03	36.78	39.35	45.00	45.91	47.29	49.65	–	–	–	–
75	1700-2800	33.34	34.00	35.03	36.78	39.35	–	–	–	–	–	–	–	–
100	2200-3300	41.68	42.51	43.79	45.98	49.20	–	–	–	–	–	–	–	–
125	2700-5250	67.91	69.27	71.35	74.91	80.16	–	–	–	–	–	–	–	–
440- or 550-Volts														
150	3200-5850	74.59	76.08	78.36	82.28	88.04	–	–	–	–	–	–	–	–
175	3600-5850	76.88	78.42	80.77	84.81	90.75	–	–	–	–	–	–	–	–
200	3600-6823	86.66	88.39	91.04	95.60	102.29	–	–	–	–	–	–	–	–

Manhours include checking out of job storage, handling, job hauling, rolling and placing or rigging and picking, setting, and aligning.

Manhours exclude installation of fasteners or supports, conduit, switches, wiring, and circuit connections. See respective tables for these time frames.

MOUNTING MOTORS

AC—25-Cycle—3- and 2-Phase—Variable Speed—220, 440, or 550-Volts

1400, 720, and 475 RPM's

MANHOURS EACH

Motor		Manhours for Mounting on												
	Weight	Floors					Walls or Machine				Ceilings			
Horse	Range	Height to					Height to				Height to			
Power	(Pounds)	Gr.	10'	15'	20'	25'	10'	15'	20'	25'	10'	15'	20'	25'
1	189-190	3.75	3.83	3.94	4.14	4.43	5.06	5.16	5.32	5.58	5.25	5.41	5.68	6.08
1-1/2	190-260	3.96	4.04	4.16	4.37	4.67	5.35	5.45	5.62	5.90	5.54	5.71	6.00	6.42
2	190-260	4.18	4.26	4.39	4.61	4.93	5.64	5.76	5.93	6.22	5.85	6.03	6.33	6.77
3	260-420	5.21	5.31	5.47	5.75	6.15	7.03	7.17	7.39	7.76	7.29	7.51	7.89	8.44
5	350-600	7.91	8.07	8.31	8.73	9.34	10.67	10.89	11.22	11.78	11.07	11.41	11.98	12.81
7-1/2	420-830	10.63	10.84	11.17	11.73	12.55	14.35	14.64	15.08	15.83	14.88	15.33	16.09	17.22
10	600-1200	14.59	14.88	15.33	16.09	17.22	19.70	20.09	20.69	21.73	20.43	21.04	22.09	23.64
15	830-1400	17.60	17.95	18.49	19.42	20.77	23.76	24.24	24.96	26.21	24.64	25.38	26.65	28.51
20	1170-1500	20.41	20.82	21.44	22.51	24.09	27.55	28.10	28.95	30.40	28.57	29.43	30.90	33.07
25	1200-2200	24.16	24.64	25.38	26.65	28.52	32.61	33.27	34.27	35.98	33.82	34.84	36.58	39.14
30	1400-2500	25.84	26.36	27.15	28.50	30.50	34.88	35.58	36.65	38.48	36.18	37.26	39.12	41.86
35	1500-2515	30.00	30.60	31.52	33.09	35.41	40.50	41.31	42.55	44.68	42.00	43.26	45.42	48.60
40	1500-2515	30.00	30.60	36.52	33.09	35.41	40.50	41.31	42.55	44.68	–	–	–	–
50	2200-3000	37.91	38.67	39.83	41.82	44.75	51.18	52.20	53.77	56.46	–	–	–	–
60	2300-3530	45.41	46.32	47.71	50.09	53.60	61.30	62.53	64.41	67.63	–	–	–	–
75	2500-3530	46.25	47.18	48.59	51.02	54.59	–	–	–	–	–	–	–	–
100	3000-5610	62.91	64.17	66.09	69.40	74.26	–	–	–	–	–	–	–	–
125	3000-6250	75.00	76.50	78.80	82.73	88.53	–	–	–	–	–	–	–	–
440- or 550-Volts														
150	3500-6250	80.84	82.46	84.93	89.18	95.42	–	–	–	–	–	–	–	–
175	4100-7033	90.00	91.80	94.55	99.28	106.23	–	–	–	–	–	–	–	–
200	4800-7033	92.91	94.77	97.61	102.49	109.67	–	–	–	–	–	–	–	–

Manhours include checking out of job storage, handling, job hauling, rolling and placing or rigging and picking, setting, and aligning.

Manhours exclude installation of fasteners or supports, conduit, switches, wiring, and circuit connections. See respective tables for these time frames.

MOUNTING MOTORS

DC—115-330-Volt—Constant Speed—Shunt Series—Compound Wound
1750, 1150, 850, 690, 575, 500, and 450 RPM's

MANHOURS EACH

Motor		Manhours for Mounting on												
Horse Power	Weight Range (Pounds)	Floors Height to				Walls or Machine Height to				Ceilings Height to			
		Gr.	10'	15'	20'	25'	10'	15'	20'	25'	10'	15'	20'	25'
1	125-320	4.19	4.27	4.40	4.62	4.95	5.66	5.77	5.94	6.24	5.87	6.04	6.34	6.79
1-1/2	170-320	4.54	4.63	4.77	5.00	5.36	6.13	6.25	6.44	6.76	6.36	6.55	6.87	7.36
2	185-500	5.63	5.74	5.91	6.21	6.65	7.60	7.75	7.99	8.38	7.88	8.12	8.52	9.12
3	185-555	6.50	6.63	6.83	7.17	7.67	8.78	8.95	9.21	9.68	9.10	9.37	9.84	10.53
5	230-720	8.25	8.42	8.67	9.10	9.74	11.14	11.36	11.70	12.29	11.55	11.90	12.49	13.37
7-1/2	320-810	10.13	10.33	10.64	11.17	11.96	13.68	13.95	14.37	15.09	14.18	14.61	15.34	16.41
10	500-1140	13.75	14.03	14.45	15.17	16.23	18.56	18.93	19.50	20.48	19.25	19.83	20.82	22.28
15	555-1995	17.05	17.39	17.91	18.81	20.12	23.00	23.48	24.18	25.39	23.87	24.59	25.82	27.62
20	720-2230	20.73	21.14	21.78	22.87	24.47	27.99	28.55	29.40	30.87	29.00	29.89	31.39	33.58
25	810-2230	21.66	22.09	22.76	23.89	25.57	29.24	29.83	30.72	32.26	30.32	31.23	32.80	35.09
30	810-2985	28.34	28.91	29.77	31.26	33.45	38.26	39.02	40.19	42.20	39.68	40.87	42.91	45.91
40	1110-3905	31.66	32.29	33.26	34.93	37.37	42.74	43.60	44.90	47.15	44.32	45.65	47.94	51.29
50	1140-3905	40.54	41.35	42.59	44.72	47.85	54.73	55.82	57.50	60.37	–	–	–	–
60	1140-3905	44.59	45.48	46.85	49.19	52.63	60.20	61.40	63.24	66.40	–	–	–	–
75	1500-5155	50.00	51.00	52.53	55.16	59.02	67.50	68.85	70.92	74.46	–	–	–	–
100	2230-6350	67.68	69.03	71.10	74.66	79.89	–	–	–	–	–	–	–	–
125	2985-7550	80.54	82.15	84.62	88.85	95.07	–	–	–	–	–	–	–	–
150	3175-7750	91.08	92.90	95.69	100.47	107.51	–	–	–	–	–	–	–	–
175	3955-8050	102.14	104.18	107.31	112.67	120.56	–	–	–	–	–	–	–	–
200	5255-8450	110.36	112.57	115.94	121.74	130.26	–	–	–	–	–	–	–	–

Manhours include checking out of job storage, handling, job hauling, rolling and placing or rigging and picking, setting, and aligning.

Manhours exclude installation of fasteners or supports, conduit, switches, wiring, and circuit connections. See respective tables for these time frames.

Section 14

POWER
TRANSFORMERS

This section is included for the installation of air and oil cooled transformers for the service of power equipment.

For other transformers see section, Outside Construction.

Manhours listed in the following tables are average of many installations for the items as outlined.

The estimator should be familiar with the introduction to this manual prior to the application or use of these units.

AIR AND OIL COOLED TRANSFORMERS
60-Cycle—Single Phase
Primary 440 or 550 Volts—Secondary 115/230 Volts

MANHOURS EACH

Type	KVA Rating	Approximate Weight (Pounds)	Approximate Gallons of Oil	Manhours for Mounting on			
				Floors	Walls or Columns Heights to		
					10'	15'	25'
Air Cooled	1	35	–	3.30	4.20	4.40	4.50
	1.5	47	–	3.85	4.70	4.95	5.10
	2	60	–	4.40	5.25	5.50	5.65
	3	90	–	4.95	5.75	6.00	6.20
	5	127	–	6.20	6.90	7.25	7.45
	7.5	180	–	6.60	7.65	8.00	8.25
	10	350	–	8.80	10.10	10.60	10.90
	15	485	–	11.20	13.85	14.50	14.95
	25	625	–	15.40	17.85	18.75	19.30
	37.5	900	–	20.90	25.20	26.45	27.25
	50	1,025	–	24.20	28.35	29.75	30.65
Oil Cooled	1	155	3.25	5.80	7.05	7.40	7.60
	1.5	165	3.75	6.15	7.30	7.65	7.85
	2	185	3.75	6.60	8.05	8.45	8.70
	3	205	4.00	7.70	8.35	8.75	9.00
	5	255	6.00	8.40	8.90	9.35	9.65
	7.5	295	9.00	8.80	10.55	11.10	11.45
	10	370	11.00	9.90	11.00	11.55	11.90
	15	520	21.00	11.20	14.50	15.20	15.65
	25	700	27.00	16.50	17.60	18.50	19.05
	37.5	990	32.00	23.10	28.60	30.00	30.90
	50	1,190	40.00	24.30	31.90	33.50	34.50

Manhours include checking out of job storage, handling, job hauling, rigging, and picking or rolling and skidding as required, setting, aligning, fastening, and connecting of transformers as outlined.

Manhours exclude installation of fasteners and supports and scaffolding. See respective tables for these time frames.

AIR AND OIL COOLED TRANSFORMERS
25-Cycle—Single Phase
Primary 550 Volts—Secondary 115/230 Volts

MANHOURS EACH

Type	KVA Rating	Approximate Weight (Pounds)	Approximate Gallons of Oil	Manhours for Mounting on			
				Floors	Walls or Columns Heights to		
					10'	15'	25'
Air Cooled	1	90	–	4.95	6.05	6.35	6.50
	1.5	140	–	6.15	7.25	7.60	7.80
	2	170	–	6.60	8.05	8.45	8.70
	3	250	–	8.35	8.90	9.35	9.65
	5	320	–	8.80	10.55	11.05	11.40
	7.5	450	–	9.55	12.30	12.90	13.30
	10	575	–	13.20	17.60	18.50	19.05
	15	630	–	15.40	18.70	19.60	20.20
	25	940	–	22.00	27.50	28.85	29 70
	37.5	1,760	–	27.50	36.30	38.10	39.25
	50	2,000	–	36.30	46.20	48.50	49.95
Oil Cooled	1	130	4.0	6.15	7.25	7.60	7.80
	1.5	180	4.5	6.60	8.05	8.45	8.70
	2	240	5.0	8.35	8.90	9.35	9.65
	3	350	6.0	8.80	10.55	11.05	11.40
	5	450	6.5	9.55	12.30	12.90	13.30
	7.5	500	11.5	11.20	14.50	15.20	15.65
	10	600	16.0	14.30	18.70	19.60	20.20
	15	850	23.0	17.60	23.10	24.25	24.95
	25	1,200	41.0	24.20	31.90	33.50	34.50
	37.5	1,600	40.0	26.40	35.20	37.00	38.10
	50	2,000	25.0	36.30	46.20	48.50	49.95

Manhours include checking out of job storage, handling, job hauling, rigging, and picking or rolling and skidding as required, setting, aligning, fastening, and connecting of transformers as outlined.

Manhours exclude installation of fasteners and supports and scaffolding. See respective tables for these time frames.

Section 15

OUTSIDE OVERHEAD SYSTEMS

This section includes manhour requirements for the installation of items as may be required for outside overhead systems such as lighting and power.

These time frames give consideration to all direct labor operations as may be required for the installation of the item involved all in accordance with the notes appearing with the tables.

For other labor operations that may be required in conjunction with the installation of these systems see other sections and tables in this manual.

OVERHEAD STREET LIGHTING
Poles, Standards, and Accessories

MANHOURS EACH

Size	Item		Manhours
	POLES		
30 feet	6"-5"-4"	Tubular Steel	20. 30
30 feet	7"-6"-5"	Tubular Steel	23. 15
35 feet	7"-6"-5"	Tubular Steel	26. 00
40 feet	7"-6"-5"-4"	Tubular Steel	29. 25
	STANDARDS		
20 feet	Steel		23. 80
25 feet	Steel		26. 25
30 feet	Steel		28. 50
35 feet	Steel		34. 05
40 feet	Steel		39. 40
20 feet	Aluminum		22. 25
25 feet	Aluminum		24. 60
30 feet	Aluminum		25. 20
35 feet	Aluminum		28. 85
40 feet	Aluminum		33. 15
20 feet	Concrete		25. 50
25 feet	Concrete		27. 80
30 feet	Concrete		30. 00
35 feet	Concrete		37. 45
40 feet	Concrete		43. 15
	POLE ACCESSORIES		
	Pole Top Pins		1. 50
	Pole Bands		3. 00

Manhours include checking out of job storage, handling, job hauling, distributing and installing poles, standards, and accessories.

Manhours exclude excavation and installation of foundations and mast arms. See respective tables for these time frames.

OVERHEAD STREET LIGHTING
Supports

MANHOURS EACH

Size	Item	Manhours
4 feet	Gooseneck Bracket	4.15
4 feet	Upsweep Bracket	4.15
6 feet	Upsweep Bracket	4.70
8 feet	Upsweep Bracket	5.20
4 feet	Straight Pipe Bracket	4.15
6 feet	Straight Pipe Bracket	4.70
8 feet	Straight Pipe Bracket	5.20
6 feet	Mast Arm	5.55
8 feet	Mast Arm	6.00
10 feet	Mast Arm	6.50
12 feet	Mast Arm	7.40
14 feet	Mast Arm	8.30
16 feet	Mast Arm	9.25
	Center Span Suspension	10.80

MISCELLANEOUS OVERHEAD SERVICE ITEMS

MANHOURS EACH

Item	Manhours
Secondary Racks	1.50
Take-Off Insulators	.75
Crossarm Bracket	.75
Angle Bracket	.75
Service Racks	1.50
Wire Holders	.75
Cable Holders	.75
#1 and Smaller — Openwire	2.25
#1/0 and Larger — Openwire	3.75
#1 and Smaller Service Drop Cable	1.50
#1/0 and Larger Service Drop Cable	3.00

Manhours include checking out of job storage, handling, job hauling, distributing and installing items as outlined.

Center Span Suspension manhours include 5/16″ guy strand, guy clamps, strain insulators, and eye bolts.

Manhours exclude installation of poles and luminaires. See respective tables for these time frames.

POLE SETTING

MANHOURS EACH

Item		Erect with:		
		Pike	Gin Pole	Boom or Crane
25 feet	Wood Pole	9.75	6.50	5.20
30 feet	Wood Pole	11.44	7.54	6.24
35 feet	Wood Pole	12.87	8.58	7.28
40 feet	Wood Pole	14.56	9.49	8.19
45 feet	Wood Pole	15.99	10.53	9.23
50 feet	Wood Pole	17.42	11.44	10.27
55 feet	Wood Pole	18.85	12.48	11.18
60 feet	Wood Pole	20.28	13.52	12.22
65 feet	Wood Pole	21.79	14.56	13.26
70 feet	Wood Pole	23.14	15.47	14.17
75 feet	Wood Pole	24.57	16.51	15.21
Roofing		-	-	.78
Gaining (on Ground)		-	-	.65
Cribbing — One 6' Log — 6 feet deep		-	-	3.25
Cribbing — Two 6' Log — 6 feet deep		-	-	7.80
Barrel		-	-	1.95

POLE ANCHORS

MANHOURS EACH

Item	Manhours
Plate Anchors	1.30
Expansion Anchors	1.30
Screw Anchors	2.60

Manhours include checking out of job storage, handling, job hauling up to 2 miles, distributing and installing items as outlined.

If pole location is hard to get to and work area is limited, additional time may be required.

For deadman anchors allow additional excavation time giving consideration to excavating methods and soil conditions.

Manhours exclude installation of supports, accessories, wiring, and luminaires. See respective tables for these time frames.

GUYS

NET MANHOURS EACH

Item	Manhours	
	5/16"	1/2"
HEAD OR STUB GUY		
No Strain Insulators	3.90	4.29
One Strain Insulators	5.20	5.85
Two Strain Insulators	6.50	7.15
ARM GUY		
No Strain Insulators	4.55	4.55
One Strain Insulators	5.85	6.50
Two Strain Insulators	7.15	7.80
DOWN GUY		
No Strain Insulators	3.25	3.90
One Strain Insulators	4.55	5.20
Two Strain Insulators	5.85	6.50
MISCELLANEOUS		
Add for Sidewalk Guy	1.95	
Add for Guy Protectors	.65	
Add for Strain Plates	.65	

Manhours include checking out of job storage, handling, job hauling up to 2 miles, distributing and placing guy strain insulator, two 3-bolt or other type guy clamp, 2 additonal clamps per strain insulator (except wood strains) thru bolts and guy hooks.

Manhours are based on ground assembly where possible.

Manhours exclude installation of poles, arms, supports and other accessories. See respective tables for these time frames.

CROSS ARMS, PINS, AND SECONDARY RACKS

MANHOURS EACH

Item	Manhours	
	8'	10'
CROSS ARMS		
Single Arm	1.69	1.69
Double Arm	3.77	3.77
Single Side or Alley Arm	3.50	3.50
Double Side or Alley Arm	5.85	5.85
PINS		
Steel Shank with Washer and Nut	.39	
Wood Shank — Nailed	.26	
Saddle Pins	.65	
Pole Top Pins	1.30	
SECONDARY RACKS		
One Spool	.65	
Two Spool	1.30	
Three Spool	1.95	
Four Spool	1.95	
Rack Extension	1.30	

Manhours include checking out of job storage, handling, job hauling, distributing, and installing items as outlined.

Time allowance for the placement of necessary thru bolts, carriage bolts, double-arming bolts, lag screws, and braces is included.

Manhours exclude installation of poles, luminaires and wire or cable. See respective tables for these time frames.

STRINGING WIRE

MANHOURS FOR UNITS LISTED

Wire Size	Item	Manhours per Thousand Feet					
		Height to:					
		25'	35'	45'	55'	65'	75'
#6		21.48	21.95	22.64	23.10	23.79	24.25
#4	Copper, Bare – Copper, TBWP –	22.74	23.23	23.96	24.45	25.18	25.67
#2	Copper, Polyethylene – ACSR, Bare –	26.78	27.36	28.22	28.80	29.66	30.24
#1/0	ACSR, TBWP & ACSR Polyethylene	33.76	34.50	35.57	36.30	37.39	38.12
#2/0		39.48	40.33	41.60	42.45	43.72	44.57
#4/0		51.06	52.16	53.80	54.90	56.55	57.65
#6		18.69	19.10	19.70	20.10	20.70	21.11
#4		19.95	20.38	21.00	21.45	22.10	22.52
#2	Aluminum, Bare – Aluminum, TBWP	23.99	24.51	25.28	25.80	26.57	27.10
#1/0	& Aluminum, Polyethylene	30.97	31.64	32.63	33.30	34.30	35.00
#2/0		36.69	37.48	38.66	39.45	40.63	41.42
#4/0		48.27	49.31	50.86	51.90	53.46	54.50
#6	Copperweld, Bare	23.58	24.09	24.85	25.36	26.12	26.63
#4	Copperweld, Bare	28.46	29.07	29.98	30.60	31.52	32.13
#6A	Copperweld – Copper, Bare	24.27	24.80	25.58	26.10	26.88	27.40
#4A	Copperweld – Copper, Bare	29.57	30.21	31.16	31.80	32.75	33.39

	Net Manhours Each
DEADENING	
Insulated Clevis	1.50
1 Suspension Insulator & Clamp	2.25
2 Suspension Insulators & Clamp	3.00
INSULATORS	
Pin	.45
Post	1.50
Preformed Arm or Rods (per set)	1.50

Manhours include checking out of job storage, handling, job hauling, stringing, splicing, sagging, and tieing of wire and installation of other items as listed.

Manhours exclude installation of poles, supports, and luminaires. See respective tables for these time frames.

AERIAL CABLE

MANHOURS FOR ITEMS LISTED

Wire Size	Item	Manhours per Hundred Feet					
		Height to:					
		25'	35'	45'	55'	65'	75'
	PRIMARY — PREASSEMBLED — SELF SUPPORTING — COPPER						
#6	Single Phase	11. 16	11. 40	11. 76	12. 00	12. 36	12. 60
#4	Single Phase	12. 56	12. 83	13. 23	13. 50	13. 90	14. 18
#2	Single Phase	13. 95	14. 25	14. 70	15. 00	15. 45	15. 75
#1/0	Single Phase	16. 04	16. 39	16. 91	17. 25	17. 77	18. 11
#2/0	Single Phase	18. 14	18. 53	19. 11	19. 50	20. 09	20. 48
#4/0	Single Phase	20. 93	21. 38	22. 05	22. 50	23. 18	23. 63
#4	Three Phase	25. 11	25. 65	26. 46	27. 00	27. 81	28. 35
#2	Three Phase	27. 90	28. 50	29. 40	30. 00	30. 90	31. 50
#1/0	Three Phase	32. 09	32. 78	33. 81	34. 50	35. 54	36. 23
#2/0	Three Phase	36. 27	37. 05	38. 22	39. 00	40. 17	40. 95
#4/0	Three Phase	41. 85	42. 75	44. 10	45. 00	46. 35	47. 25
	SECONDARY & SERVICE DROP — SELF SUPPORTING						
#6	Polyethylene	8. 37	8. 55	8. 82	9. 00	9. 27	9. 45
#4	Polyethylene	9. 77	9. 98	10. 29	10. 50	10. 82	11. 03
#2	Polyethylene	11. 16	11. 40	11. 76	12. 00	12. 36	12. 60
#1	Polyethylene	12. 56	12. 83	13. 23	13. 50	13. 91	14. 18
#1/0	Polyethylene	13. 95	14. 25	14. 70	15. 00	15. 45	15. 75
#2/0	Polyethylene	16. 74	17. 10	17. 64	18. 00	18. 54	18. 90

		Net Manhours Each
DEADENDING		
	Single Phase	4. 50
	Three Phase	15. 00
TOPPING AND SPLICING		
	Single Phase to Single Phase	3. 00
	Single Phase to Three Phase	4. 50
	Three Single Phase to Three Phase	10. 50
	Three Phase to Three Phase	13. 50

Manhours include checking out of job storage, handling, job hauling, stringing, splicing, sagging, and tieing of wire and installation of other items as listed.

Single Phase — Primary — Self Supporting Copper consists of one insulated conductor and one lashed bare neutral messenger.

Three Phase — Primary — Self Supporting Copper consists of three insulated conductors and one lashed bare neutral messenger.

Secondary and Service Drop consists of two insulated conductors and one bare neutral conductor serving as supporting member.

Manhours exclude installation of poles, supports, and luminaires. See respective tables for these time frames.

GROUNDING

MANHOURS EACH

Item	Manhours
Pole Butt Plates	.45
Pole Butt Coil	.75
Driven Ground Rod or Pipe	.75
Wire Attached to Pole 1 Connector	1.50
Wire Attached to Pole 2 Connectors	3.00
Wire Attached to Pole 3 Connectors	4.50
Attach Wood Molding	.75

PRIMARY & SECONDARY PROTECTIVE DEVICES

MANHOURS EACH

Item	Manhours
PRIMARY	
Pole Top Air Break Switch — TPST — with Operating Mechanism and Rod	45.50
Enclosed Disconnect Switches	3.00
Open Disconnect Switches	3.00
Enclosed Cutouts — One Shot	3.00
Enclosed Cutouts — Two Shot	3.00
Enclosed Cutouts — Three Shot	3.00
Open Cutouts — One Shot	3.00
Open Cutouts — Two Shot	3.00
Open Cutouts — Three Shot	3.00
Oil Circuit Reclosers — Single Phase (70 lbs.)	10.50
Oil Circuit Reclosers — Three Phase (315 lbs.)	18.00
Heavy Duty Oil Circuit Reclosers — Single Phase (170 lbs.)	15.00
Heavy Duty Oil Circuit Reclosers — Three Phase (700 lbs.)	27.00
Oil Switch — Single Phase — Remote Control	15.00
Oil Switch — One Single Phase — Manual Gang Operated	18.00
Oil Switch — Two Single Phase — Manual Gang Operated	39.00
Oil Switch — Three Single Phase — Manual Gang Operated	65.00
Oil Switch — Remote Control	65.00
Oil Switch — Manual Operated	78.00
Lightning Arresters	2.25
SECONDARY	
Fuse Cutouts	2.25
Lightning Arresters	2.25

Manhours include checking out of job storage, handling, job hauling, distributing, and installing grounding items or protective devices as outlined.

Manhours exclude installation of poles, supports, wire or cable, and luminaires. See respective tables for these time frames.

OVERHEAD STREET LIGHTING
Luminaires

MANHOURS EACH

Item	Manhours
Radial Wave Reflector	2. 60
Suburban Open	2. 60
Pendent Incandescent with Ell or Plumbizer	3. 90
Side Mounted Incandescent with Slip Fitter	3. 90
Incandescent with Built-in Control	5. 20
Mercury Vapor Unit	3. 90
Sodium Vapor Unit	5. 20
Deflectors	. 65

Control Installations

MANHOURS EACH

Item	Manhours
Contactors (Oil Switches)	7. 80
Time Switches	3. 15
Photoelectric Cell	3. 15
Pilot Wire Relay	2. 60
Control Wiring	4. 55
TRANSFORMERS	
10 KVA Constant Current	11. 15
15 KVA Constant Current	12. 75
25 KVA Constant Current	15. 10

Manhours include checking out of job storage, handling, job hauling, distributing, and installing items listed.

Luminaires manhours installation of wire or cable, fiber conduit, and other material as may be required to connect luminaires to street lighting circuit.

Control Wiring manhours include wiring from low voltage source to time switches, photoelectric cells, and relays, and to controllers and consists of low voltage wire and cable, fuse blocks, switches, etc.

Manhours exclude installation of poles, supports, accessories, and circuit wiring. See respective tables for these time frames.

POLE MOUNTED DISTRIBUTION LINE TRANSFORMERS

MANHOURS EACH

KVA Rating	Manhours	
	Single Phase	Three Phase
3	8.70	30.00
5	9.60	31.50
7-1/2	11.10	34.50
10	11.70	37.50
15	14.70	40.50
25	17.40	45.00
37-1/2	20.55	49.50
50	24.90	54.00
75	28.36	63.00
100	38.10	72.00
150	41.70	81.00
200	46.80	90.00

DISTRIBUTION LINE REGULATORS & CAPACITORS

KVA Rating	Item	Manhours
	REGULATORS	
12-1/2	Pole Mounted	18.00
19	Pole Mounted	21.00
25	Pole Mounted	24.00
12-1/2	Platform Mounted 3 Phase Bank	30.00
19	Platform Mounted 3 Phase Bank	33.00
25	Platform Mounted 3 Phase Bank	37.50
37-1/2	Platform Mounted 3 Phase Bank	41.25
50	Platform Mounted 3 Phase Bank	52.50
62-1/2	Platform Mounted 3 Phase Bank	60.00
75	Platform Mounted 3 Phase Bank	67.50
100	Platform Mounted 3 Phase Bank	82.50
125	Platform Mounted 3 Phase Bank	97.50
	CAPACITORS	
	One Single Phase Unit	7.50
	Two Single Phase Units	12.00
	Three Phase Bank	18.00
	Two Three Phase Banks	33.00

Manhours include checking out of job storage, handling, job hauling, distributing, and installing items as listed.

Manhours exclude installation of poles, supports, wire or cable, luminiares, protective devices, grounding, and switching equipment. See respective tables for these time frames.

Section 16

OUTSIDE UNDERGROUND SYSTEMS

This section includes manhour requirements for the installation of outside underground ducts and cables.

The tables throughout this section give consideration to all labor operations as may be required for the installation of the items involved all in accordance with the notes appearing with the tables.

For other labor operations that may be required in conjunction with the installation of ducts and cables, such as trenching, placing of concrete, etc., see other sections and tables included in this manual.

UNDERGROUND FIBER DUCTS

MANHOURS FOR UNITS LISTED

Item Description	Unit	Size (Inches)	Manhours required for Number of Ducts in Run				
			1	2	3	4	5 or More
Fiber Duct	Lin. Ft.	2	.087	.079	.075	.073	.072
	Lin. Ft.	2-1/2	.093	.084	.080	.079	.077
	Lin. Ft.	3	.097	.087	.083	.082	.080
	Lin. Ft.	3-1/2	.104	.094	.089	.088	.086
	Lin. Ft.	4	.109	.098	.094	.092	.090
	Lin. Ft.	4-1/2	.115	.104	.099	.097	.095
Fiber Bends	each	2-1/2	1.44	1.44	1.44	1.44	1.44
	each	3	1.56	1.56	1.56	1.56	1.56
	each	3-1/2	1.68	1.68	1.68	1.68	1.68
	each	4	1.80	1.80	1.80	1.80	1.80
	each	4-1/2	1.92	1.92	1.92	1.92	1.92
Fiber Elbows	each	2-1/2	1.13	1.13	1.13	1.13	1.13
	each	3	1.25	1.25	1.25	1.25	1.25
	each	3-1/2	1.38	1.38	1.38	1.38	1.38
	each	4	1.50	1.50	1.50	1.50	1.50
	each	4-1/2	1.63	1.63	1.63	1.63	1.63

Manhours include checking out of job storage, handling, job hauling, distributing, and installing ducts and fittings in pre-excavated, water-free trenches up to 3 feet deep.

Manhours exclude excavating, dewatering, and placing concrete and backfill. See respective tables for these time frames.

UNDERGROUND ASBESTOS CEMENT DUCTS

MANHOURS FOR UNITS LISTED

Item Description	Unit	Size (Inches)	Manhours Required for Number of Ducts in Run				
			1	2	3	4	5 or More
Asbestos Cement Duct	Lin. Ft.	2	.097	.087	.083	.082	.080
	Lin. Ft.	2-1/2	.103	.093	.088	.087	.085
	Lin. Ft.	3	.109	.098	.094	.092	.090
	Lin. Ft.	3-1/2	.116	.105	.100	.098	.096
	Lin. Ft.	4	.121	.109	.104	.102	.100
	Lin. Ft.	4-1/2	.129	.116	.110	.108	.106
Asbestos Cement Bends	each	2-1/2	1.56	1.56	1.56	1.56	1.56
	each	3	1.68	1.68	1.68	1.68	1.68
	each	3-1/2	1.80	1.80	1.80	1.80	1.80
	each	4	1.92	1.92	1.92	1.92	1.92
	each	4-1/2	2.04	2.04	2.04	2.04	2.04
Asbestos Cement Elbows	each	2-1/2	1.25	1.25	1.25	1.25	1.25
	each	3	1.38	1.38	1.38	1.38	1.38
	each	3-1/2	1.50	1.50	1.50	1.50	1.50
	each	4	1.63	1.63	1.63	1.63	1.63
	each	4-1/2	1.75	1.75	1.75	1.75	1.75

Manhours include checking out of job storage, handling, job hauling, distributing, and installing duct and fittings in pre-excavated, water-free trenches up to 3 feet deep.

Manhours exclude excavating, dewatering, and placing concrete and backfill. See respective tables for these time frames.

UNDERGROUND CLAY DUCTS

MANHOURS FOR UNITS LISTED

Item Description	Unit	Size (Inches)	Manhours Required for Number of Ducts in Run					
			1	2	3	4	6	9
Clay Duct	Lin. Ft.	3-1/4	.132	.138	.144	.150	.156	.174
	Lin. Ft.	3-1/2	.138	.144	.150	.156	.174	.192
	Lin. Ft.	4-1/4	.144	.150	.156	.174	.186	.204
	Lin. Ft.	4-1/2	.150	–	–	–	–	–
Split Clay Duct	Lin. Ft.	3-1/4	–	.156	.162	.168	.174	.192
	Lin. Ft.	3-1/2	–	.162	.168	.174	.192	.210
	Lin. Ft.	4-1/4	–	.168	.174	.192	.204	.222
Clay Duct–Single Bends	each	3-1/4	1.92	1.92	1.92	1.92	1.92	1.92
	each	3-1/2	2.16	2.16	2.16	2.16	2.16	2.16
	each	4-1/4	2.40	2.40	2.40	2.40	2.40	2.40
Clay Duct Slants	each	1 Hole	.188	.188	.188	.188	.188	.188
	each	2 Hole	.200	.200	.200	.200	.200	.200
	each	3 Hole	.213	.213	.213	.213	.213	.213
	each	4 Hole	.225	.225	.225	.225	.225	.225
	each	6 Hole	.238	.238	.238	.238	.238	.238
	each	9 Hole	.289	.289	.289	.289	.289	.289

Manhours include checking out of job storage, handling, job hauling, distributing, and installing duct and fittings in pre-excavated, water-free trenches up to 3 feet deep.

Manhours exclude excavating, dewatering, placing concrete and backfill. See respective tables for these time frames.

LEAD COVERED AND DIRECT BURIAL CABLE

MANHOURS PER HUNDRED FEET

Wire Size	Lead Covered Cable in Conduit		Direct Burial Cable	
	1 Conductor	3 Conductor	1 Conductor	3 Conductor
14	–	4.50	1.05	1.95
12	–	5.40	1.20	2.00
10	–	7.05	1.36	2.70
8	–	9.15	1.80	4.05
6	4.50	11.10	2.00	4.70
4	5.5	13.20	–	5.40
2	6.30	15.00	–	6.75
1	6.30	18.00	–	9.00
1/0	6.75	20.10	–	11.40
2/0	7.05	22.05	–	12.15
3/0	7.65	24.75	–	12.75
4/0	8.10	27.00	–	15.00
250 MCM	8.70	–	–	16.50
350 MCM	10.50	–	–	–
500 MCM	12.00	–	–	–
	14.25			

Manhours include checking out of job storage, handling, job hauling, distributing, and installing conduit and cable in pre-excavated, water-free trenches up to 3 feet deep.

Manhours exclude excavating, dewatering, placing concrete, placing backfill, splicing and hook-up. See respective tables for these time frames.

600-VOLT, TYPE RL WIRE CABLE SPLICES

MANHOURS EACH

Wire Size	1-Conductor Cable	3-Conductor Cable
14	–	3.00
12	–	3.75
10	–	4.50
8	3.30	4.25
6	3.75	7.80
4	4.20	8.25
2	4.50	10.50
1	4.95	12.00
1/0	5.25	13.50
2/0	5.63	15.00
3/0	6.00	18.00
4/0	6.75	19.50
250 MCM	7.50	–
350 MCM	8.25	–
500 MCM		–

Manhours include all operations required for make-up of splice as outlined.

Manhours exclude trench earthwork, placing cable, and prouring concrete. See respective tables for these time frames.

Section 17

COMMUNICATION AND SIGNAL SYSTEMS

This section includes manhour time frames for the installation of wire, cable, and various communication and signaling devices.

The manhours listed are for labor only. No consideration has been given to the value of materials.

Labor for checking out of project storage, handling, job hauling, and installing have been given consideration in the manhour tables for the operations as listed in accordance with the notes appearing with the tables.

These manhour units are average of many projects and should be evaluated and adjusted by the estimator for use on an individual project as outlined in the Introduction of this manual.

LOW VOLTAGE WIRE AND CABLE FOR CONTROL CIRCUITS AND COMMUNICATION AND ANNUNICIATOR SYSTEMS

MANHOURS PER LINEAR FOOT

Conductor Description	Size and Type of Conductor	Number of Conductors	Manhours Required for			
			Underfloor Ducts and Conduits	Heights to		
				10'	15'	25'
Plastic Insulated Twisted Copper Conductor	#22 Solid	2	0.009	0.010	0.011	0.011
		3	0.015	0.016	0.017	0.017
	#19 Solid	2	0.013	0.014	0.015	0.015
		3	0.019	0.021	0.021	0.022
	#18 Solid	1	0.007	0.008	0.008	0.008
		2	0.014	0.015	0.016	0.016
		3	0.021	0.023	0.024	0.024
Plastic Insulated Telephone Copper Weld Conductor	#19 Solid	2	0.012	0.013	0.013	0.014
		3	0.017	0.018	0.019	0.020
Rubber Insulated Telephone Copper Weld Conductor	#19 Solid TW Pair or Parallel	2	0.013	0.014	0.015	0.015
Rubber Insulated Telephone Bronze Conductor	#17 Solid TW Pair or Parallel	2	0.017	0.018	0.019	0.020

Manhours include checking out of job storage, handling, job hauling, placing wire coils or spools on reel stands or racks, attaching to fish tape or other pull-in means, pulling wire into conduit type raceway, ducts and underfloor ducts or pull-out and laying into wireways or channels or pulling over suspended ceilings or through bar joist or trusses, identifying or "polling out" the circuit conductors, and splicing or tagging for connection to terminal blocks or equipment.

Manhours exclude extensive identification, "ringing out", connecting to terminal blocks or equipment, and scaffolding. See respective tables for these time frames.

LOW VOLTAGE WIRE AND CABLE FOR CONTROL CIRCUITS AND COMMUNICATION AND ANNUNICIATOR SYSTEMS

Plastic Insulated, Jacket Copper Conductors

MANHOURS PER LINEAR FOOT

Size and Type of Conductor	Number of Conductors	Underfloor Ducts and Conduits	Manhours Required for		
			Heights to		
			10'	15'	25'
#22 Solid	2	0.011	0.012	0.012	0.013
	3	0.013	0 014	0.015	0.015
	4	0.014	0.015	0.016	0.016
#19 Solid	2	0.012	0.013	0.013	0.014
	3	0.013	0.014	0.015	0.015
	4	0.014	0.015	0.016	0.016
#22 Stranded	2	0.013	0.014	0.015	0.015
	3	0.013	0.014	0.015	0.015
	4	0.015	0.016	0.017	0.017
	5	0.016	0.017	0.018	0.019
	10	0.022	0.024	0.025	0.026
	12	0.023	0.025	0.026	0.027
	15	0.027	0.029	0.030	0.031
	25	0.031	0.033	0.035	0.036
#20 Stranded	2	0.014	0.015	0.016	0.016
	3	0.015	0.016	0.017	0.017
	4	0.018	0.019	0.020	0.021

Manhours include checking out of job storage, handling, job hauling, placing wire coils or spools on reel stands or racks, attaching to fish tape or other pull-in means, pulling wire into conduit type raceway, ducts and underfloor ducts or pull-out and laying into wireways or channels or pulling over suspended ceilings or through bar joist or trusses, identifying or "polling out" the circuit conductors, and splicing or tagging for connection to terminal blocks or equipment.

Manhours exclude extensive identification, "ringing out", connecting to terminal blocks or equipment, and scaffolding. See respective tables for these time frames.

LOW VOLTAGE WIRE AND CABLE FOR CONTROL CIRCUITS AND COMMUNICATIONS AND ANNUNICIATOR SYSTEMS

Plastic Insulated Jacket Copper Conductors

MANHOURS PER LINEAR FOOT

Size and Type of Conductor	Number of Conductors	Manhours Required for			
		Underfloor Ducts and Conduits	Heights to		
			10'	15'	25'
#18 Stranded	2	0.016	0.017	0.018	0.019
	3	0.018	0.019	0.020	0.021
	4	0.020	0.022	0.022	0.023
	5	0.023	0.025	0.026	0.027
	7	0.027	0.029	0.030	0.031
	9	0.030	0.032	0.034	0.035
	12	0.035	0.038	0.039	0.040
	15	0.050	0.054	0.056	0.058
#16 Stranded	2	0.019	0.021	0.021	0.022
	3	0.021	0.023	0.024	0.024
	4	0.024	0.026	0.027	0.028
	5	0.030	0.032	0.034	0.035
	7	0.043	0.046	0.048	0.050
	9	0.049	0.053	0.055	0.057
	12	0.056	0.060	0.063	0.065
	15	0.059	0.064	0.066	0.068
	19	0.062	0.067	0.070	0.072
#14 Stranded	2	0.027	0.029	0.030	0.031
	3	0.033	0.036	0.037	0.038
	4	0.051	0.055	0.057	0.059
	7	0.058	0.063	0.065	0.067
	12	0.070	0.076	0.079	0.081

Manhours include checking out of job storage, handling, job hauling, placing wire coils or spools on reel stands or racks, attaching to fish tape or other pull-in means, pulling wire into conduit type raceway, ducts and underfloor ducts or pull-out and laying into wireways or channels or pulling over suspended ceilings or through bar joist or trusses, identifying or "polling out" the circuit conductors, and splicing or tagging for connection to terminal blocks or equipment.

Manhours exclude extensive identification, "ringing out", connecting to terminal blocks or equipment, and scaffolding. See respective tables for these time frames.

LOW VOLTAGE WIRE AND CABLE FOR SOUND AND SHIELDED INSTRUMENTATION CIRCUITS

MANHOURS PER LINEAR FOOT

Conductor Description	Size and Type of Conductor	Number of Conductors	Underfloor Ducts and Conduits	Manhours Required for Heights to		
				10'	15'	25'
Plastic Insulated with Jacket—Braided Copper Shield—No Ground Wire	#22 Solid	2	0.011	0.012	0.012	0.013
	#20 Stranded	2	0.014	0.015	0.016	0.016
	#18 Stranded	2	0.01619	0.017	0.018	0.019
	#16 Stranded	2		0.021	0.021	0.022
Plastic Insulated and Jacket—Braided Copper Shield—No Ground Wire	#22 Solid	2	0.012	0.013	0.013	0.014
	#22 Stranded	2	0.013	0.014	0.014	0.015
		3	0.015	0.016	0.017	0.017
		4	0.017	0.018	0.019	0.020
	#20 Stranded	2	0.022	0.024	0.025	0.026
		3	0.023	0.025	0.026	0.027
		4	0.025	0.027	0.028	0.029
		5	0.027	0.029	0.030	0.031
		6	0.029	0.031	0.033	0.034
		7	0.031	0.033	0.035	0.036
		8	0.033	0.036	0.037	0.038
	#18 Solid	2	0.025	0.027	0.028	0.029
	#18 Stranded	2	0.026	0.028	0.029	0.030
	#16 Stranded	2	0.035	0.038	0.039	0.040
		4	0.039	0.042	0.043	0.045

Manhours include checking out of job storage, handling, job hauling, placing wire coils or spools on reel stands or racks, attaching to fish tape or other pull-in means, pulling wire into conduit type raceway, ducts and underfloor ducts or pull-out and laying into wireways or channels or pulling over suspended ceilings or through bar joist or trusses, identifying or "polling out" the circuit conductors, and splicing or tagging for connection to terminal blocks or equipment.

Manhours exclude extensive identification, "ringing out", connecting to terminal blocks or equipment, and scaffolding. See respective tables for these time frames.

LOW VOLTAGE WIRE AND CABLE FOR SOUND AND SHIELDED INSTRUMENTATION CIRCUITS

MANHOURS PER LINEAR FOOT

Conductor Description	Size and Type of Conductor	Number of Conductors	Manhours Required for			
			Underfloor Ducts and Conduits	Heights to		
				10′	15′	25′
Plastic Insulation and Jacket—Braided Copper Shield—With Ground Wire	#22 Solid	2	0.016	0.017	0.018	0.019
	#22 Stranded	2	0.017	0.018	0.019	0.020
Rubber Insulated Rubber Jacket—Braided Copper Shield—No Ground Wire	#25 Stranded	1	0.010	0.011	0.011	0.012
Plastic Insulated—Plastic Jacket—Braided Copper Shield—No Ground Wire	#25 Stranded	1	0.010	0.011	0.011	0.012

Manhours include checking out of job storage, handling, job hauling, placing wire coils or spools on reel stands or racks, attaching to fish tape or other pull-in means, pulling wire into conduit type raceway, ducts and underfloor ducts or pull-out and laying into wireways or channels or pulling over suspended ceilings or through bar joist or trusses, identifying or "polling out" the circuit conductors, and splicing or tagging for connection to terminal blocks or equipment.

Manhours exclude extensive identification, "ringing out", connecting to terminal blocks or equipment, and scaffolding. See respective tables for these time frames.

LOW VOLTAGE WIRE AND CABLE FOR SOUND AND SHIELDED INSTRUMENTATION CIRCUITS

Plastic Insulated and Jacket—Spiral Wrap Shield
No Ground Wire—Copper Conductor

MANHOURS PER LINEAR FOOT

Size and Type of Conductor	Number of Conductors	Underfloor Ducts and Conduits	Manhours Required for Heights to		
			10′	15′	25′
#22 Solid	2	0.015	0.016	0.017	0.017
	3	0.017	0.018	0.019	0.020
#22 Stranded	2	0.017	0.018	0.019	0.020
	3	0.018	0.019	0.020	0.021
	4	0.020	0.022	0.022	0.023
	6	0.024	0.026	0.027	0.028
#20 Solid	2	0.017	0.018	0.019	0.020
	3	0.018	0.019	0.020	0.021
#20 Stranded	2	0.018	0.019	0.020	0.021
	3	0.019	0.021	0.021	0.022
	4	0.022	0.024	0.025	0.026
	6	0.026	0.028	0.029	0.030
#18 Stranded	2	0.019	0.021	0.021	0.022
	3	0.020	0.022	0.022	0.023
#16 Stranded	2	0.023	0.025	0.026	0.027
	3	0.024	0.026	0.027	0.028
	4	0.027	0.029	0.030	0.031
	6	0.028	0.030	0.031	0.032

Manhours include checking out of job storage, handling, job hauling, placing wire coils or spools on reel stands or racks, attaching to fish tape or other pull-in means, pulling wire into conduit type raceway, ducts and underfloor ducts or pull-out and laying into wireways or channels or pulling over suspended ceilings or through bar joist or trusses, identifying or "polling out" the circuit conductors, and splicing or tagging for connection to terminal blocks or equipment.

Manhours exclude extensive identification, "ringing out", connecting to terminal blocks or equipment, and scaffolding. See respective tables for these time frames.

LOW VOLTAGE WIRE AND CABLE FOR SOUND AND SHIELDED INSTRUMENTATION CIRCUITS

Plastic Insulation and Jacket—Aluminum Foil Shield—with Ground Wire—Copper Conductor

MANHOURS PER LINEAR FOOT

Size and Type of Conductor	Number of Conductors	Manhours Required for			
		Underfloor Ducts and Conduits	Heights to		
			10′	15′	25′
#22 Solid	2	0.016	0.017	0.018	0.019
	3	0.018	0.019	0.020	0.021
#22 Stranded	2	0.018	0.019	0.020	0.021
	3	0.019	0.021	0.021	0.022
	4	0.021	0.023	0.024	0.024
	6	0.026	0.028	0.029	0.030
#20 Stranded	2	0.019	0.021	0.021	0.022
	3	0.021	0.023	0.024	0.024
	4	0.022	0.024	0.025	0.026
	6	0.027	0.029	0.030	0.031
#18 Stranded	2	0.023	0.025	0.026	0.027
	3	0.024	0.026	0.027	0.028
#16 Stranded	2	0.024	0.026	0.027	0.028
	3	0.026	0.028	0.029	0.030
	4	0.029	0.031	0.033	0.034
	6	0.036	0.039	0.040	0.042
#14 Stranded	2	0.029	0.031	0.033	0.034

Manhours include checking out of job storage, handling, job hauling, placing wire coils or spools on reel stands or racks, attaching to fish tape or other pull-in means, pulling wire into conduit type raceway, ducts and underfloor ducts or pull-out and laying into wireways or channels or pulling over suspended ceilings or through bar joist or trusses, identifying or "polling out" the circuit conductors, and splicing or tagging for connection to terminal blocks or equipment.

Manhours exclude extensive identification, "ringing out", connecting to terminal blocks or equipment, and scaffolding. See respective tables for these time frames.

LOW VOLTAGE CABLE
SHIELDED PAIR CABLES FOR SOUND AND INSTRUMENTATION

MANHOURS PER LINEAR FOOT

Conductor Description	Size and Type of Conductor	Pair of Conductors	Manhours Required for			
			Underfloor Ducts and Conduits	Heights to		
				10'	15'	25'
Plastic Insulation and Jacket—Individual pair Foil Shield and Ground Wire	#22 Stranded	3	0.023	0.025	0.026	0.027
		6	0.031	0.033	0.035	0.036
		9	0.039	0.042	0.043	0.045
		11	0.049	0.053	0.055	0.057
		15	0.051	0.055	0.057	0.059
		19	0.058	0.063	0.065	0.067
		27	0.065	0.070	0.073	0.075
	#20 Stranded	2	0.023	0.025	0.026	0.027
		3	0.028	0.030	0.031	0.032
		4	0.030	0.032	0.034	0.035
Plastic Insulation and Jacket—Aluminum Foil Shield Over Cabled pairs with Single Ground Wire	#22 Solid	51	0.056	0.060	0.063	0.065

Manhours include checking out of job storage, handling, job hauling, placing wire coils or spools on reel stands or racks, attaching to fish tape or other pull-in means, pulling wire into conduit type raceway, ducts and underfloor ducts or pull-out and laying into wireways or channels or pulling over suspended ceilings or through bar joist or trusses, identifying or "polling out" the circuit conductors, and splicing or tagging for connection to terminal blocks or equipment.

Manhours exclude extensive identification, "ringing out", connecting to terminal blocks or equipment, and scaffolding. See respective tables for these time frames.

LOW VOLTAGE CABLE
UNSHIELDED PAIR CABLES FOR INTERCOM AND SOUND

MANHOURS PER LINEAR FOOT

Conductor Description	Size and Type of Conductor	Pair of Conductors	Underfloor Ducts and Conduits	Heights to		
				10'	15'	25'
Plastic Insulation and Jacket	#22 Solid	1	0.007	0.008	0.008	0.008
		2	0.012	0.013	0.013	0.014
		3	0.016	0.017	0.018	0.019
		4	0.019	0.021	0.021	0.022
		5	0.021	0.023	0.024	0.024
		6	0.023	0.025	0.026	0.027
		7	0.024	0.026	0.027	0.028
		8	0.025	0.027	0.028	0.029
		9	0.026	0.028	0.029	0.030
		10	0.029	0.031	0.033	0.034
		11	0.030	0.032	0.034	0.035
		12	0.031	0.033	0.035	0.036
		13	0.035	0.038	0.039	0.040
		15	0.039	0.042	0.043	0.045
		19	0.048	0.052	0.054	0.056
		23	0.050	0.054	0.056	0.058
		27	0.051	0.055	0.057	0.059
		32	0.058	0.063	0.065	0.067
		51	0.064	0.069	0.072	0.074
		101	0.093	0.100	0.104	0.108
	#22 Stranded	6	0.029	0.031	0.033	0.034
		9	0.034	0.037	0.038	0.039
		15	0.030	0.042	0.043	0.045
		27	0.056	0.060	0.063	0.065
	#18 Stranded	3	0.031	0.033	0.035	0.036
		6	0.052	0.056	0.058	0.060
		9	0.058	0.063	0.065	0.067

Manhours include checking out of job storage, handling, job hauling, placing wire coils or spools on reel stands or racks, attaching to fish tape or other pull-in means, pulling wire into conduit type raceway, ducts and underfloor ducts or pull-out and laying into wireways or channels or pulling over suspended ceilings or through bar joist or trusses, identifying or "polling out" the circuit conductors, and splicing or tagging for connection to terminal blocks or equipment.

Manhours exclude extensive identification, "ringing out", connecting to terminal blocks or equipment, and scaffolding. See respective tables for these time frames.

MISCELLANEOUS AUDIO CABLES
COAXIAL, CLOSED CIRCUIT AND TV ANTENNA

MANHOURS PER LINEAR FOOT

Conductor Description	Size and Type of Conductor	Number of Conductors	Manhours Required for			
			Underfloor Ducts and Conduits	Heights to		
				10'	15'	25'
Plastic Jacket Copper Shielded RG/U Coaxial Cable	–	1	0.035	0.038	0.039	0.040
Closed Circuit TV Cable– Community Antenna	#22 Solid Copper Weld	1	0.028	0.030	0.031	0.032
	#20 Solid Copper Weld	1	0.028	0.030	0.031	0.032
	#14 Solid Copper	1	0.041	0.044	0.046	0.047
TV Antenna Lead-In	#22 Jacketed Stranded Copper	2	0.017	0.018	0.019	0.020
	#20 Parallel Stranded Copper	2	0.021	0.023	0.024	0.024

Manhours include checking out of job storage, handling, job hauling, distributing, attaching to fish tape or other pull-in means where required, installing and tagging for connection.

Manhours exclude connecting to outlets or equipment, and scaffolding. See respective tables for these time frames.

WIRE AND CABLE TERMINAL CONNECTIONS

MANHOURS PER HUNDRED

	Wire Gauge	Manhours Required	
		1	2
Spade Tongue	22-18	15.0	—
Spade Tongue	16-14	19.5	—
Ring Tongue	22-18	17.5	—
Ring Tongue	16-14	22.0	—
Flag Type	22-18	17.5	—
Flag Type	16-14	22.0	—
Solder Connection	22-18	—	20.0
Solder Connection	16-14	—	24.5
Terminal Screw Connection	22-18	—	12.5
Terminal Screw Connection	16-14	—	18.0
"Punch-On" Insulation Piercing Conn.	22-18	—	10.0

Code:

1–Pressure Terminal Connections to terminal Strips or Blocks.

2–Direct Conductor Connections to Terminal Strips or Blocks.

Manhours include checking out of job storage, handling, job hauling, distributing, and installing of connectors as outlined.

Manhours exclude installation of wire or cable, equipment and scaffolding. See respective tables for these time frames.

TELEPHONE AND BURGLAR AND CENTRALIZED RADIO SYSTEMS

MANHOURS EACH

Item Type and Description	Manhours
Vestibule Telephones	
Cordless Loud Speaking Telephone	4.50
Cordless Loud Speaking Telephone with Mail Box	7.20
Cord and Receiver Type Telephone	3.90
Cord and Receiver Type Telephone with Mail Box	6.60
Suite Telephones	
Loud Speaking Flush Mounted Type	2.70
Loud Speaking Surface Mounted Type	1.70
Cord and Receiver Flush Mounted Type	2.30
Cord and Receiver Surface Mounted Type	1.30
Burglar Alarm Systems	
Open or Closed Circuit Springs	0.60
Contactors	1.20
Electric Matting	1.30
Constant Ringing Drop	2.20
Relays	2.30
Centralized Radio Systems	
Master Receiver and Control	5.10
Turntable Record Player	2.70
Preamplifier	3.90

Manhours include checking out of job storage, handling, job hauling, distributing, and installating of items as outlined.

Manhours exclude installation of circuit conductors, conductor connections and scaffolding. See respective tables for these time frames.

CLOCK AND PROGRAM SYSTEMS

MANHOURS EACH

Item Type and Description	Manhours
Minute Impulse Type	
Master Controller—Flush Mounted	11.70
Master Controller—Surface Mounted	9.90
Signal Control Board—Flush Mounted—20-Signal	15.60
Signal Control Board—Flush Mounted—30-Signal	20.40
Signal Control Board—Flush Mounted—40-Signal	20.00
Signal Control Board—Surface Mounted—20-Signal	13.80
Signal Control Board—Surface Mounted—30-Signal	18.00
Signal Control Board—Surface Mounted—40-Signal	21.60
Secondary Clocks—Flush	1.90
Secondary Clocks—Surface	1.10
Secondary Clocks—Double Face	1.30
Bell or Buzzer Contained in Clock	0.40
Skeleton Dial—Set and connect Mechanism and hands	1.40
Skeleton Dial—Mount ring Type Hour Markings	1.30
Skeleton Dial—Mount Set of Individual Hour Markings	2.30
Dual Motor Type	
Switch Reset Control Unit	2.40
Automatic —Reset Control Unit—Flush Mounted	6.60
Automatic Reset Control Unit—Surface Mounted	5.70
Single Circuit Program Controller—Flush Mounted	6.30
Single Circuit Program Controller—Surface Mounted	5.10
Multiple Circuit Program Controller—Flush Mount—2-Circuit	9.00
Multiple Circuit Program Controller—Flush Mount—4-Circuit	10.50
Multiple Circuit Program Controller—Flush Mount—6-Circuit	11.40
Multiple Circuit Program Controller—Surface Mount—2-Circuit	7.80
Multiple Circuit Program Controller—Surface Mount—4-Circuit	9.10
Multiple Circuit Program Controller—Surface Mount—6-Circuit	10.80
Dual Motor Clocks—Flush	2.70
Dual Motor Clocks—Surface	1.70
Dual Motor Clocks—Double Face	1.90
Bell or Buzzer Contained in Clock	0.40
SKeleton Dial—Set and Connect Mechanism and Hands	1.50
Skeleton Dial—Mount Ring Type Hour Markings	1.30
Skeleton Dial—Mount Set of Individual Hour Markings	2.30

Manhours include checking out of job storage, handling, job hauling, distributing, and installating of items as outlined.

Manhours exclude installation of circuit conductors, conductor connections and scaffolding. See respective tables for these time frames.

FIRE ALARM AND HOSPITAL SIGNALING SYSTEMS

MANHOURS EACH

Item Type and Description	Manhours
Fire Alarm Systems	
Alarm Stations—Manual	0.60
Alarm Stations—Automatic Thermostat	1.10
Alarm Stations—Special Outlet Box	1.00
Alarm Stations—Bell	0.85
Alarm Stations—Flush Horn	1.10
Alarm Stations—Megaphone Horn	1.30
Alarm Stations—Grille Horn	1.80
Alarm Stations—Small Motor Driven Siren	1.30
Alarm Stations—Chimes	0.70
Alarm Stations—Annunicators—Mounting	1.50
Alarm Stations—Annunciators—Per Terminal	0.30
Alarm Stations—Special Rough-In Can	1.10
Control panel	4.80
City Auxiliary Station	2.40
Hospital Signaling Systems	
Elapsed Time Indicator—Flush	2.20
Elapsed Time Indicator—Surface	1.10
Elapsed Time Indicator—Control Station	1.10
In-Out Master Register—20-Name	8.40
In-Out Master Register—40-Name	15.60
In-Out Master Register—60-Name	24.00
In-Out Master Register—80-Name	31.20
In-Out Master Register—100-Name	38.40
In-Out Register Subannunciator—20-Name	7.80
In-Out Register Subannunciator—40-Name	14.40
In-Out Register Subannunciator—60-Name	20.40
In-Out Register Subannunciator—80-Name	27.60
In-Out Register Subannunciator—100-Name	34.80
In-Out Register Subannunciator with Recall—20-Name	10.50
In-Out Register Subannunciator with Recall—40-Name	19.20
In-Out Register Subannunciator with Recall—60-Name	28.80
In-Out Register Subannunciator with Recall—80-Name	38.40
In-Out Register Subannunciator with Recall—100-Name	48.00

Manhours include checking out of job storage, handling, job hauling, distributing, and installating of items as outlined.

Manhours exclude installation of circuit conductors, conductor connections and scaffolding. See respective tables for these time frames.

NURSES CALL AND PAGING SYSTEMS

MANHOURS EACH

Item Type and Description	Manhours
Nurses Call Systems	
Master Duty Station—12 Units	6.60
Master Duty Station—24 Units	12.00
Master Duty Station—36 Units	18.00
Master Duty Station—48 Units	24.00
Annunciators—12 Units	6.30
Annunciators—24 Units	12.00
Annunciators—36 Units	18.00
Annunciators—48 Units	24.00
Audible-Visible Call Bed Station	1.00
Microphone—Speaker Only	0.80
Audible-Visible Call with Microphone Speaker	1.30
Emergency Station	0.70
Explosion-Proof Station	1.80
Remote Duty Station hand Set	1.20
Pilot Lamps	0.60
Dome Lamps	0.70
Barriers	0.40
Buzzers	0.70
Bells	0.70
Night Lights	1.20
Radio Receptacle	0.90
Television Receptacle	1.00
Underpillow Speaker	0.70
Special Rough-In Can	1.00
Paging Systems	
Visual Selector 3-Digit Keyboard	9.60
Visual Control Relay Panel	6.00
Visual Annunciators—Flush Mounted	2.40
Visual Annunciators—Surface Mounted	0.90
Voice Selector Keyboard	7.20
Voice Microphone	0.60
Voice Amplifier and Power Unit	4.50
Voice Terminal Connector Panel	6.00
Voice Loud Speakers—Flush Mounted	2.10
Voice Loud Speakers—Surface Mounted	1.10

Manhours include checking out of job storage, handling, job hauling, distributing, and installating of items as outlined.

Manhours exclude installation of circuit conductors, conductor connections and scaffolding. See respective tables for these time frames.

INTER COMMUNICATION SYSTEMS

MANHOURS EACH

Item Type and Description	Manhours
Telephone Systems	
Central Switchboard—Set and Mount in Place	8.40
Central Switchboard—Connections per Pair of Terminals	0.30
Local Stations—Flush Mounted Talk Back Speaker	2.70
Local Stations—Surface Mounted Talk Back Speaker	1.80
Local Stations—Suspended Wall Handset	1.20
Local Stations—Suspended Desk Handset	1.20
Local Stations—Cradle Desk Set	1.10
Desk Type Intercom Systems	
Master Station—Per Terminal	0.60
Substations	1.30
Public Address Systems	
Amplifiers—Mounting	5.10
Amplifiers—Connecting Per Terminal	0.30
Microphone—Outlet	0.80
Microphone—Mounting and Connecting	0.60
Speakers—Flush Wall—Mounting and Connecting	1.80
Speakers—Flush Ceiling—Mounting and Connecting	2.30
Speakers—Surface Wall—Mounting and Connecting	1.00
Trumpet Speakers—Indoor Wall—Mounting and Connecting	2.40
Trumpet Speakers—Outdoor Wall—Mounting and Connecting	3.60
Trumpet Speakers—Outdoor Pole—Mounting and Connecting	4.80

Manhours include checking out of job storage, handling, job hauling, distributing, and installing items as outlined.

Manhours exclude installation of circuit conductors, conductor connection except where indicated, and scaffolding. See respective tables for these time frames.

MISCELLANEOUS SIGNAL SYSTEM DEVICES

MANHOURS EACH

Item Type and Description	Manhours
Door Chimes	
Mechanical	1.30
Suspended Tubes	1.40
Wall Mounted	0.70
Directory Boards	
Directory—Medium Size	0.70
Door Openers	
Mortise Type	2.30
Rim Type	1.20
Push Buttons and Contacts	
Single Push Button	0.40
Push Button Panels with 1-5 Buttons	3.30
Pushbutton Panels with 6-10 Buttons	4.50
Pushbutton Panels with 11-15 Buttons	5.70
Large Panels—Mounting Only	2.30
Push Buttons on Large panels—Per Button	0.30
Desk Block—6 Button	0.50
Floor Tread	0.60

Manhours include checking out of job storage, handling, job hauling, distributing, and installating of items as outlined.

Manhours exclude installation of circuit conductors, conductor connections and scaffolding. See respective tables for these time frames.

POWER SUPPLY ITEMS

MANHOURS EACH

Item Type and Description	Manhours
Transformers	
Door Bell Type	0.70
Signaling Type–50 VA	1.20
Signaling Type–100 VA	1.40
Signaling Type–250 VA	1.80
Signaling Type–500 VA	2.30
Signaling Type–750 VA	3.30
Signaling Type–1000 VA	4.50
Batteries	
Dry Type	0.40
Storage Type	0.80
Battery Charges	
Copper Oxide Rectifier Automatic type–12-Cell	4.80
Copper Oxide Rectifier Automatic type–24-Cell	6.00
Copper Oxide Rectifier Automatic Type–60-Cell	7.50
Automatic Unit with Generator and Control panel–5 HP	14.70
Automatic Unit with Generator and Control panel–7.5 HP	16.80
Automatic Unit with Generator and Control panel–15 HP	18.90
Rectifier	
Solenium–Telephone Type	2.30
Solenium–Small Capacity Type	2.30
Solenium–Single Voltage Type	4.50
Solenium–Dual Voltage Type	5.70

Manhours include checking out of job storage, handling, job hauling, distributing, and installating of items as outlined.

Manhours exclude installation of circuit conductors, conductor connections and scaffolding. See respective tables for these time frames.

Section 18

ELECTRICAL INSTRUMENT INSTALLATION

This section covers manhours required for installing electrical instrumentation for monitoring various process systems. The manhours listed are for installing the instruments only and do not include time required for the electrical or mechanical tie-in to the process systems.

Time has been allowed for unloading and storage at the project warehouse, checking job storage, calibrating, hauling to the installation site, installing, testing, and final check.

LOCAL-MOUNTED FLOW INSTRUMENTS

MANHOURS EACH

Item Description	Manhours Each
FT Flow Transmitter, Electronic. D/P Cell. Foxboro Model: 823DP— For static pressure to 3,000 psig. Range Capsule: 5–30 inches water, adjustable span. Body material: Cadmium-plated carbon steel. Process Connection: ¼ or ½ NPT female of ½-inch Sch. 80 welding neck. Output Signal: 0 to 10 V dc with Spec 200 input component control system. Electric Code: Class I, Groups B, C and D, Division 1. Enclosure: NEMA 4, watertight. Mounting: Transmitter in any orientation; input component in a Spec 200 Nest.	7.8
FT Flow Transmitter, Electronic. D/P Cell. Foxboro Model: E13DM— For static pressure to 2,000 psig. Range Capsule: 20–205 or 200–850 inches water, adjustable span. Body material: Cadmium-plated carbon steel, 2-wire system; power supply located in control panel. Electric Code: Class I, Group C & D, Division 1. Enclosure: NEMA 4, watertight. Mounting: Direct to process, or by bracket for 2-inch pipe.	7.6
FT Flow Transmitter, Electronic. D/P Cell. Foxboro Model: E13DH— For static pressure to 6,000 psi. Range Capsule: 20–205 or 200–850 inches water, adjustable span. Body material: Cadmium-plated carbon steel. Process Connection. ¼ or ½ NPT female or body machined for 9/16–18 Aminco fittings. Output Signal: 4 to 20 ma d-c, 2-wire system; power supply located in control panel. Electric Code: Class I, Groups C & D, Division 1. Enclosure: NEMA 4, watertight. Mounting: Direct to process, or by bracket for 2-inch pipe.	7.9

Manhours include removing from storage, calibrating, hauling to erection site, installing, testing, and final check out.

Manhours do not include conduit or wiring and connecting to process system. See other electrical accounts for these time frames.

Manhours do not include piping connection time. See *Piping Manhour Manual* for these time frames.

FLANGED ELECTRICAL LIQUID-LEVEL TRANSMITTERS

MANHOURS EACH

Item Description	Manhours Each
LT Level Transmitter, Electronic. D/P Cell. Foxboro Model: E17DM. Range Capsule: 20–205 or 200–850 inches water, adjustable span. Body & Flange: Cadmium-plated carbon steel. Process Connection: High pressure—ANSI raised face modified flange; low pressure— ½ NPT. Flange: 3-inch, 150-lb with 5-inch flange extension. Output Signal: 4 to 20 ma d-c, 2-wire system; power supply located in control panel. Electric Code: Class I, Group C & D, Division 1. Enclosure: NEMA 4, watertight. Mounting: by flange process connection.	9.0
LT Level Transmitter, Electronic. D/P Cell. Foxboro Model: E17DL. Range Capsule: 5-25 inches water, adjustable span. Body & Flange: Cadmium-plated carbon steel. Process Connection: High pressure— ANSI raised face modified flange; low pressure—1/2 NPT. Flange: 6-inch, 150-lb with 5-inch flange extension. Output Signal: 4 to 20 ma d-c, 2-wire system; power supply located in control panel. Electric Code: Class I, Group C & D, Division 1. Enclosure: NEMA 4, watertight. Mounting: By flange process connection.	9.2
LT Liquid Level Transmitter, Electronic. D/P Cell. Foxboro Model: E17DEMK. Range Capsule: 20–205 or 200–850 inches water, adjustable span. Body & Flange: Cadmium-plated carbon steel. Process Connection: High pressure—ANSI raised face modified flange; low pressure—1/2 NPT. Flange: 4-inch, 150-lb with 5-inch flange extension. Diaphragm Extension: 2-inch, 3.695-inch diameter for 4-inch Sch. 80 unlined nozzle only. Output Signal: 4 to 20 ma d-c, 2-wire system; power supply located in control panel. Electric Code: Class I, Group C & D, Division 1. Enclosure: NEMA 4, watertight. Mounting: By flange process connection.	9.4

Manhours include removing from storage, calibrating, hauling to erection site, installing, testing, and final check.

Manhours do not include installing conduit or wiring and connecting to process system. See other electrical accounts for these time frames.

Manhours do not include piping connection time. See *Piping Manhour Manual* for these time frames.

ELECTRICAL LOCAL-MOUNTED LIQUID-LEVEL INSTRUMENTS

MANHOURS EACH

Displacer Length Inches	Manhours Each		
	Type 1	Type 2	Type 3
14	3.9	3.9	3.9
32	5.1	5.1	5.1
48	6.4	6.4	6.4
60	7.6	7.6	7.6
72	8.9	8.9	8.9
84	10.1	10.1	10.1
96	11.4	11.4	11.4
108	12.3	12.3	12.3
120	14.0	14.0	14.0

Note: LT or LC, Level Transmitter or Controller, Electronic, Side Mounted, External Displacement Type, Fabricated Steel Cage.

Fisher Type 2340 electronic transmitter direct or reverse acting; 4–20 ma or 10–50 ma output signal; construction suitable for Class I, Group D, Divisions 1 and 2 hazardous locations mounted on a fabricated steel displacer cage. 316 SS trim; 304 SS displacer; K-Monel torque tube. DC power supply is required.

Type 1: 300# flanged connections, 1½″ or 2″, fabricated steel cage. Fisher Type 249B. Side-side connected.
Type 2: 600# flanged connections, 1½″ or 2″, fabricated steel cage. Fisher Type 249B. Side-side connected.
Type 3: 600# screwed connections, 1½″ or 2″, fabricated steel cage. Top and bottom connections. Fisher Type 249B.

Manhours include removing from storage, calibrating, hauling to erection site, installing, testing, and final check of the transmitters or controllers, complete with air supply filter-regulator and output gauges.

Manhours do not include installing conduit or wiring and connecting to process system. See other electrical accounts for these time frames.

Manhours do not include piping connection time. See *Piping Manhour Manual* for these time frames.

LOCAL-MOUNTED LIQUID-LEVEL INSTRUMENTS

MANHOURS EACH

Displacer Length	Manhours Each			
Inches	Type 1	Type 2	Type 3	Type 4
14	5.1	5.1	5.2	5.7
13	5.7	5.7	5.9	6.4
48	6.4	6.4	6.6	7.1
60 thru 96	8.2	8.2	8.6	9.3
108	8.3	8.3	8.7	9.4
120	9.0	9.0	9.2	9.9

Note: LT or LC, Level Transmitter or Controller, Electronic, Top-Mounted, Internal Displacer, Cast Iron Heads.

Fisher Type 2340 electronic transmitter direct or reverse acting; 4–20 ma or 10–50 ma output signal; construction suitable for Class I, Group D, Divisions 1 and 2 hazardous locations mounted on a top-mounted displacer assembly. 316 SS trim; 304 SST displacer; K-Monel torque tube. DC power supply is required.

Type 1: 4" 125# flanged, cast iron head. Fisher Type 249P.
Type 2: 4" 250# flanged, cast iron head. Fisher Type 249P.
Type 3: 8" 125# flanged, cast iron head. Fisher Type 249P.
Type 4: 8" 250# flanged, cast iron head. Fisher Type 259P.

Manhours include removing from storage, calibrating, hauling to erection site, installing, testing and final check of the transmitters or controllers, complete with air supply filter-regulator and output gauges.

Manhours do not include installing conduit or wiring and connecting to process system. See other electrical accounts for these time frames.

Manhours do not include piping connecting time. See *Piping Manhour Manual* for these time frames.

LOCAL-MOUNTED LIQUID-LEVEL INSTRUMENTS

MANHOURS EACH

Displacer Length Inches	Manhours Each				
	Type 1	Type 2	Type 3	Type 4	Type 5
14	5.1	5.1	5.4	5.2	5.4
32	5.7	5.7	6.1	5.8	6.2
48	6.4	6.4	6.7	6.5	6.8
60 thru 96	6.9	6.9	7.4	7.1	7.4
108	7.6	7.6	7.9	7.7	8.1
120	8.3	8.3	8.6	8.4	8.9

Note: LT or LC, Level Transmitter or Controller, Electronic, Top-Mounted, Internal Displacer, Fabricated Steel Heads.

Fisher Type 2340 electronic transmitter direct or reverse acting; 4–20 ma or 10–50 ma output signal; construction suitable for Class I, Group D, Divisions 1 and 2. Hazardous locations mounted on a top-mounted displacer assembly. 316 SS trim; 304 SS Displacer, K-Monel torque tube. DC power supply is required.

Type 1: 4″ 150# flanged, fabricated steel head. Fisher Type 249BP.
Type 2: 4″ 300# flanged, fabricated steel head. Fisher Type 249BP.
Type 3: 4″ 600# flanged, fabricated steel head. Fisher Type 249BP.
Type 4: 6″ 150# flanged, fabricated steel head. Fisher Type 249BP.
Type 5: 6″ 300# flanged, fabricated steel head. Fisher Type 249BP.

Manhours include removing from storage, calibrating, hauling to erection site, installing, testing, and final check of the transmitters or controllers complete with air supply filter-regulator and output gauges.

Manhours do not include installing conduit or wiring and connecting to process system. See other electrical accounts for these time frames.

Manhours do not include piping connecting time. See *Piping Manhour Manual* for these time frames.

LOCAL-MOUNTED LIQUID-LEVEL INSTRUMENTS

MANHOURS EACH

Displacer Length Inches	Manhours Each	
	Type 1	Type 2
14	5.3	5.6
32	5.9	6.2
48	6.6	6.8
60 thru 96	7.3	7.5
108	7.8	8.1
120	8.5	9.0

Note: LT or LC, Level Transmitter or Controller, Electronic, Top-Mounted, Internal Displacer, Fabricated Steel Heads.

Type 1: 8″ 150# flanged, fabricated steel head. Fisher Type 249 BP.
Type 2: 8″ 300# flanged, fabricated steel head. Fisher Type 249 BP.

Manhours include removing from storage, calibrating, hauling to erection site, installing, testing, and final check of the transmitters or controllers.

Manhours do not include installing conduit or wiring and connecting to process system. See other electrical accounts for these time frames.

Manhours do not include piping connecting time. See *Piping Manhour Manual* for these time frames.

LOCAL-MOUNTED PRESSURE INSTRUMENTS

MANHOURS EACH

Item Description	Manhours Each
PT Gauge Pressure Transmitter, Electronic. Foxboro Model: E11GM. Range Capsule: 7.5 psi to 2000 psi adjustable span. Materials: 316 SS body and capsule. Process Connection: ¼, ½ NPT female or connection block machined to accept 9/16–18 Aminco fittings. Output Signal: 4 to 20 ma d-c, 2-wire system; power supply located in control panel. Electric Code: Class I, Groups B, C & D, Division 1. Mounting: Bracket for 2-inch pipe. With mounting bracket.	7.6
PT Gauge Pressure Transmitter, Electronic. Foxboro Model: 821GM. Range Capsule: 10 psi to 600 psi adjustable span. Materials: 316 SS body. Process Connection: ¼, ½ NPT or connection block machined to accept 9/16–18 Aminco fittings. Output Signal: 4 to 20 ma d-c, 2-wire system. Electric Code: Class I, Groups B, C & D, Division 1. Mounting: Bracket for 2-inch pipe.	7.6
PT Absolute Pressure Transmitter, Electronic. Foxboro Model: E11AH. Range Capsule 20–200 or 40–400 psi, adjustable span maximum overrange to 350 and 750 psi respectively. Body Material: 316 SS. Process Connection: ½ NPT female. Output Signal: 4 to 20 ma d-c, 2-wire system; power supply located in control panel. Electric Code: Class I, Groups B, C & D, Division 1. Enclosure: NEMA 4, watertight. Mounting: Bracket for 2-inch pipe. With Mounting Bracket.	7.6

Manhours include removing from storage, calibrating, hauling to erection site, installing, testing, and final check of the transmitter pressure gauges.

Manhours do not include installing conduit or wiring and connecting to process system. See other electrical accounts for these time frames.

Manhours do not include piping connecting time. See *Piping Manhour Manual* for these time frames.

LOCAL-MOUNTED TEMPERATURE INSTRUMENTS

MANHOURS EACH

Transmitter Model	Manhours Each
Basic Package:	
E 93-A	7.6
E93-B	7.6
Add for:	
Surface Mounting	1.2
Element Mounting	1.8

Note: TT Temperature Transmitter, Blind, Electronic. Foxboro Model: E93 Series. Electric Code: Class I, Division 2. Output Signal: 4 to 20 ma d-c proportional to the measured temperature or temperature difference. Output is linear with millivolt input signal from thermocouple temperature sensors of other d-c mV sources. 2-wire system; power supply located in control panel.

The transmitter is available in two versions; the E93-A is non-isolated and the E93-B provides electrical isolation between input and output.

Manhours include removing from storage, calibrating, hauling to erection site, installing, testing, and final check.

Manhours do not include conduit or wiring and connecting to process system. See other electrical accounts for these time frames.

LOCAL-MOUNTED TEMPERATURE INSTRUMENT

MANHOURS EACH

Transmitter Description	Manhours Each
TTI Temperature Indicating Transmitter, Electronic. Foxboro Model: E45P Series, Electric Code: Class I, Division 2. Output Signal: 4 to 20 ma d-c, 2-wire system; power supply located in control panel. Mounting: Universal bracket for surface or 2-inch pipe. Scale: Eccentric. Thermal System: Class IIIB, gas pressure. Bulb: 316 SS, fixed union with 8″ bendable extension. Bushing: Plain, 316 SS, ¾ NPT. Tubing: 5 feet, ⅛-inch OD 316 SS. Temperature Ranges: 450° F to + 1000° F (270° C to + 535°)	11.4

Manhours include removing from storage, calibrating, hauling to erection site, installing, testing, and final check.

Manhours do not include conduit or wiring and connecting to process system. See other electrical accounts for these time frames.

CONTROL PANEL INSTALLATION

Control panels are usually fabricated by a subcontractor who specializes in this type of work. The instruments that are to be installed on the panelboard or cabinet are usually furnished to the subcontractor by the general contractor.

Panelboards are usually fabricated in sections up to approximately 12'0" in length.

To unload control panel from carrier, move into position, and set on foundation.

Per linear foot of control board length .... 1.5 manhours

If more than one section of control panel is required, <u>add additional time for each connection</u> .... 2.5 manhours

CONNECTING ELECTRICAL PANELBOARD INSTRUMENTS

MANHOURS EACH

Description	Manhours Each
Hand Control Station with auto./man. sw. Auto. position from remote source man. position local adjustment	2.2
Recorder/Hand Control Station with auto/man. sw., 4″ chart,	
one pen	7.7
two pens	8.8
three pens	9.9
Indicator/Hand Control Station with auto./man. sw.,	
single point	3.3
two points	4.4
three points	5.5
Indicator, single point	1.7
two points	2.8
three points	3.9
Indicator Controllers	
Mechanical Setpoint for local adjustment with integral front access mode adjustment, auto./man. transfer sw., and 3″ output meter, single point.	
proportional plus fast reset	6.6
proportional plus reset plus rate	6.6
for two points, <u>add</u>	1.7
Electrical Setpoint for both local and remote adjustment with integral front access mode adjustment, auto./man. transfer sw., and 3″ output meter, single point.	
proportional plus fast reset	6.6
proportional plus reset plus rate	6.6
for two points, <u>add</u>	2.2
Ratio Setpoint with ratio indicating pointer for local ratio adjustment with integral front access mode adjustment, auto./man. transfer sw., and 3″ output meter, single point.	
proportional plus fast reset	6.6
proportional plus reset plus rate	6.6
for two points, <u>add</u>	2.2
Recorder 4″ chart, speed ¾″/hour, one pen	2.2
two pens	3.3
three pens	5.0
Panel-Mounted Electronic Accessories.	
Thermocouple amplifier accepts thermocouple inputs and converts them to 1–5 volts DC, up or down scale burn out protection, automatic cold junction compensation complete with rear panel mounting case	2.2
Panel Meter Scale 2.7″ long, horizontal or vertical, range 4–20 ma DC or 10–50 ma DC, 0–100 Linear, or 1–1000 square root.	2.2

Manhours include making electrical terminations at rear of panel, calibrating, checking, adjusting, testing, and commissioning of electrical instruments.

Manhours for installing air supply piping and air signal lines from control panel to remote instruments are not included. See *Piping Manhour Manual* for these manhours.

Manhours do not include installing conduit, wire, and wiring from control panel to remote instruments. See other electrical accounts for these time frames.

No manufacturer's representatives are included in the above manhours.

Section 19

ANCHORS, FASTENERS, HANGERS, AND SUPPORTS

This section provides manhour units for the installation of anchors, fasteners, hangers, and supports for stabilization or holding in place various items described throughout this manual.

These manhour units are averages, compiled from many projects and are generally representative for the type of construction that the item is normally used in or with. For example, beam clamps are used in conjunction with structural steel framing, toggle bolts with hollow masonry, drywall or plaster, etc. We therefore caution the estimator to adjust and apply these units with discretion.

In applying the manhour units that appear in the following tables consideration must be given to the various items that may affect the overall productivity of the chosen crew and location for the individial project. A method that may be used to obtain this evaluation is outlined in the Introduction of this manual for the convenience of the estimator.

MISCELLANEOUS ANCHORS AND BOLTS

MANHOURS EACH

Item Description	Size	Manhours for Height to							
		10'		15'		20'		25'	
		H	V	H	V	H	V	H	V
Toggle Bolts	1/8"	.16	.19	.17	.21	.18	.22	.19	.23
	3/16"	.16	.19	.17	.21	.18	.22	.19	.23
	1/4"	.20	.24	.22	.26	.23	.27	.24	.28
	5/16"	.20	.24	.22	.26	.23	.27	.24	.28
	3/8"	.23	.27	.25	.29	.26	.30	.27	.31
	1/2"	.23	.27	.25	.29	.26	.30	.27	.31
Hollow Wall Anchors	1/4"	.15	.18	.16	.19	.17	.20	.18	.21
	5/16"	.15	.18	.16	.19	.17	.20	.18	.21
Machine Screw Anchors	# 6	.21	.25	.23	.27	.24	.28	.25	.29
	# 8	.21	.25	.23	.27	.24	.28	.25	.29
	#10	.24	.28	.26	.30	.27	.31	.28	.32
	#12	.24	.28	.26	.30	.27	.31	.28	.32
	1/4"	.28	.33	.30	.36	.31	.37	.32	.38
Lead Anchors	# 6	.22	.26	.24	.28	.25	.29	.26	.30
	#10	.24	.28	.26	.30	.27	.31	.28	.32
	#16	.28	.33	.30	.36	.31	.37	.32	.38
	#20	.31	.37	.33	.40	.35	.42	.36	.43
Plastic or Fiber Anchors	# 6	.16	.19	.17	.21	.18	.22	.19	.23
	# 8	.16	.19	.17	.21	.18	.22	.19	.23
	#10	.19	.22	.21	.24	.22	.25	.23	.26
	#12	.19	.22	.21	.24	.22	.25	.23	.26
	#14	.22	.26	.24	.28	.25	.29	.26	.30
	#16	.22	.26	.24	.28	.25	.29	.26	.30
	#20	.25	.30	.27	.32	.28	.34	.29	.35

H—Horizontal, V—Vertical

Manhours are for wall and ceiling construction and are average for the type of construction that the item is normally used in. For example, Toggle Bolts are for use in drywall, plsater or suspended ceilings, etc.

Manhours include checking out of job storage, handling, job hauling, and installing of items as listed.

For items inserted in floors, use 75% of horizontal manhours.

Manhours exclude scaffolding. See respective table for this time frame.

EXPANSION AND SELF-DRILL ANCHORS

MANHOURS EACH

Item Description	Size	Manhours for Height to							
		10'		15'		20'		25'	
		H	V	H	V	H	V	H	V
Lag Expansion Shields	1/4"	.27	.32	.29	.35	.30	.36	.31	.37
	3/8"	.39	.46	.42	.50	.44	.52	.45	.53
	1/2"	.49	.58	.53	.63	.55	.65	.57	.67
	5/8"	.49	.58	.53	.63	.55	.65	.57	.67
	3/4"	.55	.65	.59	.70	.62	.73	.64	.75
	7/8"	.69	.81	.75	.87	.78	.91	.80	.94
Single or Double Expansion Shields	1/4"	.27	.32	.29	.35	.30	.36	.31	.37
	3/8"	.39	.46	.42	.50	.44	.52	.45	.53
	1/2"	.49	.58	.53	.63	.55	.65	.57	.67
	5/8"	.49	.58	.53	.63	.55	.65	.57	.67
	3/4"	.55	.65	.59	.70	.62	.73	.64	.75
	7/8"	.69	.81	.75	.87	.78	.91	.80	.94
Self-Drill Anchors	1/4"	.14	.17	.15	.18	.16	.19	.17	.20
	5/16"	.14	.17	.15	.18	.16	.19	.17	.20
	3/8"	.21	.25	.23	.27	.24	.28	.25	.29
	1/2"	.21	.25	.23	.27	.24	.28	.25	.29
	5/8"	.27	.32	.29	.35	.30	.36	.31	.37
	3/4"	.27	.32	.29	.35	.30	.36	.31	.37
	7/8"	.32	.38	.35	.41	.36	.43	.37	.44

H—Horizontal, V—Vertical

Manhours are average for solid masonry or concrete wall and ceiling construction.

Manhours include checking out of job storage, handling, job hauling, and installing of items as listed.

For items inserted in floors, use 75% of horizontal manhours.

Manhours exclude installation of scaffolding. See respective table for this time frame.

CONCRETE INSERTS AND POWDER ACTUATED PINES AND STUDS

MANHOURS EACH

Item Description	Size	Manhours for Height to							
		10'		15'		20'		25'	
		H	V	H	V	H	V	H	V
Concrete Inserts	1/4"	.29	.34	.31	.37	.33	.38	.34	.40
	3/8"	.29	.34	.31	.37	.33	.38	.34	.40
	1/2"	.32	.38	.35	.41	.36	.43	.37	.44
	5/8"	.32	.38	.35	.41	.36	.43	.37	.44
	3/4"	.35	.41	.38	.45	.39	.46	.40	48
Powder Actuated Drive Pins	1/8"	.09	.11	.10	.12	.11	.13	.12	.14
	5/32"	.09	.11	.10	.12	.11	.13	.12	.14
	11/64"	.09	.11	.10	.12	.11	.13	.12	.14
	3/16"	.09	.11	.10	.12	.11	.13	.12	.14
Powder Actuated External Threaded Studs	5/32"	.14	.17	.15	.18	.16	.19	.17	.20
	3/16"	.14	.17	.15	.18	.16	.19	.17	.20
	9/64"	.14	.17	.15	.18	.16	.19	.17	.20
	1/4"	.14	.17	.15	.18	.16	.19	.17	.20
	5/16"	.14	.17	.15	.18	.16	.19	.17	.20
	3/4"	.14	.17	.15	.18	.16	.19	.17	.20
Powder Actuated Internal Threaded Studs	3/16"	.14	.17	.15	.18	.16	.19	.17	.20
	11/64"	.14	.17	.15	.18	.16	.19	.17	.20
	1/4"	.14	.17	.15	.18	.16	.19	.17	.20
	5/16"	.14	.17	.15	.18	.16	.19	.17	.20
	3/4"	.14	.17	.15	.18	.16	.19	.17	.20

H–Horizontal, V–Vertical

Manhours are average for solid masonry or concrete wall and ceiling construction.

Manhours include checking out of job storage, handling, job hauling, and installing of items as listed.

For items inserted in floors use 75% of horizontal manhours.

Manhours exclude installation of scaffolding. See respective table for this time frame.

CHANNEL INSERTS, CONTINUOUS CHANNEL INSERTS, AND CHANNEL SUPPORTS
CHANNEL INSERTS AND SUPPORTS

MANHOURS EACH

Item Description	Channel Size (Inches)	Manhours for Height to							
		10'		15'		20'		25'	
		H	V	H	V	H	V	H	V
	3/4 x 1-1/2 x 12	0.26	0.31	0.28	0.33	0.29	0.35	0.30	0.36
	3/4 x 1-1/2 x 24	0.37	0.41	0.40	0.44	0.42	0.46	0.43	0.47
	3/4 x 1-1/2 x 36	0.52	0.61	0.56	0.66	0.58	0.69	0.60	0.71
	3/4 x 1-1/2 x 48	0.77	0.91	0.83	0.98	0.86	1.02	0.89	1.05
Channel Inserts	3/4 x 1-1/2 x 60	1.02	1.20	1.10	1.30	1.15	1.35	1.18	1.39
	1-1/2 x 1-1/2 x 12	0.26	0.31	0.28	0.33	0.29	0.35	0.30	0.36
	1-1/2 x 1-1/2 x 24	0.37	0.41	0.40	0.44	0.42	0.46	0.43	0.47
	1-1/2 x 1-1/2 x 36	0.52	0.61	0.56	0.66	0.58	0.69	0.60	0.71
	1-1/2 x 1-1/2 x 48	0.77	0.91	0.83	0.98	0.86	1.02	0.89	1.05
	1-1/2 x 1-1/2 x 60	1.02	1.20	1.10	1.30	1.15	1.35	1.18	1.39
Channel "U" Support	–	0.22	0.26	0.24	0.28	0.25	0.29	0.26	0.30
Channel Angle Support	–	0.22	0.26	0.24	0.28	0.25	0.29	0.26	0.30

Continuous Channel

MANHOURS PER LINEAR FOOT

Item Description	Channel Size (Inches)	Manhours for Height to							
		10'		15'		20'		25'	
		H	V	H	V	H	V	H	V
	3/4 x 1-1/2	.28	.33	.30	.36	.31	.37	.32	.38
Continuous Channel Inserts	1 x 1-1/2	.28	.33	.30	.36	.31	.37	.32	.38
	1-1/4 x 1-1/2	.28	.33	.30	.36	.31	.37	.32	.38
	1-1/2 x 1-1/2	.28	.33	.30	.36	.31	.37	.32	.38

H—Horizontal, V—Vertical

Manhours are average for concrete wall and ceiling construction.

Manhours include checking out of job storage, handling, job hauling, and installing items as listed.

For items inserted in floors use 80% of horizontal manhours.

Manhours exclude installation of scaffolding. See respective table for this time frame.

CONDUIT STRAP SUPPORTS

MANHOURS EACH

Item Description	Conduit Size	Manhours for Height to							
		10′		15′		20′		25′	
		H	V	H	V	H	V	H	V
1-Hole Strap Supports	1/2″	.02	.03	.02	.03	.03	.04	.03	.04
	3/4″	.02	.03	.02	.03	.03	.04	.03	.04
	1″	.02	.03	.02	.03	.03	.04	.03	.04
	1-1/4″	.03	.04	.03	.04	.04	.05	.04	.05
	1-1/2″	.03	.04	.03	.04	.04	.05	.04	.05
	2″	.04	.05	.04	.05	.05	.06	.05	.06
	2-1/2″	.04	.05	.04	.05	.05	.06	.05	.06
	3″	.06	.07	.06	.07	.07	.08	.07	.08
	3-1/2″	.08	.09	.08	.09	.08	.10	.09	.10
	4″	.11	.12	.12	.13	.12	.13	.13	.14
	5″	.16	.19	.17	.21	.18	.22	.19	.23
	6″	.21	.25	.23	.27	.24	.28	.25	.29
2-Hole Strap Supports	1/2″	.03	.04	.03	.04	.04	.05	.04	.05
	3/4″	.03	.04	.03	.04	.04	.05	.04	.05
	1″	.03	.04	.03	.04	.04	.05	.04	.05
	1-1/4″	.04	.05	.04	.05	.05	.06	.05	.06
	1-1/2″	.04	.05	.04	.05	.05	.06	.05	.06
	2″	.05	.06	.05	.06	.06	.07	.06	.07
	2-1/2″	.05	.06	.05	.06	.06	.07	.06	.07
	3″	.08	.09	.08	.09	.09	.10	.09	.10
	3-1/2″	.10	.12	.11	.13	.11	.13	.12	.14
	4″	.14	.17	.15	.18	.16	.19	.17	.20
	5″	.20	.24	.22	.26	.23	.27	.24	.28
	6″	.26	.31	.28	.33	.29	.35	.30	.36

H—Horizontal, V—Vertical

Manhours include checking out of job storage, handling, job hauling, and installing strap support anchors for the size conduit as listed.

Manhours exclude installation of conduit and scaffolding. See respective tables for these time requirements.

STRUCTURAL STEEL BEAM CLAMPS AND BOLTS

MANHOURS EACH

Item Description	Size	Manhours for Height to			
		10'	15'	20'	25'
Adjustable or Hinged Beam Clamps	1/2"	0.28	0.30	0.31	0.32
	3/4"	0.28	0.30	0.31	0.32
	1"	0.35	0.38	0.39	0.40
	1-1/4"	0.42	0.45	0.47	0.49
	1-1/2"	0.42	0.45	0.47	0.49
	2"	0.55	0.59	0.62	0.64
	2-1/2"	0.55	0.59	0.62	0.64
	3"	0.65	0.70	0.73	0.75
	3-1/2"	0.65	0.70	0.73	0.75
	4"	0.73	0.79	0.82	0.84
	6"	0.88	0.95	0.99	1.02
	8"	1.03	1.11	1.16	1.19
	12"	1.18	1.27	1.33	1.37
Rod Beam Clamps	1/4"	0.35	0.38	0.39	0.40
	3/8"	0.43	0.46	0.48	0.50
	1/2"	0.43	0.46	0.48	0.50
	5/8"	0.50	0.54	0.56	0.58
Structural Steel Clamps	3/8"	0.42	0.45	0.47	0.49
	1/4"	0.42	0.45	0.47	0.49
	1/2"	0.42	0.45	0.47	0.49
	5/8"	0.48	0.52	0.54	0.56
	3/4"	0.48	0.52	0.54	0.56
	1"	0.50	0.54	0.56	0.58
Drill and Bolt Through Steel	3/16"	0.35	0.38	0.39	0.40
	1/4"	0.35	0.38	0.39	0.40
	3/8"	0.43	0.46	0.48	0.50
	1/2"	0.43	0.46	0.48	0.50
	5/8"	0.50	0.54	0.56	0.58
	3/4"	0.50	0.54	0.56	0.58
Drill and Tap Steel and Bolt	3/8"	0.50	0.54	0.56	0.58
	1/2"	0.55	0.59	0.62	0.64
	5/8"	0.65	0.70	0.73	0.75
	3/4"	0.76	0.82	0.85	0.88

Manhours include checking out of job storage, handling, job hauling, and installing item as outlined.

Manhours exclude installation of scaffolding and other items. See respective tables for these time frames.

HANGER CHANNELS AND RODS

MANHOURS EACH

Item Description	Channel or Rod Length	Manhours for Height to			
		10'	15'	20'	25'
	6"	.12	.13	.14	.15
	12"	.13	.14	.15	.16
3/4" x 1-1/2" Channel	18"	.14	.15	.16	.17
	24"	.16	.17	.18	.19
	30"	.18	.19	.20	.21
	6"	.14	.15	.16	.17
	12"	.15	.16	.17	.18
1-1/2" x 1-1/2" Channel	18"	.16	.17	.18	.19
	24"	.18	.19	.20	.21
	30"	.21	.23	.24	.25
	6"	.11	.12	.13	.14
	12"	.12	.13	.14	.15
3/8" or 1/2" Round	18"	.13	.14	.15	.16
Hanger Rods	24"	.14	.15	.16	.17
	30"	.16	.17	.18	.19
	36"	.18	.19	.20	.21
	6"	.14	.15	.16	.17
	12"	.15	.16	.17	.18
5/8" or 3/4" Round	18"	.16	.17	.18	.19
Hanger Rods	24"	.17	.18	.19	.20
	30"	.20	.22	.23	.24
	36"	.22	.24	.25	.26

Manhours include checking out of job storage, handling, job hauling, and installing items as outlined.

Manhours exclude installation of conduits, ducts, and scaffolding. See respective tables for these time frames.

CONDUIT AND PIPE CLAMPS

MANHOURS EACH

Item Description	Conduit Size	Manhours for Height to			
		10'	15'	20'	25'
Right Angle, Parallel	1/2"	0.28	0.30	0.31	0.32
and Edge Conduit	3/4"	0.28	0.30	0.31	0.32
Clamps	1"	0.31	0.33	0.35	0.36
	1-1/4"	0.35	0.38	0.39	0.40
	1-1/2"	0.35	0.38	0.39	0.40
	2"	0.50	0.54	0.56	0.58
	2-1/2"	0.50	0.54	0.56	0.58
	3"	0.70	0.76	0.79	0.81
	3-1/2"	0.70	0.76	0.79	0.81
	4"	0.95	1.03	1.07	1.10
	1-1/4"	0.30	0.32	0.34	0.35
	1-1/2"	0.30	0.32	0.34	0.35
	2"	0.41	0.44	0.46	0.47
	2-1/2"	0.41	0.44	0.46	0.47
Riser Pipe Clamps	3"	0.47	0.51	0.53	0.54
	3-1/2"	0.47	0.51	0.53	0.54
	4"	0.61	0.66	0.69	0.71
	5"	0.61	0.66	0.69	0.71
	6"	0.77	0.83	0.86	0.89

Manhours include checking out of job storage, handling, job hauling, and installing clamps as outlined.

Manhours exclude installation of conduits and scaffolding. See respective tables for these time frames.

CONDUIT, CABLE, AND RING HANGERS

MANHOURS EACH

Item Description	Conduit Size	Manhours for Height to			
		10'	15'	20'	25'
Conduit and Cable hangers	1/2"	0.28	0.30	0.31	0.32
	3/4"	0.28	0.30	0.31	0.32
	1"	0.31	0.33	0.35	0.36
	1-1/4"	0.35	0.38	0.39	0.40
	1-1/2"	0.35	0.38	0.39	0.40
	2"	0.50	0.54	0.56	0.58
	2-1/2"	0.50	0.54	0.56	0.58
	3"	0.70	0.76	0.79	0.81
	3-1/2"	0.70	0.76	0.79	0.81
Ring and Adjustable Ring Hangers	1-1/4"	0.28	0.30	0.31	0.32
	1-1/2"	0.28	0.30	0.31	0.32
	2"	0.35	0.38	0.39	0.40
	2-1/2"	0.35	0.38	0.39	0.40
	3"	0.50	0.54	0.56	0.58
	3-1/2"	0.50	0.54	0.56	0.58
	4"	0.70	0.76	0.79	0.81
	5"	0.70	0.76	0.79	0.81
	6"	1.09	1.18	1.22	1.26

Manhours include checking out of job storage, handling, job hauling, and installing hangers as outlined.

Manhours exclude installation of conduit or cable and scaffolding. See respective tables for these time frames.

CONDUIT SUPPORTS, STRAPS, AND HANGERS

MANHOURS EACH

Item Description	Conduit Size or Hanger Length	Manhours for Height to			
		10'	15'	20'	25'
1-Conduit Support	1/2"	.28	.30	.31	.32
	3/4"	.28	.30	.31	.32
	1"	.35	.38	.39	.40
2-Conduit Support	1/2"	.42	.45	.47	.49
	3/4"	.42	.45	.47	.49
	1"	.51	.55	.57	.59
4-Conduit Support	1/2"	.64	.69	.72	.74
	3/4"	.64	.69	.72	.74
Conduit Channel Straps	1/2"	.02	.02	.03	.03
	3/4"	.02	.02	.03	.03
	1"	.02	.02	.03	.03
	1-1/4"	.03	.03	.04	.04
	1-1/2"	.03	.03	.04	.04
	2"	.04	.04	.05	.05
	2-1/2"	.04	.04	.05	.05
	3"	.06	.06	.07	.07
	3-1/2"	.08	.08	.09	.09
	4"	.11	.12	.13	.14
	5"	.16	.17	.18	.19
	6"	.21	.23	.24	.25
Slotted Conduit Hangers	6"	.03	.03	.04	.04
	8"	.04	.04	.05	.05
	12"	.06	.06	.07	.07
	16"	.07	.07	.08	.08
	20"	.09	.09	.10	.10
	24"	.11	.12	.13	.14

Manhours include checking out of job storage, handling, job hauling and installing supports, straps, and hangers as outlined.

Manhours exclude installation of conduit and scaffolding. See respective tables for these time requirements.

BUS DUCT WALL BRACKETS

MANHOURS REQUIRED EACH

Item Description	Bracket Size	Manhours for Height to			
		10'	15'	20'	25'
Channel Wall Brackets	6"	.23	.25	.26	.27
	8"	.26	.28	.29	.30
	10"	.28	.30	.31	.32
	12"	.30	.32	.34	.35
	14"	.33	.36	.37	.38
	16"	.35	.38	.39	40
	18"	.38	.41	43	44
	20"	40	.43	45	.46
	24"	43	46	.48	.50
Gusset Wall Brackets	6"	.26	.28	.29	.30
	8"	.28	.30	.31	.32
	10"	.30	.32	.34	.35
	12"	.33	.36	.37	.38
	14"	.35	.38	.39	40
	16"	.38	.41	.43	.44
	18"	.40	43	45	46
	20"	.43	.46	.48	.50
	24"	.49	.53	.55	.57

Manhours include checking out of job storage, handling, job hauling, and installing brackets as outlined.

Manhours exclude installation of duct and scaffolding. See respective tables for these time frames.

BUS DUCT HANGERS AND SUPPORTS

MANHOURS EACH

Item Description	Amp, Size or Weight	Manhours for Height to			
		10'	15'	20'	25'
Bus Duct Hangers	225-AMP	0.26	0.28	0.29	0.30
	400-AMP	0.28	0.30	0.31	0.32
	600-AMP	0.30	0.32	0.34	0.35
	800-AMP	0.33	0.36	0.37	0.38
	1000-AMP	0.35	0.38	0.39	0.40
	1350-AMP	0.38	0.41	0.43	0.44
	1600-AMP	0.40	0.43	0.45	0.46
	2000-AMP	0.43	0.46	0.48	0.50
	2500-AMP	0.48	0.52	0.54	0.56
	3000-AMP	0.51	0.55	0.57	0.59
	4000-AMP	0.56	0.60	0.62	0.65
	5000-AMP	0.59	0.64	0.66	0.68
Hanger Adapter	–	0.33	0.36	0.37	0.38
Threaded Clevis	–	0.33	0.36	0.37	0.38
Vertical Supports	8" Long	0.33	0.36	0.37	0.38
	12" Long	0.44	0.48	0.49	0.51
	16" Long	0.55	0.59	0.62	0.64
	20" Long	0.66	0.71	0.74	0.76
	24" Long	0.77	0.83	0.86	0.89
Horizontal Spring Supports	8" Long	0.73	0.79	0.82	0.84
	12" Long	1.06	1.14	1.19	1.23
	16" Long	1.39	1.50	1.56	1.61
	20" Long	1.79	1.93	2.01	2.07
	24" Long	2.13	2.30	2.39	2.46
Vertical Spring Supports	50# Load	1.46	1.58	1.64	1.69
	100# Load	1.79	1.93	2.01	2.07
	200# Load	2.12	2.29	2.38	2.45
	300# Load	2.52	2.72	2.83	2.92
	400# Load	2.92	3.15	3.28	3.38

Manhours include checking out of job storage, handling, job hauling, and installing hangers and supports as listed.

Manhours exclude installation of ducts and scaffolding. See respective tables for these time frames.

FIELD CUT KNOCKOUTS FOR PANELBOARDS AND CONTROL CENTERS

MANHOURS FOR ITEMS LISTED

Conduit or Connector Size (Inches)	Manual Cutter Manhours	Hydraulic	
		Equipment Set-up Manhours	Each Knockout Manhours
1/2	0.55	0.85	0.30
3/4	0.55	0.85	0.30
1	0.55	0.85	0.30
1-1/4	0.85	0.85	0.35
1-1/2	0.85	0.85	0.35
2	1.10	0.85	0.40
2-1/2	1.10	0.85	0.40
3	1.10	0.85	0.40
3-1/2	1.25	0.85	0.45
4	1.25	0.85	0.45
5	1.45	0.85	0.45
6	1.65	0.85	0.45

Manhours include locating and layout of knockout, guide hole drilling, changing the cutter heads knockout, cutting, deburring and cleaning. Manhours for equipment set-up include checking equipment out of storage, handling, hauling, set-up, take-down, cleaning and returning to storage.

Manhours exclude removing or installing panelboards, control center modulars, pull box covers, and raceway or cable terminations. See respective tables for these time frames.

SUPPORTS AND FASTENERS FOR DISTRIBUTION CABINETS

MANHOURS FOR EACH OPERATION

Cabinet Back Box Weight (Pounds)	Weld Cabinet to Steel Column	Thru-Bolts to 5/8" x 12"	
		Concrete	Conc. or Cinder Block
50	4.80	1.80	0.60
75	7.20	1.80	0.60
100	9.60	1.80	0.60
125	11.50	2.20	0.75
150	13.80	2.20	0.75
175	16.10	2.30	0.80
200	17.60	2.30	0.85
225	19.80	2.40	0.85
250	22.00	2.40	0.90

GROUTING IN STEEL CHANNELS FOR EQUIPMENT

MANHOURS EACH

4" Channel Length in Feet	Manhours Each	6" Channel Length in Feet	Manhours Each
5	0.90	5	1.15
6	1.10	6	1.35
7	1.30	7	1.60
8	1.45	8	1.80
9	1.60	9	2.00
10	1.80	10	2.25
12	2.15	12	2.70

Manhours include checking out of job storage or fabrication shop where required, handling, job hauling, distributing, and layout.

DISTRIBUTION CABINET manhours include welding or placing bolts.

GROUTING manhours include placing and grouting in channels.

Manhours exclude setting of cabinets or fabrication of channels. See respective tables for these time frames.

Section 20

DEMOLITION, EXCAVATION, AND CONCRETE

This section covers labor in manhours with reference to channeling, cutting holes through walls, excavation and duct encasement and manhole concrete construction.

First of all we cover channeling and cutting holes through various sized walls for the installation of conduits and cables through same. Next we include manhour tables for machine and hand excavation of duct encasements and manholes.

Before an estimate is made on excavation it is well to know the kind of soil that may be encountered. For this reason, we have divided soil into five groups according to the difficulty experienced in excavating it. Soils vary greatly in charter and no two are exactly alike.

GROUP 1: Light Soil — Earth which can be shoveled easily and requires no loosening, such as sand.

GROUP 2: Medium or Ordinary Soil — Type of earth easily loosened by pick. Preliminary loosening is not required when power excavating equipment such as shovels, dragline scrapers and backholes are used. This type of earth is usually classified as ordinary soil and loam.

GROUP 3: Heavy or Hard Soil — This type of soil can be loosened by pick but is sometimes very hard to do. It may be excavated by sturdy power shovels without preliminary loosening. Hard and compacted loam containing gravel, small stones and boulders, stiff clay or compacted gravel are good examples of this type.

GROUP 4: Hard Pan or Shale — A soil that has hardened and is very difficult to loosen with picks. Light blasting is often required when excavating with power equipment.

GROUP 5: Rock — Requires blasting before removal and transporting. (May be divided into different grades such as hard, soft or medium.)

Also covered are forms and ready-mixed concrete for duct entrenchments and manholes.

Usually, excavation and concrete work of this nature are subcontracted. However, a complete direct labor estimate can be obtained from the following manhour tables if desired.

CHANNELING CONCRETE, BRICK & TILE CONSTRUCTION

MANHOURS PER FOOT

Size	Concrete Construction	Brick Construction	Tile Construction
1/2"	1.20	.55	.08
3/4"	1.43	.70	.12
1"	1.50	.80	.18
2"	2.63	.90	.23
3"	3.60	1.36	.30
4"	4.88	1.58	.38

Manhours include groving with power tools and is average for all heights of walls for above type construction.

Add 75% to manhours for overhead channeling.

Manhours do not include scaffolding. See respective table for this charge.

CUTTING HOLES IN WALL

Heights to Ten Feet

MANHOURS EACH

Size	8" Wall		12" Wall		16" Wall		20" Wall		24" Wall	
	Conc.	Brick	Conc.	Brick	Conc.	Brick	Conc.	Brick	Conc.	Brick
1/2"	1.20	.48	1.68	.56	2.40	.68	2.96	.80	3.52	.96
3/4"	1.52	.56	1.76	.72	2.48	.76	3.20	.96	3.84	1.12
1"	1.76	.64	2.24	.80	2.80	.88	3.52	1.04	4.16	1.52
1-1/4"	1.96	.68	2.64	.96	3.00	1.00	3.92	1.12	4.48	1.60
1-1/2"	2.16	.72	2.80	1.04	3.20	1.20	4.16	1.32	4.76	1.72
2"	2.40	.84	3.20	1.20	4.00	1.32	4.72	1.44	4.96	1.92
2-1/2"	2.52	.92	3.28	1.28	4.24	1.40	4.80	1.56	5.20	2.08
3"	2.88	.96	3.84	1.44	4.80	1.56	5.12	1.64	5.60	2.32
3-1/2"	3.04	1.04	4.16	1.52	5.12	1.68	5.52	1.80	5.80	2.48
4"	3.28	1.20	4.80	1.60	5.76	2.00	5.84	2.08	6.40	2.96

Heights from Ten to Fifteen Feet

Size	8" Wall		12" Wall		16" Wall		20" Wall		24" Wall	
	Conc.	Brick	Conc.	Brick	Conc.	Brick	Conc.	Brick	Conc.	Brick
1/2"	1.50	.60	2.10	.70	3.00	.85	3.70	1.00	4.40	1.20
3/4"	1.90	.70	2.20	.90	3.10	.95	4.00	1.20	4.80	1.40
1"	2.20	.80	2.80	1.00	3.50	1.10	4.40	1.30	5.20	1.90
1-1/4"	2.45	.85	3.30	1.20	3.75	1.25	4.90	1.40	5.60	2.00
1-1/2"	2.70	.90	3.50	1.30	4.00	1.50	5.20	1.65	5.95	2.15
2"	3.00	1.05	4.00	1.50	5.00	1.65	5.90	1.80	6.20	2.40
2-1/2"	3.15	1.15	4.10	1.60	5.30	1.75	6.00	1.95	6.50	2.60
3"	3.60	1.20	4.80	1.80	6.00	1.95	6.40	2.05	7.00	2.90
3-1/2"	3.80	1.30	5.20	1.90	6.40	2.10	6.90	2.25	7.25	3.10
4"	4.10	1.50	6.00	2.00	7.20	2.50	7.30	2.60	8.00	3.70

Manhours include proper set-up for use of power tools with star drill and the removal of excess material.

If koredrill is used on brick, deduct 20%, and on concrete, deduct 10% from manhours.

Manhours do not include scaffolding. See respective table for this charge.

CUTTING HOLES IN WALL

Heights from Fifteen to Twenty Feet

NET MANHOURS EACH

Size	8" Wall		12" Wall		16" Wall		20" Wall		24" Wall	
	Conc.	Brick	Conc.	Brick	Conc.	Brick	Conc.	Brick	Conc.	Brick
1/2"	1.73	.69	2.42	.81	3.45	.98	4.26	1.15	5.06	1.38
3/4"	2.19	.81	2.53	1.04	3.57	1.09	4.60	1.38	5.52	1.61
1"	2.53	.92	3.22	1.15	4.03	1.27	5.06	1.50	5.98	2.19
1-1/4"	2.82	.98	3.80	1.38	4.31	1.44	5.64	1.61	6.44	2.30
1-1/2"	3.11	1.04	4.03	1.50	4.60	1.73	5.98	1.90	6.84	2.47
2"	3.45	1.21	4.60	1.73	5.75	1.90	6.79	2.07	7.13	2.76
2-1/2"	3.62	1.32	4.72	1.84	6.10	2.01	6.90	2.24	7.48	2.99
3"	4.14	1.38	5.52	2.07	6.90	2.24	7.36	2.36	8.05	3.34
3-1/2"	4.37	1.50	5.98	2.19	7.36	2.42	7.94	2.59	8.34	3.57
4"	4.72	1.73	6.90	2.30	8.28	2.88	8.40	2.99	9.20	4.26

Heights from Twenty to Twenty-Five Feet*

Size	8" Wall		12" Wall		16" Wall		20" Wall		24" Wall	
	Conc.	Brick	Conc.	Brick	Conc.	Brick	Conc.	Brick	Conc.	Brick
1/2"	1.80	.72	2.52	.84	3.60	1.02	4.44	1.20	5.28	1.44
3/4"	2.28	.84	2.64	1.08	3.72	1.14	4.80	1.44	5.76	1.68
1"	2.64	.96	3.36	1.20	4.20	1.32	5.28	1.56	6.24	2.28
1-1/4"	2.94	1.02	3.96	1.44	4.50	1.50	5.88	1.68	6.72	2.40
1-1/2"	3.24	1.08	4.20	1.56	4.80	1.80	6.24	1.98	7.14	2.58
2"	3.60	1.26	4.80	1.80	6.00	1.98	7.08	2.16	7.44	2.88
2-1/2"	3.78	1.38	4.92	1.92	6.36	2.10	7.20	2.34	7.80	3.12
3"	4.32	1.44	5.76	2.16	7.20	2.34	7.68	2.46	8.40	3.48
3-1/2"	4.56	1.56	6.24	2.28	7.68	2.52	8.28	2.70	8.70	3.72
4"	4.92	1.80	7.20	2.40	8.64	3.00	8.76	3.12	9.60	4.44

Manhours include proper set-up for use of power tools with star drill and the removal of excess material.

If koredrill is used on brick, deduct 20%, and on concrete, deduct 10% from manhours.

Manhours do not include scaffolding. See respective table for this charge.

*For heights above 25 feet, add 1% per foot of rise to above manhours.

EXCAVATION FOR POLES

NET MANHOURS EACH

Hole Depth	Hand Excavation				Power Excavation
	Sand or Loam	Wet Sticky Clay	Slate	Rock	
4' Hole Depth	1.95	3.90	8.45	9.10	.91
5' Hole Depth	2.60	5.20	9.88	10.40	1.04
6' Hole Depth	4.55	9.10	12.74	13.78	1.30
7' Hole Depth	5.20	10.40	15.60	16.25	1.56
8' Hole Depth	6.24	12.48	18.20	19.50	1.95
9' Hole Depth	7.80	15.60	20.80	23.40	2.34
10' Hole Depth	10.40	20.80	24.70	27.30	2.60
Dynamiting (Av. per Hole in Rock)	-	-	-	-	8.5
Dewatering (Av. per Hole)	-	-	-	-	8.5

Manhours are for all labor involved in the above types of excavation under average conditions and in accordance with the following typical crews.

Typical Hand Digging Crew—Foreman, truck driver, and 4 workers.

Typical Power Digger Crew—Foreman, truck driver, and 2 workers.

Manhours are for maximum hole diameter of 2'0".

Manhours do not include equipment usage.

If hole locations are hard to get to and work area is small and limited, add for these factors.

MACHINE EXCAVATION FOR DUCTS & MANHOLES

Light, Medium and Heavy Soil*

MANHOURS PER 100 CUBIC YARDS

Equipment	LIGHT SOIL			MEDIUM SOIL			HEAVY SOIL		
	Oper. Engr.	Oiler	La- borer	Oper. Engr.	Oiler	La- borer	Oper. Engr.	Oiler	La. borer
Power Shovel:									
1 cu. yd. Dipper	1.1	1.1	1.1	2.0	2.0	2.0	2.7	2.7	2.7
3/4 cu. yd. Dipper	1.5	1.5	1.5	2.8	2.8	2.8	3.7	3.7	3.7
1/2 cu. yd. Dipper	2.0	2.0	2.0	3.7	3.7	3.7	4.9	4.9	4.9
Backhoe:									
1 cu. yd. Bucket	1.4	1.4	1.4	2.6	2.6	2.6	3.5	3.5	3.5
3/4 cu. yd. Bucket	1.5	1.5	1.5	3.8	3.8	3.8	3.7	3.7	3.7
1/2 cu. yd. Bucket	2.0	2.0	2.0	3.7	3.7	3.7	4.9	4.9	4.9
Dragline:									
2 cu. yd. Bucket	0.7	0.7	0.7	1.3	1.3	1.3	1.7	1.7	1.7
1 cu. yd. Bucket	1.1	1.1	1.1	2.0	2.0	2.0	2.7	2.7	2.7
1/2 cu. yd. Bucket	2.0	2.0	2.0	3.7	3.7	3.7	4.9	4.9	4.9
Trenching Machine	-	-	-	3.8	-	7.5	4.8	-	9.4

Hard Pan and Rock*

MANHOURS PER 100 CUBIC YARDS

Equipment	HARD PAN			ROCK		
	Oper. Engr.	Oiler	La- borer	Oper. Engr.	Oiler	La- borer
Power Shovel						
1 cu. yd. Dipper	3.4	3.4	3.4	3.4	3.4	3.4
3/4 cu. yd. Dipper	4.6	4.6	4.6	4.6	4.6	4.6
1/2 cu. yd. Dipper	6.1	6.1	6.1	6.1	6.1	6.1
Backhoe:						
1 cu. yd. Bucket	4.4	4.4	4.4	4.4	4.4	4.4
3/4 cu. yd. Bucket	4.6	4.6	4.6	4.6	4.6	4.6
1/2 cu. yd. Bucket	6.1	6.1	6.1	6.1	6.1	6.1
Dragline:						
2 cu. yd. Bucket	-	-	-	-	-	-
1 cu. yd. Bucket	-	-	-	-	-	-
1/2 cu. yd. Bucket	-	-	-	-	-	-
Trenching Machine	-	-	-	-	-	-

Manhours are for operational procedures only and do not include equipment rental or depreciation. This must be added in all cases.

Operation includes excavating and dumping on side line or into trucks for hauling but does not include hauling. See sheets on hauling for this charge.

For excavations deeper than 6 feet, add 25% to manhours.

*For descriptions of various soils, see Section Introduction.

HAND EXCAVATION FOR DUCTS & MANHOLES

LABORER MANHOURS PER CUBIC YARD

Type of Soil*	Excavation	First Lift	Second Lift	Third Lift
Light	General Dry	1.07	1.42	1.89
Light	General Wet	1.60	2.13	2.83
Light	Special Dry	1.34	1.78	2.37
Medium	General Dry	1.60	2.13	2.83
Medium	General Wet	2.14	2.85	3.79
Medium	Special Dry	1.07	2.49	3.31
Hard or Heavy	General Dry	2.67	3.55	4.72
Hard or Heavy	General Wet	3.21	4.27	5.68
Hard or Heavy	Special Dry	2.94	3.91	5.20
Hard Pan	General Dry	3.74	4.97	6.61
Hard Pan	General Wet	4.28	5.69	7.57
Hard Pan	Special Dry	4.01	5.33	7.09

Manhours include picking and loosening where necessary and placing on bank out of way of excavation or loading into trucks or wagons for hauling away. Manhours do not include hauling and unloading.

*For description of various soils, see Section Introduction.

ROCK EXCAVATION FOR DUCTS AND MANHOLES

Operation	Labor Hours per Cubic Yard		
	Soft	Medium	Hard
Hand Drill, Plug and Feathers	15.0	21.0	30.0
Hand Drill, Blasting	13.0	16.0	22.0
Machine Drill, Plug and Feathers	8.0	11.0	14.0
Machine Drill, Blasting	4.0	6.0	7.0

Manhours are for drilling, blasting, and loading per cubic yard of rock in place in ground.

For hauling see respective manhour table.

Equipment and materials must be added in all cases.

SHORING & BRACING TRENCHES

NET MANHOURS per 100 SQUARE FEET

Operation	Laborers	Carpenters	Truck Drivers
Placing	3.0	3.0	0.4
Removing	2.5	-	0.4

Manhours include hauling, erecting and stripping.

DISPOSAL OF EXCAVATED MATERIALS

MANHOURS PER HUNDRED (100) CUBIC YARDS

Truck Capacity and Length of Haul	Manhours								
	Average Speed 10 mph			Average Speed 15 mph			Average Speed 20 mph		
	Truck Driver	Laborer	Total	Truck Driver	Laborer	Total	Truck Driver	Laborer	Total
3 Cu Yd Truck:									
1 Mile Haul	15.0	2.8	17.8	11.6	2.8	14.4	10.5	2.8	13.3
2 Mile Haul	21.8	2.8	24.6	16.2	2.8	19.0	14.0	2.8	16.8
3 Mile Haul	28.2	3.0	31.2	20.6	3.0	23.6	17.3	3.0	20.3
4 Mile Haul	36.0	3.0	39.0	26.8	3.0	29.8	21.0	3.0	24.0
5 Mile Haul	41.7	2.5	44.2	31.00	2.5	33.5	25.5	2.5	28.0
4 Cu Yd Truck:									
1 Mile Haul	11.3	2.1	13.4	8.8	2.0	10.8	7.9	2.1	9.0
2 Mile Haul	16.2	2.1	18.3	12.0	2.0	14.0	10.4	2.1	12.5
3 Mile Haul	21.6	2.0	23.6	15.8	2.3	18.1	13.2	2.2	15.4
4 Mile Haul	26.4	2.0	28.4	18.7	2.3	21.0	15.6	2.2	17.8
5 Mile Haul	31.3	1.3	32.6	22.2	1.6	23.8	18.5	1.5	20.0
5 Cu Yd Truck:									
1 Mile Haul	9.0	1.7	10.7	7.0	1.7	8.7	6.3	1.6	7.9
2 Mile Haul	13.0	1.7	14.7	9.7	1.7	11.4	8.3	1.7	10.0
3 Mile Haul	17.1	1.8	18.9	12.3	1.8	14.1	10.4	1.7	12.1
4 Mile Haul	21.0	2.0	23.0	15.0	2.0	17.0	12.4	1.7	14.1
5 Mile Haul	25.0	1.7	26.7	17.9	1.7	19.6	14.8	1.6	16.4
8 Cu Yd Truck:									
1 Mile Haul	5.6	1.0	6.6	4.8	1.0	5.8	4.0	1.0	5.0
2 Mile Haul	8.2	1.0	9.2	6.0	1.0	7.0	5.2	1.0	6.2
3 Mile Haul	10.5	1.1	11.6	7.8	1.1	8.9	6.5	1.0	7.5
4 Mile Haul	13.2	1.1	14.3	9.2	1.1	10.3	7.6	1.0	8.6
5 Mile Haul	15.6	1.3	16.9	10.9	1.3	12.2	9.0	1.1	10.1

Manhours include round trip for truck driver, spotting at both ends, unloading, and labor for minor repairs.

Manhours do not include labor for excavation or loading of trucks. See respective tables for these charges.

DISPOSAL OF EXCAVATED MATERIALS

MANHOURS PER HUNDRED (100) CUBIC YARDS

Truck Capacity and Length of Haul	MANHOURS								
	Average Speed 20 mph			Average Speed 25 mph			Average Speed 30 mph		
	Truck Driver	Laborer	Total	Truck Driver	Laborer	Total	Truck Driver	Laborer	Total
3 Cu Yd Truck:									
6 Mile Haul	26.0	2.5	28.5	21.4	2.5	23.9	17.6	2.4	20.0
7 Mile Haul	27.1	2.3	29.4	22.3	2.3	24.6	18.3	2.2	20.5
8 Mile Haul	28.7	2.3	31.0	23.6	2.3	25.9	19.4	2.2	21.6
9 Mile Haul	30.8	2.1	32.9	25.3	2.1	27.4	20.8	2.0	22.8
10 Mile Haul	33.4	2.1	35.5	27.5	2.1	29.6	22.6	2.0	24.6
4 Cu Yd Truck:									
6 Mile Haul	19.3	1.5	20.8	16.0	2.0	18.0	13.4	1.9	15.3
7 Mile Haul	21.0	1.5	22.5	17.4	2.0	19.4	14.8	1.9	16.7
8 Mile Haul	23.5	1.5	25.0	19.5	1.8	21.3	16.6	1.8	18.4
9 Mile Haul	26.9	1.3	28.2	22.3	1.8	24.1	19.0	1.8	20.8
10 Mile Haul	31.1	1.3	32.4	25.8	1.6	27.4	21.9	1.5	23.4
5 Cu Yd Truck:									
6 Mile Haul	15.6	1.4	17.0	14.1	1.6	15.7	11.7	1.5	13.2
7 Mile Haul	17.3	1.4	18.7	15.5	1.5	17.0	12.9	1.4	14.3
8 Mile Haul	19.8	1.3	21.1	17.5	1.5	19.0	14.5	1.3	15.8
9 Mile Haul	23.2	1.3	24.5	20.3	1.2	21.5	16.8	1.1	17.9
10 Mile Haul	27.4	1.2	28.6	23.7	1.2	24.9	19.7	1.1	20.8
8 Cu Yd Truck:									
6 Mile Haul	9.8	1.2	11.0	9.2	1.0	10.2	7.7	1.0	8.7
7 Mile Haul	11.5	1.2	12.7	10.6	1.0	11.6	8.9	1.0	9.9
8 Mile Haul	14.1	1.1	15.2	12.7	1.0	13.7	10.6	0.9	11.5
9 Mile Haul	17.4	1.1	18.5	15.4	0.9	16.3	12.9	0.9	13.8
10 Mile Haul	21.7	1.0	22.7	19.0	0.9	19.9	15.9	0.8	16.7

Manhours include round trip for truck driver, spotting at both ends, unloading, and labor for minor repairs.

Manhours do not include labor for excavation or loading of trucks. See respective tables for these charges.

MACHINE & HAND BACKFILL
Average for Sand or Loam, Ordinary Soil, Heavy Soil and Clay

MANHOURS PER UNITS LISTED

Item	Unit	Manhours			
		Laborer	Oper. Engr.	Oiler	Total
Hand Place	cu yd	.55	—	—	.55
Bulldoze Loose Material	100 cu yds	—	3.32	—	3.32
Clamshell					
1 cubic yard bucket	100 cu yds	—	1.60	1.60	3.20
¾ cubic yard bucket	100 cu yds	—	2.00	2.00	4.00
½ cubic yard bucket	100 cu yds	—	2.75	2.75	5.50
Hand Spread					
Stone or gravel fill	cu yd	.40	—	—	.40
Sand fill	cu yd	.35	—	—	.35
Cinder fill	cu yd	.40	—	—	.40
Tamp by Hand	cu yd	.60	—	—	.60
Pneumatic Tamping	cu yd	.25	—	—	.25

Hand Place units include placing by hand with shovels loose earth within hand-throwing distance of stockpiles. This unit does not include compaction.

Bulldoze Loose Material units include the moving of prestockpiled loose earth over an area.

Clamshell units include the placement of materials from reachable stockpiles.

Stone, Sand and Cinder Spread units include the hand placing, with shovels, these materials from strategically located stockpiles.

Tamp By Hand and Penumatic Tamping units include the compacting of prespread materials in 6″ layers. Manhours for this type work are shown as laborer hours. Should air tool operator be required for this work, substitute his time for laborer hours.

Manhours do not include trucking or fine grading. See respective tables for these charges.

DUCT ENCASEMENT & MANHOLE FORMS

MANHOURS PER HUNDRED SQUARE FEET

Description	Manhours			
	Carpenter	Laborer	Truck Driver	Total
WOOD FORMS				
Duct Encasement or Manhole Footings:				
Fabricate	2.50	.75	-	3.25
Erect	2.00	1.00	.08	3.08
Strip & Clean	.50	1.50	.08	2.08
TOTAL	5.00	3.25	.16	8.41
Manhole Walls:				
Fabricate	2.50	.75	-	3.25
Erect	3.50	1.00	.10	4.60
Strip & Clean	1.00	2.50	.10	3.60
TOTAL	7.00	4.25	.20	11.45
METAL FORMS				
Metal Wall Forms:				
Erect & Brace	4.50	1.75	.15	6.40
Strip & Clean	.50	2.00	.25	2.75
TOTAL	5.00	3.75	.40	9.15
FACTORS FOR RE-USE OF WOOD FORMS				
Repairs 1st Re-Use	1.25	.50	-	1.75
Repairs 2nd Re-Use	1.50	.50	-	2.00
Repairs 3rd Re-Use	2.00	.75	-	2.75
Oiling after each Re-Use	-	1.00	-	1.00

Manhours include all operations as outlined above.

For reuse of wood forms add factors as appear above.

Manhours do not include excavation, the installation of ducts or the placement of concrete. See respective tables for this charges.

DUCT ENCASEMENT & MANHOLE CONCRETE

MANHOURS PER CUBIC YARD

Description	Manhours		
	Laborer	Carpenter	Total
Duct Encasement & Manhole Footings:			
Ready-Mix from Truck	.50	-	.50
Ready-Mix with Wood Chute	.60	.10	.70
Manhole Walls:			
Ready-Mix from Truck	.60	-	.60
Ready-Mix with Wood Chute	.75	.12	.87

Manhours include the placement of ready-mix concrete, vibrating, and puddling.

If wood chutes are used, ample carpenter time has been allowed for the fabrication and repairs or minor chutes.

Manhours do not include mixing or hauling or finishing of concrete.

Section 21

TECHNICAL INFORMATION

As was stated in the Preface of this manual, its intention is solely for the estimation of labor and is not intended for the design of electrical installations. Therefore, this section is held to a minimum.

Included in this section are a few tables, formulas, and definitions which may be of some use to the estimator. These and many other tables and definitions may be found in the *National Electrical Code,* published by the National Board of Fire Underwriters, as well as any number of textbooks.

DEFINITIONS

Centimeter is the unit of length and equals .3937 inch, or .000000001 of a quadrant of the earth.

Gram is the unit of mass, and is equal to 15.432 grains, the mass of a cubic centimeter of water at 4°C.

Second is the unit of time and is the time of one swing of a pendulum, swinging 86464.09 times per day, or the 1·86400th part of a mean solar day.

Volt is the unit of electro-motive force (E).

Electro-motive force, which is the force that moves electricity, is usually written E.M.F. (in formulae E) and various writers use it to express potential difference of potential, electrical pressure and electric force.

One Volt will force an ampere of current through one ohm of resistance. Its value is purely arbitrary, but fixed.

Milli-Volt is one thousandth of a volt, 0.001 volt.

Ohm is the unit of resistance (R), and it is equal to the resistance of a column of pure mercury 1 square millimeter in section and 106 centimeters long at the temperature of melting ice.

One Ohm is that resistance through which one ampere of current will flow at a pressure of one volt of E.M.F.

Megohm is 1,000,000 ohms.

Ampere is the unit of current strength (C). Its value may be defined as that quantity of electricity which flows through one Ohm of resistance when impelled by one volt of E.M.F.

One Ampere of current flowing through a bath will deposit 0.017253 grain of silver of 0.005085 grain of copper per second.

Milli-Ampere is one thousandth of an ampere.

Coulomb is the unit of quantity (Q), and is the quantity of electricity passing per second when the current is one ampere.

Farad is the unit of capacity (K), and is that capacity that will contain one coulomb at a potential of one volt. A condenser of one farad capacity, if charged to two volts, will contain two coulombs; if to 100 volts, 100 coulombs, etc.

Joule is the unit of work (W). It is the work done or heat generated by 1 watt in a second. It is equal to .7373 foot-pound.

Watt is the unit of electrical power (P), and is the energy contained in a current of one ampere with an electro-motive force of one volt. 746 Watts is one horsepower. A current of 10 amperes and 74.6 volts will do the work of one horsepower.

Horsepower in a steam engine or other prime mover is 550 lbs. raised one foot per second, or 33,000 lbs. one foot per minute.

Kilowatt (KW) is 1,000 watts.

Kilo-Volt Ampere (KVA) is one thousand volt amperes.

E.M.F. is distributed according to the resistance of the various parts on one circuit, except where there is counter E.M.F

Counter E.M.F is like back pressure in hydraulics. Thus, to find the available E.M.F., or the resulting current against a resistance where there is a counter E.M.F., the counter E.M.F must be de-ducted. For Example: Suppose a storage battery with a resistance of .02 ohm and a C.E.M.F of 15 volts, and you wish to charge it with a dynamo which gives an E.M.F. of 20 Volts at the battery binding posts. There are 20 - 15 = 5 volts working through a resistance of .02 of an ohm with consequently a current of 250 amperes. The fall of potential is, however, virtually 20 volts, and the power is 20 × 250 = 5000 watts, and not 5 × 250 = 1250 watts, as might perhaps be supposed. It is obvious that the C.E.M.F. has acted as a true resistance. In the above case 5 × 250 = 1250 watts were wasted in overcoming the resistance of the storage battery and the remaining 3750 watts were stored up in the chemical changes which they brought about in the active material of the storage battery.

Mils. is thousandths of an inch.

d^1 is circular mils.

Circular Mil. (d²) is now generally used as the unit of area when considering the cross-section of electrical conductors, the resistance being inversely, and weight of copper directly, proportional to the circular mils.

Hertz is the preferred terminology for "cycles per second."

MENSURATION UNITS

Diameter of a circle x 3.1416 = Circumference
Radius of a circle x 6.283185 = Circumference

Square of the radius of a circle x 3.1416 = Area
Square of the diameter of a circle x 0.7854 = Area
Square of the circumference of a circle x 0.07985 = Area
Half the circumference of a circle x half its diameter = Area

Circumference of a circle x 0.159155 = Radius
Square root of the area of a circle x 0.56419 = Radius

Circumference of a circle x 0.31831 = Diameter
Square root of the area of a circle x 1.12838 = Diameter

Diameter of a circle x 0.86 = Side of an inscribed equilateral triangle
Diameter of a circle x 0.7071 = Side of an inscribed square
Circumference of a circle x 0.225 = Side of an inscribed square
Circumference of a circle x 0.282 = Side of an equal square
Diameter of a circle x 0.8862 = Side of an equal square

Base of a triangle x 1/2 the altitude = Area
Multiplying both diameters and .7854 together = Area of an Elipse

Surface of a sphere x 1/6 of its diameter = Solidity

Circumference of a sphere x its diameter = Surface
Square of the diameter of a sphere x 3.1416 = Surface
Square of the circumference of a sphere x 0.3183 = Surface

Cube of the diameter of a sphere x 0.5236 = Solidity
Cube of the radius of a sphere x 4.1888 = Solidity
Cube of the circumference of a sphere x 0.016887 = Solidity

Square of the surface of a sphere x 0.56419 = Diameter
Square root of the surface of a sphere x 1.772454 = Circumference

Cube root of the solidity of a sphere x 1.2407 = Diameter
Cube root of the solidity of a sphere x 3.8978 = Circumference

Radius of a sphere x 1.1547 = Side of inscribed cube
Square root of (1/3 of the square of) the diameter of a sphere = Side of inscribed cube

Area of its base x 1/3 of its altitude = Solidity of a cone or pyramid, whether round, square or triangular
Area of one of its sides x 6 = Surface of a cube
Altitude of trapezoid x 1/2 the sum of its parallel sides = Area

CONVERSION FACTORS

AREA

1 sq. mile = 640 acres
1 acre = 4840 sq. yards
1 acre = 43,560 sq. ft.
1 sq. foot = 144 sq. inches
1 sq. yard = .836 sq. meters
1 sq. meter = 1.196 sq. yards

1 sq. ft. = .0929 sq. meters
1 cir. mil = 7.854 x 10^7 sq. inch.
1 cir. mil = .7854 sq. mils
1 sq. mil = 1.273 cir. mils
1 sq. inch = 6.452 sq. cm.
1 sq. cm. = .155 sq. inch

ANGLE

1 quadrant = 90 degrees
1 quadrant = 1.57 radians
1 radian = 57.3 degrees

1 degree = .0175 radian
1 minute = .01667 degree
1 minute = 2.9 x 10^4 radian

LENGTH

1 mile = 5280 feet
1 mile = 1.609 kilometers
1 kilometer = .621 miles
1 yard = .9144 meters
1 meter = 3.28 feet
1 meter = 39.37 inches
1 meter = 1.094 yards

1 foot = 12 inches
1 foot = .3048 meters
1 inch = 2.54 centimeters
1 centimeter = .394 inch
1 fathom = 6 feet
1 rod = 5·1/2 yards

WEIGHT

1 short ton = 2000 pounds
1 short ton = 907.2 kilograms
1 kilogram = 2.205 pounds

1 pound = 453.6 grams
1 ounce = 28.35 grams
1 gram = .0353 ounces

DRY VOLUME

1 cu. meter = 1.308 cu. yards
1 cu. yard = .7646 cu. meters

1 cu. meter = 35.31 cu. feet
1 cu. foot = .0283 cu. meters

LIQUID VOLUME

1 U.S. gallon = 3.785 liters
1 liter = .2642 U.S. gallons

1 U.S. quart = .9463 liters
1 liter = 1.057 U.S. quarts

POWER

1 horsepower = 746 watts
1 horsepower = 33000 ft·lbs/min
1 horsepower = 550 ft·lbs/sec.

1 BTU/hour = .293 watts
1 BTU = 252 gram-calories
1 BTU = 778.3 ft·lbs.

DECIMAL AND METRIC EQUIVALENTS OF COMMON FRACTIONS OF AN INCH

Fraction		Decimal	Mm	Fraction		Decimal	Mm
	1/64	0.01562	0.397		23/64	0.51562	13.097
1/32		0.03125	0.794	17/32		0.53125	13.494
	3/64	0.04688	1.191		35/64	0.54688	13.891
1/16		0.06250	1.588	9/16		0.56250	14.288
	5/64	0.07812	1.984		37/64	0.57812	14.684
3/32		0.09375	2.381	19/32		0.59375	15.081
	7/64	0.10938	2.778		39/64	0.60938	15.478
1/8		0.12500	3.175	3/8		0.62500	15.875
	9/64	0.14062	3.572		41/64	0.64062	16.272
3/32		0.15625	3.696	21/32		0.65625	16.669
	11/64	0.17188	4.366		43/64	0.67188	17.066
3/16		0.18750	4.763	11/16		0.68750	17.463
	13/64	0.20312	5.159		45/64	0.70312	17.859
7/32		0.21875	5.556	23/32		0.71875	18.256
	15/64	0.23438	5.953		47/64	0.73438	18.653
1/4		0.25000	6.350	3/4		0.75000	19.050
	17/64	0.26562	6.747		49/64	0.76562	19.447
9/32		0.28125	7.144	25/32		0.78125	19.844
	19/64	0.29688	7.541		51/64	0.79688	20.241
5/16		0.31250	7.938	13/16		0.81250	20.638
	21/64	0.32812	8.334		53/64	0.81812	21.034
11/32		0.34375	8.731	27/32		0.84375	21.431
	23/64	0.35938	9.128		55/64	0.85938	21.828
3/8		0.37500	9.525	7/8		0.87500	22.225
	25/64	0.39062	9.922		57/64	0.89062	22.622
13/32		0.40625	10.319	29/32		0.90625	23.019
	27/64	0.42188	10.716		59/64	0.92188	23.416
7/16		0.43750	11.113	15/16		0.93750	23.813
	29/64	0.45312	11.509		61/64	0.95312	24.209
15/32		0.46875	11.906	31/32		0.96875	24.606
	31/64	0.48438	12.303		63/64	0.98438	
1/2		0.50000	12.700	1/1		1.00000	25.400

NATURAL TRIGONOMETRIC FUNCTIONS

Angle	Sin	Tan	Cot	Cos	Deg
0	0.0000	0.0000	–	1.0000	90
1	0.0175	0.0175	57.2900	0.9998	89
2	0.0349	0.0349	28.6363	0.9994	88
3	0.0523	0.0524	19.0811	0.9986	87
4	0.0698	0.0699	14.3007	0.9976	86
5	0.0672	0.0875	11.4300	0.9962	85
6	0.1045	0.1051	9.5144	0.9945	84
7	0.1219	0.1228	8.1443	0.9925	83
8	0.1392	0.1405	7.1154	0.9903	82
9	0.1564	0.1584	6.3138	0.9877	81
10	0.1736	0.1763	5.6713	0.9848	80
11	0.1908	0.1944	5.1446	0.9816	79
12	0.2079	0.2126	4.7046	0.9781	78
13	0.2250	0.2309	4.3315	0.9744	77
14	0.2419	0.2493	4.0108	0.9703	76
15	0.2588	0.2679	3.7321	0.9659	75
16	0.2756	0.2867	3.4874	0.9613	74
17	0.2924	0.3057	3.2709	0.9563	73
18	0.3090	0.3249	3.0777	0.9511	72
19	0.3256	0.3443	2.9042	0.9455	71
20	0.3420	0.3640	2.7475	0.9397	70
21	0.3584	0.3839	2.6051	0.9336	69
22	0.3746	0.4040	2.4751	0.9272	68
23	0.3907	0.4245	2.3559	0.9205	67
24	0.4067	0.4452	2.2460	0.9135	66
25	0.4226	0.4663	2.1445	0.9063	65
26	0.4384	0.4877	2.0503	0.8988	64
27	0.4540	0.5095	1.9626	0.8910	63
28	0.4695	0.5317	1.8807	0.8829	62
29	0.4848	0.5543	1.8040	0.8746	61
30	0.5000	0.5774	1.7321	0.8660	60
31	0.5150	0.6009	1.6643	0.8572	59
32	0.5299	0.6249	1.6003	0.8480	58
33	0.5446	0.6494	1.5399	0.8387	57
34	0.5592	0.6745	1.4826	0.8290	56
35	0.5736	0.7002	1.4281	0.8192	55
36	0.5878	0.7265	1.3764	0.8090	54
37	0.6018	0.7536	1.3270	0.7986	53
38	0.6157	0.7813	1.2799	0.7880	52
39	0.6293	0.8098	1.2349	0.7771	51
40	0.6428	0.8391	1.1918	0.7660	50
41	0.6561	0.8693	1.1504	0.7547	49
42	0.6691	0.9004	1.1106	0.7431	48
43	0.6820	0.9325	1.0724	0.7314	47
44	0.6947	0.9657	1.0355	0.7193	46
45	0.7071	1.0000	1.0000	0.7071	45
Deg	Cos	Cot	Tan	Sin	Angle

MECHANICAL, ELECTRICAL AND HEAT EQUIVALENTS

Unit	Equivalent Value in Other units	Unit	Equivalent Value in Other Units

1 HP =	746 watts
	.746 KW
	33,000 ft-lbs per minute
	550 ft-lbs per second
	2,545 heat-units per hour
	42.4 heat-units per minute
	.707 heat-units per second
	.175 lb carbon oxidized per hr
	2.64 lbs water evaporated
	  per hour from and at 212°F
		1 Heat-unit =	1,055 watt seconds
			778 ft-lbs
			107.6 kilogram metres
			.000293 KW hour
			.000393 HP hour
			.001036 lbs water evaporated
			  from and at 212°F

		1 Heat-unit per sq ft per min =	.122 watts per sq in .0176 KW per sq ft .0236 HP per sq ft.

1 HP Hour =	746 KW hours
	1,980,000 ft-lbs
	2,545 heat-units
	273,740 kgm
	175 lb carbon oxidized
	  with perfect efficiency
	2.64 lbs water evaporated
	  from and at 212°F
	17.0 lbs water raised from
	  62° to 212°F
		1 Watt =	1 joule per second
			.00134 HP
			3,412 heat-units per hour
			.7373 ft-lbs per second
			.0035 lb water evaporated
			  per hour
			44.24 ft-lbs per minute

1 Kilowatt =	1,000 watts
	1.34 HP
	2,654,200 ft-lbs per minute
	44,240 ft-lbs per minute
	737.3 ft-lbs per second
	3,412 heat-units per hour
	56.9 heat-units per minute
	.948 heat-units per second
	.2275 lbs carbon oxidized
	  per hour
	3.53 lbs water evaporated per
	  hour from and at 212°F
		1 KW Hour =	1,000 watt hours
			1.34 HP hours
			2,654,200 ft-lbs
			3,600,000 joules
			3,412 heat-units
			367,000 kilogram metres
			.235 lb carbon oxidized with
			  perfect efficiency
			3.53 lbs water evaporated
			  from and at 212°F
			22.75 lbs of water raised from
			  62° to 212°F

1 Watt per sq in =	8.9 heat-units per sq ft per min 6371 ft-lbs per sq ft per min .193 HP per sq ft	1 Joule =	1 watt second .000000278 KW hour .102 kgm .0009477 heat-units .7373 ft-lb
1 Kilogram Metre =	7.233 ft-lbs .00000365 HP hour .00000272 KW hour .0093 heat-units	1 ft-lb =	1.356 joules .1383 kgm .000000377 KW hours .001285 heat-units .0000005 HP hour
1 lb Water Evaporated from and at 212°F =	.283 KW hour .379 HP hour 965.7 heat-units 103,900 kgm 1,019,000 joules 751,300 ft-lbs .0664 lb of carbon oxidized	1 lb Carbon Oxidized with Perfect Efficiency =	14,544 heat-units 1.11 lb anthracite coal oxidized 2.5 lbs dry wood oxidized 21 cu ft illuminating gas 4.26 KW hours 5.71 HP hours 11,315,000 ft-lbs 15 lbs of water evaporated from and at 212°F

ELECTRICAL FORMULA FOR DETERMINING AMPERES, HORSEPOWER, KILOWATTS AND KILOVOLT-AMPERES

Desired Date	Alternating Current			Direct Current
	1-Phase	2-Phase 4-Wire*	3-Phase	
Kilowatts	$\dfrac{1 \times E \times PF}{1000}$	$\dfrac{1 \times E \times 2 \times PF}{1000}$	$\dfrac{1 \times E \times 1.73 \times PF}{1000}$	$\dfrac{1 \times E}{1000}$
KVA	$\dfrac{1 \times E}{1000}$	$\dfrac{1 \times E \times 2}{1000}$	$\dfrac{1 \times E \times 1.73}{1000}$	–
Horsepower Output	$\dfrac{1 \times E \times \% \text{ Eff} \times PF}{746}$	$\dfrac{1 \times E \times 2 \times \% \text{ Eff} \times PF}{746}$	$\dfrac{1 \times E \times 1.73 \times \% \text{ Eff} \times PF}{746}$	$\dfrac{1 \times E \times \% \text{ Eff}}{746}$
Amperes when HP is known	$\dfrac{HP \times 746}{E \times \% \text{ Eff} \times PF}$	$\dfrac{HP \times 746}{2 \times E \times \% \text{ Eff} \times PF}$	$\dfrac{HP \times 746}{1.73 \times E \times \% \text{ Eff} \times PF}$	$\dfrac{HP \times 746}{E \times \% \text{ Eff}}$
Amperes when KW is known	$\dfrac{KW \times 1000}{E \times PF}$	$\dfrac{KW \times 1000}{2 \times E \times PF}$	$\dfrac{KW \times 1000}{1.73 \times E \times PF}$	$\dfrac{KW \times 1000}{E}$
Amperes when KVA is known	$\dfrac{KVA \times 1000}{E}$	$\dfrac{KVA \times 1000}{2 \times E}$	$\dfrac{KVA \times 1000}{1.73 \times E}$	–

*In three-wire, two-phase circuits the current in the common conductor is 1.41 times that in either conductor.

E = Volts
1 = Amperes
% Eff = Percent Efficiency
PF = Power Factor

ALLOWABLE AMPACITIES OF INSULATED COPPER CONDUCTORS

Not More Than 3 conductors in Raceway
or Cable or Direct Burial

(Based on Ambient Temperature of 30°C, 86°F)

Size	Temperature Rating of Conductor.							
AWG MCM	60°C (140°F)	75°C (167°F)	85°C (185°F)	90°C (194°F)	110°C (230°F)	125°C (257°F)	200°C (392°F)	250°C (482°F)
	TYPES RUW (14-2), T, TW, UF	TYPES RH, RHW, RUH, (14-2), THW, THWN, XHHW, USE	TYPES V, MI	TYPES TA, TBS, SA, AVB, SIS, FEP, FEPB, RHH, THHN, XHHW*	TYPES AVA, AVL	TYPES AI (14-8), AIA	TYPES A (14-8), AA, FEP* FEPB*	TYPE TFE (Nickel or nickel-coated copper only)
18				21				
16			22	22				
14	15	15	25	25*	30	30	30	40
12	20	20	30	30*	35	40	40	55
10	30	30	40	40*	45	50	55	75
8	40	45	50	50	60	65	70	95
6	55	65	70	70	80	85	95	120
4	70	85	90	90	105	115	120	145
3	80	100	105	105	120	130	145	170
2	95	115	120	120	135	145	165	195
1	110	130	140	140	160	170	190	220
1/0	125	150	155	155	190	200	225	250
2/0	145	175	185	185	215	230	250	280
3/0	165	200	210	210	245	265	285	315
4/0	195	230	235	235	275	310	340	370
250	215	255	270	270	315	335		
300	240	285	300	300	345	380		
350	260	310	325	325	390	420		
400	280	335	360	360	420	450		
500	320	380	405	405	470	500		
600	355	420	455	455	525	545		
700	385	460	490	490	560	600		
750	400	475	500	500	580	620		
800	410	490	515	515	600	640		
900	435	520	555	555				
1000	455	545	585	585	680	730		
1250	495	590	645	645				
1500	520	625	700	700	785			
1750	545	650	735	735				
2000	560	665	775	775	840			

For ambient temperatures over 30°C, and derating factors for more conductors see following tables.

*The ampacities for types FEP, FEPB, RHH, THHN, and XHHW conductors for 14, 12, and 10 shall be the same as designated for 75°C conductors in this table.

ALLOWABLE AMPACITIES OF INSULATED COPPER CONDUCTORS

Single Conductor in Free Air

(BASED ON AMBIENT TEMPERATURE OF 30°C, 86°F)

Size	Temperature Rating of Conductor								
AWG MCM	60°C (140°F)	75°C (167°F)	85°C (185°F)	90°C (194°F)	110°C (230°F)	125°C (257°F)	200°C (392°F)	250°C (482°F)	
	TYPES RUW (14-2), T, TW	TYPES RH, RHW, RUH (14-2), THW, THWN, XHHW	TYPES V, MI	TYPES TA, TBS, SA, AVB, SIS, FEP, FEPB, RHH, THHN, XHHW*	TYPES AVA, AVL	TYPES AI (14-8), AIA	TYPES A (14-B), AA, FEP* FEPB*	TYPE TFE (Nickel or nickel-coated copper only)	Bare and Covered Conductors
18				25					
16			27	27					
14	20	20	30	30*	40	40	45	60	30
12	25	25	40	40*	50	50	55	80	40
10	40	40	55	55*	65	70	75	110	55
8	55	65	70	70	85	90	100	145	70
6	80	95	100	100	120	125	135	210	100
4	105	125	135	135	160	170	180	285	130
3	120	145	155	155	180	195	210	335	150
2	140	170	180	180	210	225	240	390	175
1	165	195	210	210	245	265	280	450	205
1/0	195	230	245	245	285	305	325	545	235
2/0	225	265	285	285	330	355	370	605	275
3/0	260	310	330	330	385	410	430	725	320
4/0	300	360	385	385	445	475	510	850	370
250	340	405	425	425	495	530			410
300	375	445	480	480	555	590			460
350	420	505	530	530	610	655			510
400	455	545	575	575	665	710			555
500	515	620	660	660	765	815			630
600	575	690	740	740	855	910			710
700	630	755	815	815	940	1005			780
750	655	785	845	845	980	1045			810
800	680	815	880	880	1020	1085			845
900	730	870	940	940					905
1000	780	935	1000	1000	1165	1240			965
1250	890	1065	1130	1130					
1500	980	1175	1260	1260	1450				1215
1750	1070	1280	1370	1370					
2000	1155	1385	1470	1470	1715				1405

*The ampacities for types FEP, FEPB, RHH, THHN, and XHHW conductors for 14, 12 and 10 shall be the same as designated for 75°C conductors in this table.

For ambient temperatures over 30°C, and derating factors for more conductors see following tables on page

ALLOWABLE AMPACITIES OF INSULATED ALUMINUM AND COPPER-CLAD ALUMINUM CONDUCTORS

No More Than Three Conductors in Raceway or Cable or Direct Buria

(BASED ON AMBIENT TEMPERATURE OF 30°C, 86°F)

Size	Temperature Rating of Conductor						
AWG MCM	60°C (140°F)	75°C (167°F)	85°C (185°F)	90°C (194°F)	110°C (230°F)	125°C (257°F)	200°C (392°F)
	TYPES RUW (12-2), T, TW, UF	TYPES RH, RHW, RUH (12-2), THW THWN XHHW, USE	TYPES V, MI	TYPES TA, TBS, SA, AVB, SIS, RHH THHN XHHW*	TYPES AVA, AVL	TYPES AI (12-8), AIA	TYPES A (12-8), AA
12	15	15	25	25*	25	30	30
10	25	25	30	30*	35	40	45
8	30	40	40	40	45	50	55
6	40	50	55	55	60	65	75
4	55	65	70	70	80	90	95
3	65	75	80	80	95	100	115
2	75	90	95	95	105	115	130
1	85	100	110	110	125	135	150
1/0	100	120	125	125	150	160	180
2/0	115	135	145	145	170	180	200
3/0	130	155	165	165	195	210	225
4/0	155	180	185	185	215	245	270
250	170	205	215	215	250	270	
300	190	230	240	240	275	305	
350	210	250	260	260	310	335	
400	225	270	290	290	335	360	
500	260	310	330	330	380	405	
600	285	340	370	370	425	440	
700	310	375	395	395	455	485	
750	320	385	405	405	470	500	
800	330	395	415	415	485	520	
900	355	425	455	455			
1000	375	445	480	480	560	600	
1250	405	485	530	530			
1500	435	520	580	580	650		
1750	455	545	615	615			
2000	470	560	650	650	705		

*The ampacities for types RHH, THHN, and XHHW conductors for sizes 12, and 10 shall be the same as designated for 75°C conductors in this table.

For ambient temperatures over 30°C, and derating factors for more conductors see following tables on page

ALLOWABLE AMPACITIES OF INSULATED ALUMINUM AND COPPER-CLAD ALUMINUM CONDUCTORS

Single Conductor in Free Air

(BASED ON AMBIENT TEMPERATURE OF 30°C., 86°F)

Size	Temperature Rating of Conductor							
AWG MCM	60°C (140°F)	75°C (167°F)	85°C (185°F)	90°C (194°F)	110°C (230°F)	125°C (257°F)	200°C (392°F)	
	TYPES RUW (12-2), T, TW	TYPES RH, RHW, RUH (12-2), THW THWN XHHW	TYPES V, MI	TYPES TA, TBS, SA, AVB, SIS, RHH, THHN XHHW*	TYPES AVA, AVL	TYPES AI (12-8), AIA	TYPES A (12-8), AA	Bare and Covered Conductors
12	20	20	30	30*	40	40	45	30
10	30	30	45	45*	50	55	60	45
8	45	55	55	55	65	70	80	55
6	60	75	80	80	95	100	105	80
4	80	100	105	105	125	135	140	100
3	95	115	120	120	140	150	165	115
2	110	135	140	140	165	175	185	135
1	130	155	165	165	190	205	220	160
1/0	150	180	190	190	220	240	255	185
2/0	175	210	220	220	255	275	290	215
3/0	200	240	255	255	300	320	335	250
4/0	230	280	300	300	345	370	400	290
250	265	315	330	330	385	415		320
300	290	350	375	375	435	460		360
350	330	395	415	415	475	510		400
400	355	425	450	450	520	555		435
500	405	485	515	515	595	635		490
600	455	545	585	585	675	720		560
700	500	595	645	645	745	795		615
750	515	620	670	670	775	825		640
800	535	645	695	695	805	855		670
900	580	700	750	750				725
1000	625	750	800	800	930	990		770
1250	710	855	905	905				
1500	795	950	1020	1020	1175			985
1750	875	1050	1125	1125				
2000	960	1150	1220	1220	1425			1165

*The ampacities for types RHH, THHN and XHHW, conductors for sizes 12 and 10 shall be the same as designated for 75°C conductors in this table.

For ambient temperatures over 30°C, and derating factors for more conductors see following tables on page

CORRECTION FACTORS FOR ROOM TEMPERATURES OVER 30°C (86°F)

Conductor Temperature Rating

C.	F.	60°C (140°F)	75°C (167°F)	85°C (185°F)	90°C (194°F)	110°C (230°F)	125°C (257°F)	200°C (392°F)	250°C (482°F)
40	104	.82	.88	.90	.90	.94	.95		
45	113	.71	.82	.85	.85	.90	.92		
50	122	.58	.75	.80	.80	.87	.89		
55	131	.41	.67	.74	.74	.83	.86		
60	140		.58	.67	.67	.79	.83	.91	.95
70	158		.35	.52	.52	.71	.76	.87	.91
75	167			.43	.43	.66	.72	.86	.89
80	176			.30	.30	.61	.69	.84	.87
90	194					.50	.61	.80	.83
100	212						.51	.77	.80
120	248							.69	.72
140	284							.59	.59
160	320								.56
180	356								.50
200	392								.43
225	437								.30

DERATING FACTORS FOR MORE CONDUCTORS

Conductors	% Value	Conductors	% Value
4-6	80	25-42	60
7-24	70	43 & over	50

For ampacities on insulated copper and aluminum conductors see preceding tables.

MAXIMUM NUMBER OF CONDUCTORS IN TRADE SIZE OF CONDUIT OR TUBING

Type Letters	Conductor Size AWG, MCM	1/2	3/4	1	1-1/4	1-1/2	2	2-1/2	3	3-1/2	4	4-1/2	5	6
TW, T, RUH,	14	9	15	25	44	60	99	142						
RUW,	12	7	12	19	35	47	78	111	171					
XHHW (14 thru 8)	10	5	9	15	26	36	60	85	131	176				
	8	2	4	7	12	17	28	40	62	84	108			
RHW and RHH	14	6	10	16	29	40	65	93	143	192				
(without outer	12	4	8	13	24	32	53	76	117	157				
covering),	10	4	6	11	19	26	43	61	95	127	163			
THW	8	1	3	5	10	13	22	32	49	66	85	106	133	
TW,	6	1	2	4	7	10	16	23	36	48	62	78	97	141
T,	4	1	1	3	5	7	12	17	27	36	47	58	73	106
THW,	3	1	1	2	4	6	10	15	23	31	40	50	63	91
RUH (6 thru 2),	2	1	1	2	4	5	9	13	20	27	34	43	54	78
RUW (6 thru 2),	1		1	1	3	4	6	9	14	19	25	31	39	57
FEPB (6 thru 2),	0		1	1	2	3	5	8	12	16	21	27	33	49
RHW and	00		1	1	1	3	5	7	10	14	18	23	29	41
RHH (without	000		1	1	1	2	4	6	9	12	15	19	24	35
outer covering)	0000			1	1	1	3	5	7	10	13	16	20	29
	250			1	1	1	2	4	6	8	10	13	16	23
	300			1	1	1	2	3	5	7	9	11	14	20
	350				1	1	1	3	4	6	8	10	12	18
	400				1	1	1	2	4	5	7	9	11	16
	500				1	1	1	1		4	6	7	9	14
	600					1	1	1	3	4	5	6	7	11
	700					1	1	1	2	3	4	5	7	10
	750					1	1	1	2	3	4	5	6	9
THWN,	14	13	24	39	69	94	154							
	12	10	18	29	51	70	114	164						
	10	6	11	18	32	44	73	104	160					
	8	3	5	9	16	22	36	51	79	106	136			
THHN,	6	1	4	6	11	15	26	37	57	76	98	125	154	
	4	1	2	4	7	9	16	22	35	47	60	75	94	137
FEP (14 thru 2),	3	1	1	3	6	8	13	19	29	39	51	64	80	116
FEPB (14 thru 8).	2	1	1	3	5	7	11	16	25	33	43	54	67	97
	1		1	1	3	3	8	12	18	25	32	40	50	72
XHHW (4 thru	0		1	1	3	4	7	10	15	21	27	33	42	61
500MCM)	00		1	1	2	3	6	8	13	17	22	28	35	51
	000		1	1	1	2	5	7	11	14	18	23	29	42
	0000		1	1	1	2	4	6	9	12	15	19	24	35
	250			1	1	1	3	4	7	10	12	16	20	28
	300			1	1	1	3	4	6	8	11	13	17	24
	350			1	1	1	2	3	5	7	9	12	15	21
	400				1	1	1	3	5	6	8	10	13	19
	500				1	1	1	2	4	5	7	9	11	16
	600				1	1	1	1	3	4	5	7	9	13
	700					1	1	1	3	4	5	6	8	11
	750					1	1	1	2	3	4	6	7	11

(continued on next page)

(continued)

Type Letters	Conductor Size AWG, MCM	1/2	3/4	1	1-1/4	1-1/2	2	2-1/2	3	3-1/2	4	4-1/2	5	6
XHHW	6	1	3	5	9	13	21	30	47	63	81	102	128	185
	600					1	1	1	3	4	5	7	9	13
	700					1	1	1	3	4	5	6	7	11
	750					1	1	1	2	3	4	6	7	10
RHW,	14	3	6	10	18	25	41	58	90	121	155			
	12	3	5	9	15	21	35	50	77	103	132			
	10	2	4	7	13	18	29	41	64	86	110	138		
	8	1	2	4	7	9	16	22	35	47	60	75	94	137
RHH (with outer covering)	6	1	1	2	5	6	11	15	24	32	41	51	64	93
	4	1	1	1	3	5	8	12	18	24	31	39	50	72
	3	1	1	1	3	4	7	10	16	22	28	35	44	63
	2		1	1	3	4	6	9	14	19	24	31	38	56
	1			1	1	3	5	7	11	14	18	23	29	42
	0		1	1	1	2	4	6	9	12	16	20	25	37
	00			1	1	1	3	5	8	11	14	18	22	32
	000			1	1	1	3	4	7	9	12	15	19	28
	0000			1	1	1	2	4	6	8	10	13	16	24
	250				1	1	1	3	5	6	8	11	13	19
	300				1	1	1	3	4	5	7	9	11	17
	350				1	1	1	2	4	5	6	8	10	15
	400				1	1	1	1	3	4	6	7	9	14
	500				1	1	1	1	3	4	5	6	8	11
	600					1	1	1	2	3	4	5	6	9
	700					1	1	1	1	3	3	4	6	8
							1	1	1	3	3	4	5	8

CONDUIT SPACINGS
Spacings in Inches between Centers of Conduits

The top figures are the minimum dimensions to provide clearance between locknuts. The bottom spacings should be used whenever possible.

Size	1/2	3/4	1	1-1/4	1-1/2	2	2-1/2	3	3-1/2	4	4-1/2	5	8
1/2	1-3/16												
	1-3/16												
3/4	1-5/16	1-7/16											
	1-1/2	1-7/8											
1	1-1/3	1-3/8	1-3/4										
	1-3/4	1-7/8	2										
1-1/4	1-3/4	1-7/8	2	2-1/4									
	2	2-1/8	2-1/4	2-1/2									
1-1/2	1-15/16	2-1/16	2-3/16	2-7/16	2-9/16								
	2-1/8	2-1/8	2-3/8	2-3/8	2-3/4								
2	2-3/16	2-5/16	2-1/2	2-3/4	2-7/8	3-1/8							
	2-3/8	2-1/2	2-3/4	3	3-1/8	3-3/8							
2-1/2	2-7/16	2-9/16	2-3/4	3	3-1/8	3-3/8	3-5/8						
	2-5/8	2-3/4	3	3-1/4	3-3/8	3-5/8	4						
3	2-13/16	2-15/16	3-1/16	3-5/16	3-7/16	3-3/8	4	4-3/16					
	3	3-1/8	3-3/8	3-5/8	3-3/8	4	4-3/8	4-3/4					
3-1/2	3-1/8	3-1/4	3-3/8	3-5/8	3-3/8	4-1/16	4-5/16	4-5/8	4-15/16				
	3-3/8	3-1/2	3-5/8	3-7/8	4	4-3/8	4-5/8	5	5-3/8				
4	3-7/16	3-9/16	3-11/16	3-17/16	4-1/16	4-3/8	4-5/8	4-13/16	5-1/8	5-7/16			
	3-3/4	3-7/8	4	4-1/4	4-7/8	4-3/8	5	4-3/8	5-5/8	6			
4-1/2	3-3/4	3-7/8	4	4-1/4	4-3/8	4-5/8	4-7/8	5-1/4	5-1/16	5-7/8	6-1/8		
	4	4-1/8	4-1/4	4-1/2	4-3/4	5	5-1/4	5-1/8	6	6-1/4	6-1/2		
5	4-1/8	4-1/4	4-3/8	4-5/8	4-3/4	5	5-1/4	5-9/16	5-7/8	6-3/16	6-1/8	6-13/16	
	4-3/8	4-1/2	4-5/8	4-7/8	5	5-3/8	5-5/8	6	6-1/4	6-5/8	7	7-1/4	
6	4-3/4	4-7/8	5	5-1/4	5-3/8	5-5/8	5-7/8	6-3/16	6-1/2	6-13/16	7-1/8	7-3/16	8-1/8
	5	5-1/4	5-1/4	5-1/2	5-5/8	6	6-1/4	6-3/8	7	7-1/4	7-3/8	8	8-5/8

Approximate Diameters

	1/2	3/4	1	1-1/4	1-1/2	2	2-1/2	3	3-1/2	4	4-1/2	5	8
Locknut	1-1/8	1-3/8	1-11/16	2-5/16	2-7/16	3	3-7/16	4-7/16	4-13/16	5-3/8	6	6-13/16	7-15/16
bushing	1	1-1/4	1-1/2	1-7/8	2-1/8	2-5/8	3-3/16	3-7/8	4-7/16	5	5-7/16	6-7/8	7-3/8
Conduit	7/8	1-1/16	1-3/8	1-11/16	1-15/16	2-3/8	2-7/8	3-1/2	4	4-1/2	5	5-7/16	6-5/8

Number per 100 Lbs.

	1/2	3/4	1	1-1/4	1-1/2	2	2-1/2	3	3-1/2	4	4-1/2	5	8
Locknuts	5882	3845	2500	1250	833	625	500	333	278	208	133	151	53
Bushings	4166	2858	1110	833	667	500	333	250	179	125	83	67	435

ELECTRICAL SYMBOLS

Standard Symbols for Wiring Plans and Electrical Apparatus Diagrams

In order to understand plans and diagrams, a set of standard graphical symbols should be used. Any arbitrary symbols with an appropriate key may be used, however the following standard symbols have been accepted and are more widely used, recognized and understood by the electrical industry. The following symbols as outlined are preferred types based upon the recommendations of the American Standard Association.

GRAPHICAL SYMBOLS FOR ONE LINE SWITCHGEAR AND APPARATUS DIAGRAMS

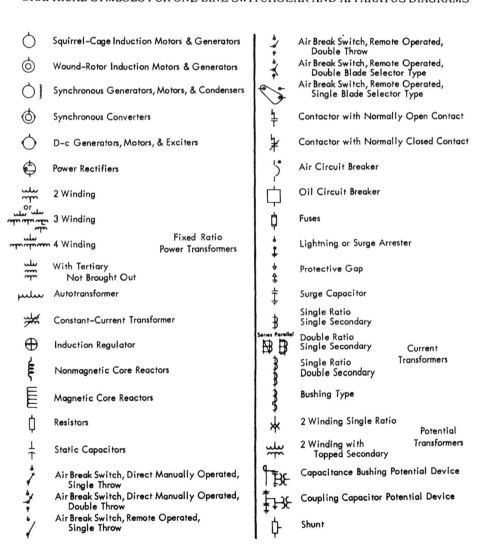

Squirrel-Cage Induction Motors & Generators

Wound-Rotor Induction Motors & Generators

Synchronous Generators, Motors, & Condensers

Synchronous Converters

D-c Generators, Motors, & Exciters

Power Rectifiers

2 Winding

3 Winding

4 Winding — Fixed Ratio Power Transformers

With Tertiary Not Brought Out

Autotransformer

Constant-Current Transformer

Induction Regulator

Nonmagnetic Core Reactors

Magnetic Core Reactors

Resistors

Static Capacitors

Air Break Switch, Direct Manually Operated, Single Throw

Air Break Switch, Direct Manually Operated, Double Throw

Air Break Switch, Remote Operated, Single Throw

Air Break Switch, Remote Operated, Double Throw

Air Break Switch, Remote Operated, Double Blade Selector Type

Air Break Switch, Remote Operated, Single Blade Selector Type

Contactor with Normally Open Contact

Contactor with Normally Closed Contact

Air Circuit Breaker

Oil Circuit Breaker

Fuses

Lightning or Surge Arrester

Protective Gap

Surge Capacitor

Single Ratio Single Secondary

Double Ratio Single Secondary — Series Parallel — Current Transformers

Single Ratio Double Secondary

Bushing Type

2 Winding Single Ratio

2 Winding with Topped Secondary — Potential Transformers

Capacitance Bushing Potential Device

Coupling Capacitor Potential Device

Shunt

ELECTRICAL SYMBOLS

GRAPHICAL ELECTRICAL SYMBOLS FOR ARCHITECTURAL PLANS

Ceiling Wall   **GENERAL OUTLETS**

Ceiling	Wall	
O	─O	Outlet
Ⓑ	─Ⓑ	Blanked Outlet
Ⓓ	─Ⓓ	Drop Cord
Ⓔ	─Ⓔ	Electrical Outlet, for use only when circle used alone might be confused with columns, plumbing symbols, etc.
Ⓕ	─Ⓕ	Fan Outlet
Ⓙ	─Ⓙ	Junction Box
Ⓛ	─Ⓛ	Lamp Holder
Ⓛ$_{PS}$	─Ⓛ$_{PS}$	Lamp Holder with Pull Switch
Ⓢ	─Ⓢ	Pull Switch
Ⓥ	─Ⓥ	Outlet for Vapor Discharge Lamp
Ⓧ	─Ⓧ	Exit Light Outlet
Ⓒ	─Ⓒ	Clock Outlet (Specify Voltage)

CONVENIENCE OUTLETS

⊐Ⓔ	Duplex Convenience Outlet
⊐Ⓔ$_{1,3}$	Convenience Outlet Other than Duplex 1 = Single, 3 = Triplex, etc.
⊐Ⓔ$_{WP}$	Weatherproof Convenience Outlet
⊐Ⓔ$_R$	Range Outlet
⊐Ⓔ$_S$	Switch and Convenience Outlet
⊐Ⓔ-Ⓡ	Radio and Convenience Outlet
⬤	Special Purpose Outlet (Des. in Spec.)
⊙	Floor Outlet

SWITCH OUTLETS

S	Single Pole Switch
S$_2$	Double Pole Switch
S$_3$	Three Way Switch
S$_4$	Four Way Switch
S$_D$	Automatic Door Switch
S$_E$	Electrolier Switch
S$_K$	Key Operated Switch
S$_P$	Switch and Pilot Lamp
S$_{CB}$	Circuit Breaker
S$_{WCB}$	Weatherproof Circuit Breaker
S$_{MC}$	Momentary Contact Switch
S$_{RC}$	Remote Control Switch
S$_{WP}$	Weatherproof Switch
S$_F$	Fused Switch
S$_{WF}$	Weatherproof Fused Switch

SPECIAL OUTLETS

O a,b,c,etc.
⊐Ⓔ a,b,c,etc.
s a,b,c,etc.

Any Standard Symbol as given above with the addition of a lower case subscript letter may be used to designate some special variation of Standard Equipment of particular interest in a specific set of Architectural Plans.

When used they must be listed in the Key of Symbols on each drawing and if necessary further described in the specifications.

PANELS, CIRCUITS & MISCELLANEOUS

▦	Lighting Panel
▨	Power Panel
──	Branch Circuit; Concealed in Ceiling or Wall
─ ─ ─	Branch Circuit; Concealed in Floor
-----	Branch Circuit; Exposed
─→	Home Run to Panel Board. Indicate number of circuits by number of arrows. Note: Any circuit without further designation indicates a 2-wire circuit. For a greater number of wires indicate as follows: (3 wires) (4 wires), etc.
──	Feeders. Note: Use heavy lines and designate by number corresponding to listing in Feeder Schedule.
⊏▣⊐	Underfloor Duct & Junction Box Triple System. Note: For double or single systems eliminate one or two lines. This symbol is equally adaptable to auxiliary system layouts.
Ⓖ	Generator
Ⓜ	Motor
Ⓘ	Instrument
Ⓣ	Power Transformer (or draw to scale)
⊠	Controller
⊏⊐	Isolating Switch

AUXILIARY SYSTEMS

⊡	Push Button
⊏⟩	Buzzer
⊏⟩	Bell
◇	Annunciator
◀	Outside Telephone
⋈	Interconnecting Telephone
⋈	Telephone Switchboard
⊕	Bell Ringing Transformer
Ⓓ	Electric Door Opener
Ⓕ⟩	Fire Alarm Bell
Ⓕ	Fire Alarm Station
⊠	City Fire Alarm Station
⒡Ⓐ	Fire Alarm Central Station
⒡Ⓢ	Automatic Fire Alarm Device
Ⓦ	Watchman's Station
〚Ⓦ〛	Watchman's Central Station
Ⓗ	Horn
Ⓝ	Nurse's Signal Plug
Ⓜ	Maid's Signal Plug
Ⓡ	Radio Outlet
〚SC〛	Signal Central Station
▭	Interconnection Box
⊪⊪⊪	Battery
─··─	Auxiliary System Circuits.

Note: Any line without further designation indicates a 2-Wire System. For a greater number of wires designate with numerals In manner similar to −12-No. 18W − ¾"C, or designate by number corresponding to listing in Schedule.

▢ a,b,c Special Auxiliary Outlets
Subscript letters refer to notes on plans or detailed description in specifications.

CONVERSION TABLE

Minutes to Decimal Hours

Minutes	Hours	Minutes	Hours
1	.017	31	.517
2	.034	32	.534
3	.050	33	.550
4	.067	34	.567
5	.084	35	.584
6	.100	36	.600
7	.117	37	.617
8	.135	38	.634
9	.150	39	.650
10	.167	40	.667
11	.184	41	.684
12	.200	42	.700
13	.217	43	.717
14	.232	44	.734
15	.250	45	.750
16	.267	46	.767
17	.284	47	.784
18	.300	48	.800
19	.317	49	.817
20	.334	50	.834
21	.350	51	.850
22	.368	52	.867
23	.384	53	.884
24	.400	54	.900
25	.417	55	.917
26	.434	56	.934
27	.450	57	.950
28	.467	58	.967
29	.484	59	.984
30	.500	60	1.000

JOB ESTIMATING FORM

COMPANY

PROJECT

DESCRIPTION OF WORK

LOCATION

COMPOSITE CREW RATE

ESTIMATOR

CHECKED BY

SHEET NO. ___ OF ___

ESTIMATE NO

DATE IN

DATE DUE

No	Description	Unit	Quantity	Weight		Unit Man-Hours	Total Man-Hours	Unit Labor Cost	Unit Material Cost	Total Cost		
				Unit	Total					Labor	Material	Total

Gulf Publishing Company, Houston　Form 310

This Job Estimating Form is ideal for use when working with the Estimating Man-Hour Manuals. Prices and further information on this form available from Gulf Publishing Co., P.O. Box 2608, Houston, Texas 77252-2608.

Printed and bound by CPI Group (UK) Ltd, Croydon, CR0 4YY

08/05/2025

01864834-0002